Mathematics 9

ADDISON-WESLEY SECONDARY MATHEMATICS

Senior Authors

Brendan Kelly
*Co-ordinator of Pure
and Applied Sciences
Halton Board of Education
Burlington, Ontario*

Bob Alexander
*Assistant Co-ordinator
of Mathematics
Toronto Board of Education
Toronto, Ontario*

Authors

Paul Atkinson
*Vice Principal
Bluevale Collegiate Institute
Waterloo, Ontario*

Earle Warnica
*Director of Curriculum
and Instruction
Lethbridge School District 51
Lethbridge, Alberta*

ADDISON-WESLEY PUBLISHERS

Don Mills, Ontario • Reading, Massachusetts •
Menlo Park, California • London • Amsterdam • Sydney

Layout and Illustrations: Acorn Technical Art Inc.

Special Features: Glyphics Inc., Designers and Consultants

Photo Credits

Abitibi Paper Company Ltd, 167
Addison-Wesley Photo Library, 1, 27, 79, 92, 257
David Andrew, 334
Boy Scouts of Canada, 121
British Airways, 173
Byron Bush, 148, 203
Canadian Pacific Rail, x, 263, 334
Canon Optics and Business Machines Canada Ltd, 113
CCM, 170
Chrysler Canada Ltd, 71
Convention and Tourist Bureau of Metropolitan Toronto, 262
Dr. W. Aubrey Crich, 146
De Havilland Aircraft of Canada Ltd, 58
Environment Canada Ltd, 160, 197, 216
Fiat Motors of Canada Ltd, 96
Government of Manitoba, Department of Tourism, 2
Government of Quebec, Tourist Branch, 2
Government of Yukon Territory, Tourism and Information Branch, 334
IBM Canada Ltd, 250
Imperial Oil Ltd, 143, 385
Jaguar Rover Triumph Ltd, 104
V. Last, 181
Manitoba Government Travel, 2, 119
Miller Services Ltd, ix, 11, 12, 106, 109, 334, 395
NASA, 166, 257, 293
National Gallery of Art, Washington, D.C., 199
NFB-Photothèque, 169, 263
New Brunswick Department of Tourism, 50
New York Mets, 187
Ontario Ministry of Culture and Education, 78, 320
Ontario Ministry of Industry and Tourism, 32, 119, 159, 169, 201, 251, 309, 406
P.E.I. Department of Tourism, 2
Queensway General Hospital, Etobicoke, Ontario, 100
Royal Canadian Mint, 173
Royal Observatory, Edinburgh, 142
Royal Ontario Museum, 199
Singer Company of Canada Ltd, 2
A. Strunk, 199
Swiss National Tourist Office, 31, 33, 145, 191
Toronto Institute of Medical Technology, 131, 141
Toronto Transit Commission, 276
Travelways, 271
U.S. Naval Observatory, 199
Burk Uzzle (Magnum Photos Inc), 207
Volkswagen Canada Inc., 59, 87
Xerox Canada Inc., 193

Written, printed, and bound in Canada

B C D E F - ASP -86 85 84 83 82   ISBN 0-201-18600-4

Contents

From Arithmetic to Algebra   **1**

Integers   **2**

Rational Numbers   **3**

Statistics and Probability 7

Polynomials 8

Relations 9

Prologue

Mathematics, a product of the human mind,...

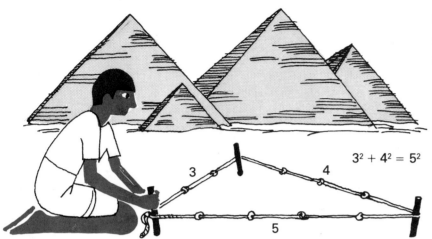

How can a square corner be constructed?

$$3^2 + 4^2 = 5^2$$

More than 4000 years ago, the Babylonians and the Egyptians knew that a triangle with side lengths in the ratio 3 : 4 : 5 is right-angled. They used knotted ropes to lay out a square corner for a field or the base of a pyramid. Much later, Pythagoras proved the theorem which states how the sides of any right triangle are related.

...has been used to solve significant problems of the past,...

Before the sixteenth century, people believed that the heavenly bodies moved around the Earth. In 1543, Nicolas Copernicus suggested instead that Earth and the other planets revolved around the Sun. The discovery that Earth was not the centre of the solar system had a profound effect on civilization and raised many new mathematical problems.

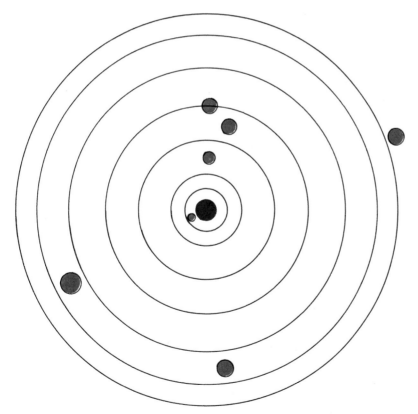

Is Earth or Sun the centre of the solar system?

...and of the present.

A ship's position at sea is determined by its latitude and longitude. Early navigators understood how to measure a ship's latitude, but it took more than 2000 years to solve the problem of finding its longitude. This problem was so important to exploration and trade in the sixteenth and seventeenth centuries that several countries offered substantial prizes for its solution. It was eventually solved by the combined work of mathematicians, astronomers, and clock makers.

How did explorers find their way?

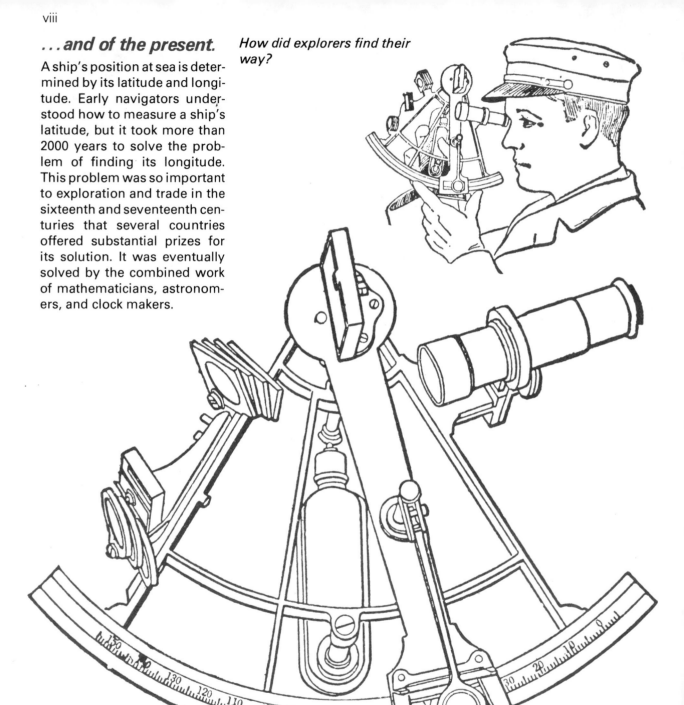

What is the best way to make a map of Earth?

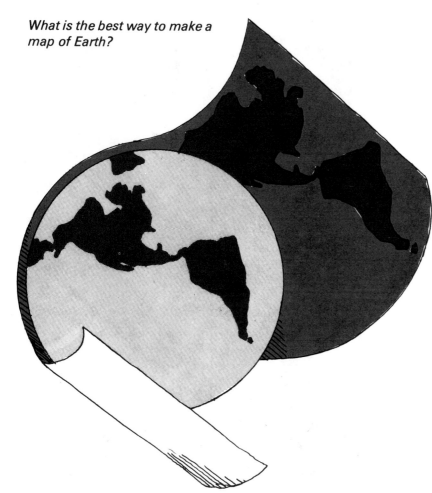

Can matter be totally destroyed?

Mathematics, and mathematicians,...

The surface of a sphere cannot be flattened out without cutting or stretching. Since Earth is a sphere, this means that every flat map cannot truly represent its surface. Using geometric principles, many different kinds of maps can be made. Each kind is designed for a specific purpose.

...will continue to play...

In 1905, Albert Einstein formulated his famous equation: $E = mc^2$ which expresses the amount of energy, E, released when a mass, m, is totally destroyed. In 1945, the atomic bomb verified the mass-energy equivalence which Einstein had predicted by his equation. Einstein and other scientists worried deeply when they saw their work being applied to weapons of destruction.

...an increasingly more important role...

To be able to guarantee his products, a manufacturer must have a quality-control program. Samples of the product are thoroughly tested and the results used to predict the total number of articles that are likely to be defective. In this way the manufacturer can make improvements in product design, and can also estimate the cost of providing replacements.

How can a manufacturer guarantee his products?

...in solving the complex problems...

A communications satellite, high in orbit above Earth, can relay a telephone or television signal to a receiver many thousands of kilometres away. Because it must always be above the same point on Earth, its orbit must be synchronized with Earth's rotation. Mathematical formulas have been developed to determine both the altitude of the satellite and its velocity.

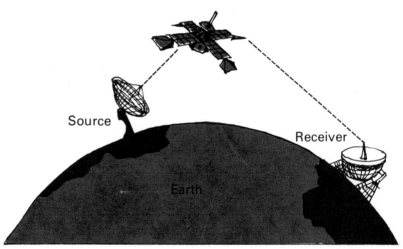

How high should a communications satellite be?

...of the future.

Genetics is the study of inherited characteristics in plants and animals. This important field of science depends heavily on a branch of mathematics called probability. Using genetics, scientists, for example, have succeeded in creating a new grain called triticale which combines the high yield of wheat with the disease-resistant strength of rye.

How can more food be produced to feed Earth's starving people?

1

From Arithmetic To Algebra

History was made in August 1978 when three Americans made the first successful attempt to cross the Atlantic Ocean by balloon. The height of the balloon was approximately 33 m and its diameter was approximately 19.6 m. About how much helium was needed to fill the balloon? (See *Example 3* of Section 1-2.)

1-1 Using Mathematics

People in all walks of life use mathematics every day, often without realizing it. Think about the mathematics that might be needed for these activities:

- Redecorating a bedroom
- Planning a vacation
- Giving a party
- Starting a vegetable garden
- Taking a survey of time spent on homework
- Going on a diet
- Being a disk jockey
- Raising tropical fish
- Making a dress
- Framing a picture

In the examples of this section, you will see mathematics used to solve a variety of problems. You will have the opportunity to solve several on your own. Try to solve them any way you can. Methods and techniques will be developed as you work through the book.

Example 1. The model of a locomotive is made to a scale of 1 cm to 1.6 m.

   a) About how long is the full-sized locomotive if the model is 9.7 cm long?

   b) About how wide is the model if the locomotive is 2.4 m wide?

Solution. a) Since 1 cm corresponds to 1.6 m, 9.7 cm corresponds to 9.7 × 1.6 m, or 15.52 m. The full-sized locomotive is about 15.5 m long.

   b) Since 1.6 m corresponds to 1 cm, 2.4 m corresponds to $\frac{2.4}{1.6}$ cm, or 1.5 cm.

The model is about 1.5 cm wide.

Example 2. The outside dimensions of the border of a rectangular picture are 45 cm by 35 cm. If the border is 5 cm wide, what are the dimensions of the picture?

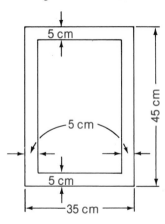

Solution. Picture's length: 45 cm $-$ 2 $\times$ 5 cm
                                     = 45 cm $-$ 10 cm, or 35 cm
                         width: 35 cm $-$ 2 $\times$ 5 cm
                                     = 35 cm $-$ 10 cm, or 25 cm
                 The picture is 35 cm long by 25 cm wide.

Example 3. Eva drove 100 km on 8.5 L of gasoline. At this rate, how much gasoline will she use on a trip of 320 km?

Solution. A trip of 100 km uses 8.5 L of gasoline.

A trip of 320 km will use $\frac{320}{100}$ $\times$ 8.5 L of gasoline.

Eva will use 3.2 $\times$ 8.5 L, or 27.2 L, of gasoline on a trip of 320 km.

Exercises 1 - 1

A   1. 25 people have equal shares in a $10 lottery ticket that wins $100 000.

  a) How much did each person pay for a share?

  b) What is each person's share of the prize?

2. A golf ball is travelling at 30 m/s. How far does it travel in 5 s.?

3. How long does it take a golf ball to travel 240 m if its average speed is 30 m/s?   40 m/s?

4. Dale drives 120 km on 6 L of gasoline.

  a) How far would she drive on 3 L?   5 L?

  b) How much gasoline would she need to drive 240 km? 140 km?

B   5. A model car, 12 cm long by 5 cm wide, is made to a scale of 1 : 32.

  a) What are the length and width of the full-sized car?

  b) If the wheelbase (the distance between the front and rear wheels) of the car is 288 cm, what is the wheelbase of the model?

6. Two towns are 7.5 cm apart on a map. If the scale of the map is 1 cm to 48 km, what is the actual distance between the two towns?

7. Twelve articles are bought at 3 for $1 and sold at 2 for 75¢. What was the profit?

8. A microscope magnifies an organism to 2000 times its normal size. What is the actual width of an organism that appears to be 3 cm wide when viewed under the microscope?

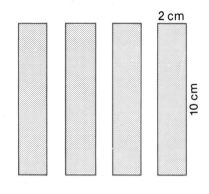

2 cm

10 cm

9. A concrete patio, 1 m wide, surrounds a rectangular ornamental pool 5 m long by 4 m wide.
   a) What is the outside perimeter of the patio?
   b) What is the area of the patio?

10. Four strips of colored cardboard, each measuring 2 cm by 10 cm, are used to make the border for a picture. If none of the cardboard is wasted, what are the dimensions of the picture?

11. Nine dozen baseballs, valued at $891, were stolen. What was the value of each baseball?

12. The winner of a 73-lap race at the Fuji International Speedway circuit in Japan averaged 207.83 km/h. He completed the race in 1 h 31 min 51.68 s.
    a) What distance did the winner drive?
    b) How long is a lap?

13. A slapshot travelled 10 m in 0.25 s. What speed is this in kilometres per hour?

14. Karen's car used 22.5 L of gasoline in travelling 180 km.
    a) What was the total cost of this gasoline at 39.7¢/L?
    b) What was the car's rate of fuel consumption in litres per 100 km?

Food	Energy Value
Chocolate milk shake	2200 kJ
Fried egg	460
Hamburger	1550
Glass of skim milk	350
Strawberry shortcake	1400

Activity	Energy Consumption
Walking	25 kJ/min
Cycling	35
Swimming	50
Running	80

15. The tables give the approximate energy value of some foods, in kilojoules (kJ), and the energy requirements of some activities. How long would it take to use up the energy from
    a) a fried egg by swimming?
    b) a glass of skim milk by walking?
    c) a piece of strawberry shortcake by running?
    d) a hamburger and a chocolate milkshake by cycling?

C 16. One person guessed that a jar in a store window contained 475 jelly beans. A second person guessed 455 beans, and a third person guessed 510. One guess was wrong by 20, another by 15, and another by 40. How many jelly beans were in the jar?

THE MATHEMATICAL MIND

PROBLEMS AND THEIR SOLVERS OF TIMES GONE BY

Some problems involving mathematics have required far more than correct arithmetic and the application of the right formula, they have required the discovery of a new principle and the invention of special mathematical techniques. Here are three of the world's greatest mathematicians and the kinds of problems they solved.

Why does a small rock sink and a large block of wood float?

ARCHIMEDES 287-212 B.C.

Archimedes is regarded as the greatest problem solver of the ancient world. Apparently his powers of concentration were so deep that, when working on a problem, he became unaware of his surroundings. The story is told that he was in his bathtub when he discovered the principle of buoyancy. So great was his excitement that he leaped from his tub and ran through the streets naked shouting: "Eureka! Eureka!" (I have found it! I have found it!)

What holds up the moon?

SIR ISAAC NEWTON 1642-1727 A.D.

Before Isaac Newton, no one understood the idea of gravity. No one knew why the moon travels in an orbit around the Earth instead of hurtling off into space or crashing to the Earth. By the time Newton was 25 years old, he had formulated the law of gravitation and cracked the problem – a problem that had baffled scientists from the beginning of time.

Is there a way to send messages around the world instantaneously?

CARL FRIEDRICH GAUSS 1777-1855 A.D.

The greatest pure mathematician of all time was Carl Friedrich Gauss. In addition to his computer-like skill in performing mental calculations, he possessed an almost superhuman ability to solve problems. Though his achievements were mainly in pure mathematics, he is also known for his invention of the telegraph. This invention was a giant step forward in communications and led the way to the development of the telephone and radio.

1-2 Substituting Into Formulas

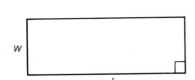

The scope of mathematics extends well beyond its everyday applications. In fact, many of the important ideas in science, economics, and engineering are expressed mathematically in formulas. For example, the area of a rectangle is found by multiplying the length by the width:

$$\text{Area} = \text{length} \times \text{width}$$

or written as a **formula**: $A = lw$

Letters such as A, l, and w, which represent numbers, are called **variables**.

Formulas are the result of combining variables and numbers using the basic operations of arithmetic.

Addition:	$x + 7, \quad y + z, \quad a + b + c$
Subtraction:	$n - 5, \quad 4 - p, \quad r - s$
Multiplication:	$7n, \quad bh, \quad 13gh$
Division:	$\dfrac{t}{5}, \quad \dfrac{6x}{y}$

The signs, $\times$ and $\div$, are usually omitted with variables.

Formulas are designed to solve particular kinds of problems. By substituting given values for all but one of the variables in a formula, we can calculate the value of the remaining unknown variable.

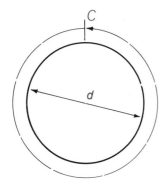

Example 1. The circumference, C, of a circle is given by the formula $C = \pi d$, where $\pi \doteq 3.14$ and d is the diameter. What is the circumference, to the nearest centimetre, of a bicycle wheel with a diameter of 70 cm?

Solution. When $d = 70$, $C \doteq (3.14)(70)$
$$\doteq 219.8$$
The circumference is 220 cm to the nearest centimetre.

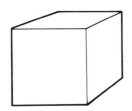

Example 2. The formula relating the number of faces *(F)*, vertices *(V)*, and edges *(E)* of any polyhedron is:
$$E = F + V - 2.$$
a) For a cube, $F = 6$ and $V = 8$. How many edges does a cube have?

b) A dodecahedron has 12 faces and 30 edges. How many vertices does it have?

Solution. a) If $F = 6$ and $V = 8$,
$$E = 6 + 8 - 2$$
$$= 12$$

A cube has 12 edges.

b) If $F = 12$ and $E = 30$,
$$30 = 12 + V - 2$$
$$30 = V + 10$$
$$V = 20$$

A dodecahedron has 20 vertices.

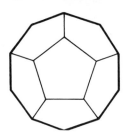

Example 3. The approximate dimensions of the balloon that made the first successful crossing of the Atlantic are: height 33 m, diameter 19.6 m. How much helium was needed to fill the balloon?

Solution. The balloon has the shape of a hemisphere on a cone. The formulas for the volumes, V, of these shapes are:

$$V(\text{sphere}) = \frac{4}{3}\pi r^3 \qquad V(\text{cone}) = \frac{1}{3}\pi r^2 h$$

where r is the radius of the sphere and base of the cone, and h is the height of the cone.

$$r = 19.6 \div 2 \qquad\qquad h = 33 - 9.8$$
$$= 9.8 \qquad\qquad\qquad = 23.2$$

Volume of hemisphere: Volume of cone:

$$V \doteq \frac{1}{2}\left(\frac{4}{3}\right)(3.14)(9.8)^3 \qquad V \doteq \frac{1}{3}(3.14)(9.8)^2(23.2)$$

$$\doteq 1970.2 \qquad\qquad\qquad \doteq 2332.1$$

$$V(\text{hemisphere}) + V(\text{cone}) \doteq 1970.2 + 2332.1$$
$$\doteq 4302.3$$

About 4300 m³ of helium were needed to fill the balloon.

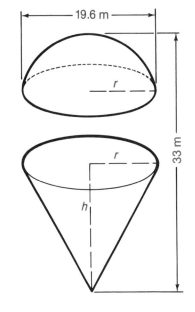

19.6 m

33 m

Example 4. A survey shows that grade 9 students of the same height, h cm, have an average mass of M kg, where mass and height are related by the formula: $M = \frac{3}{4}h - 72$. What is the average mass of students who are 150 cm tall?

Solution. If $h = 150$, then $M = \frac{3}{4}(150) - 72$
$$= 112.5 - 72$$
$$= 40.5$$

The average mass of students 150 cm tall is 40.5 kg.

Exercises 1 - 2

A   1. An old recipe book gives temperatures in degrees Fahrenheit.
They can be converted to degrees Celsius by the formula:
$$C = \tfrac{5}{9}(F - 32).$$
What are these temperatures in degrees Celsius?
a) 68°F   b) 176°F   c) 350°F   d) 32°F

2. In the formulas below, A is the area, P the perimeter, and C
the circumference. For each of the figures, substitute the
given values for the variables in the formulas and complete the
calculations.

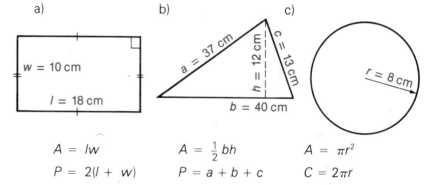

a)   b)   c)

$w = 10$ cm $a = 37$ cm $h = 12$ cm $c = 13$ cm $r = 8$ cm

$l = 18$ cm $b = 40$ cm

$A = lw$ $A = \tfrac{1}{2}bh$ $A = \pi r^2$

$P = 2(l + w)$ $P = a + b + c$ $C = 2\pi r$

3. Use the formulas given in Exercise 2 to find the area of the
shaded part of each diagram.

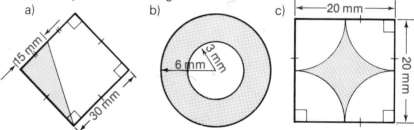

a)   b)   c) $\leftarrow$―20 mm―$\rightarrow$

15 mm   30 mm   6 mm   3 mm   20 mm

4. Typing speed, S, in words per minute, is calculated by the
formula $s = \dfrac{w - 10e}{5}$ where w is the number of words typed
in 5 min and e is the number of errors in 5 min. Calculate the
typing speeds for these numbers of words and errors in 5 min:
a) 300 words, 5 errors   b) 280 words, 2 errors
c) 420 words, 8 errors   d) 490 words, 7 errors

5. In September 1979, an East German family crossed the heavily
guarded frontier into West Germany in a homemade hot-air
balloon. This was the first escape by this means. The balloon
was a sphere approximately 22 m in diameter. Use the formula
in *Example 3* to calculate its volume.

6. The profit, P, in dollars, that a firm makes in the manufacture of windows and doors is given by the formula $P = 7w + 10d$, where d is the number of doors sold and w the number of windows sold. What is the firm's profit from these sales:

    a) 11 windows, 9 doors?   b) 29 windows, 18 doors?

    c) 16 windows, 8 doors?   d) 428 windows, 186 doors?

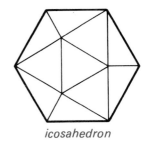

icosahedron

7. Use the formula of *Example 2* to answer the following:

    a) A regular tetrahedron has 4 faces and 6 edges. How many vertices does it have?

    b) A regular icosahedron has 12 vertices and 30 edges. How many faces does it have?

8. The rate at which a cricket chirps can be used to find an approximate value for the temperature. The relationship is $t = \frac{n}{7} + 4$, where t is the temperature in degrees Celsius, and n is the number of chirps in 1 min.

    a) What is the temperature if the number of chirps per minute is: i) 70?   ii) 98?   iii) 140?

    b) What should be the rate of chirping at 16°C?

9. The fare charged by a taxi company is an initial $0.80 plus $0.50/km or fraction of a kilometre. Written as a formula: $F = 0.80 + 0.50d$, where F is the fare in dollars and d is the distance in kilometres.

    a) Calculate the fare for a trip of
       i) 2.7 km;   ii) 13.5 km;   iii) 8.2 km.

    b) For what length of trip is the fare $1.80?

B 10. A ski club's net income, in dollars, I, is given by the formula:
$$I = 9A + 6C - 725n,$$
where A is the number of adults, C the number of children, and n the number of tows and lifts in operation.

    a) Find the net income when there are 4 tows and lifts operating and 425 adults and 210 children pay to use them.

    b) If there were 375 children, 3 tows and lifts operating, and a net income of $975, how many adults were there?

C 11. A regular polygon has equal sides and equal angles. The size of each angle, A, in degrees, is given by $A = 180 - \frac{360}{n}$, where n is the number of sides.

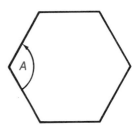

    a) Find the size of an angle of
       i) a decagon ($n = 10$);   ii) a dodecagon ($n = 12$).

    b) If a regular polygon has angles of 120°, how many sides does it have?

1-3 Introduction to Algebra

One use of the language of algebra is to express and solve problems of arithmetic. In order to use it for this purpose, we need to know the words of the language.

Consider the formula $A = 1 + x - 2z + 7xy$

These are **terms**.

$1 + x - 2z + 7xy$

These are **coefficients**.

$1 + 1x - 2z + 7xy$

These are **variables**.

$1 + x - 2z + 7xy$

This is an **expression**.

$1 + x - 2z + 7xy$

Example 1. Copy and complete:

Expression	Variables	Terms	Coefficients
$2m - 9n$			
$35x + 17y$			
$5a - 4b + 6c$			

Solution.

Expression	Variables	Terms	Coefficients
$2m - 9n$	m, n	$2m, 9n$	2, 9
$35x + 17y$	x, y	$35x, 17y$	35, 17
$5a - 4b + 6c$	a, b, c	$5a, 4b, 6c$	5, 4, 6

Example 2. Copy and complete:

a)

+	3	$\frac{2}{5}$	x	w	d
5	8	$5\frac{2}{5}$	$5+x$		
1				$1+w$	
7					

b)

×	3	$\frac{2}{5}$	x	w	$3d$
5	15	2	$5x$		
1				w	
7					

When the coefficient is 1, it is omitted.

Solution. a) b)

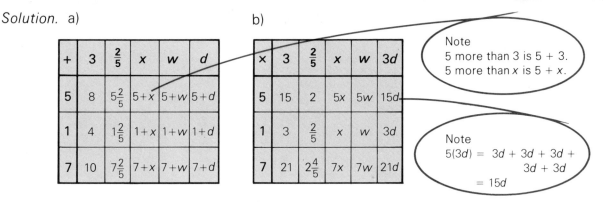

+	3	$\frac{2}{5}$	x	w	d
5	8	$5\frac{2}{5}$	5+x	5+w	5+d
1	4	$1\frac{2}{5}$	1+x	1+w	1+d
7	10	$7\frac{2}{5}$	7+x	7+w	7+d

×	3	$\frac{2}{5}$	x	w	3d
5	15	2	5x	5w	15d
1	3	$\frac{2}{5}$	x	w	3d
7	21	$2\frac{4}{5}$	7x	7w	21d

Note
5 more than 3 is 5 + 3.
5 more than x is 5 + x.

Note
$5(3d) = 3d + 3d + 3d + 3d + 3d$
$= 15d$

Example 3. Victoria Falls on the Zambesi River is twice as high as Niagara Falls. The world's highest waterfall is Angel Falls in Venezuela. It is nine times as high as Victoria Falls. If the height of Niagara Falls is x metres write an expression for the height of

a) Victoria Falls;   b) Angel Falls;

c) a waterfall three times as high as Niagara Falls.

Victoria Falls

Solution. a) If Niagara Falls is x metres high, Victoria Falls is 2x metres high.

b) If Victoria Falls is 2x metres high, Angel Falls is 9(2x) metres high, or 18x metres high.

c) A falls three times as high as Niagara Falls would be 3x metres high.

In the last example, we cannot find the height of Victoria Falls until we are given a value for x. Substituting a particular value for a variable in an expression is called *evaluating* the expression.

Example 4. Evaluate:

a) 5x + 9 when x = 6;

b) 3a − 7b when a = 9 and b = 2;

c) 2.6 m when m = 2.5.

Solution. a) When x = 6, $5x + 9 = 5(6) + 9$
$= 30 + 9$
$= 39$

b) When $a = 9$ and $b = 2$,
$$3a - 7b = 3(9) - 7(2)$$
$$= 27 - 14$$
$$= 13$$

c) When $m = 2.5$, $2.6\,m = (2.6)(2.5)$
$$= 6.5$$

Exercises   1 - 3

A 1. Copy and complete:

	Expression	Variables	Terms	Coefficients
a)	$6a - 2b$			
b)	$a - 2b + 9c$			
c)	$1.8C + 32$			
d)	$2\,\pi r$			

2. Copy and complete:

a)

+	3	7	q	$2b$	ab
4					
9					
x					
y					
xy					

b)

×	3	7	a	$2b$	ab
4					
9					
x					
y					
xy					

3. Students at the John Cabot Secondary School write their examinations in the gymnasium. How many students can write at one time if
   a) there are 11 rows and 32 desks in each row?
   b) there are 11 rows and d desks in each row?
   c) there are r rows and 32 desks in each row?
   d) there are r rows and d desks in each row?

4. At a track and field meet, points are awarded as follows: first place—5 points, second place—3 points, third place—1 point. How many points would a school be awarded for
   a) 4 firsts, 2 seconds, and 6 thirds?
   b) x firsts?   c) y seconds?   d) z thirds?
   e) x firsts, y seconds, and z thirds?

B 5. Evaluate:

a) $2x + 7$ for $x = 1, 3, 5, 8, 12$;

b) $28 - 5m$ for $m = 1, 2, 3, 4, 5$;

c) $9x - 4y$ for $x = 8$ and $y = 7$;

d) $8a - 19b$ for $a = 28$ and $b = 8$;

e) $8a - 4b - c$ for $a = 12, b = 18$, and $c = 3$;

6. Evaluate:

a) $2.3x + 0.7y$ for $x = 4$ and $y = 8$;

b) $0.27j - 3k$ for $j = 2.3$ and $k = 0.09$;

c) $3.7a - 2.1b$ for $a = 4.8$ and $b = 3.7$;

d) $5m - 9.2n$ for $m = 2.8$ and $n = 0.6$;

e) $8.3r - 1.27s + 0.6t$ for $r = 0.8, s = 5$, and $t = 0.5$.

7. Evaluate:

a) $\frac{3}{4}c + \frac{5}{7}d$ for $c = 12$ and $d = 14$;

b) $\frac{5}{6}m - \frac{2}{9}n$ for $m = \frac{2}{5}$ and $n = \frac{3}{8}$;

c) $\frac{2}{5}x + \frac{1}{3}y$ for $x = \frac{3}{4}$ and $y = \frac{4}{5}$;

d) $\frac{2}{5}p + \frac{2}{3}q$ for $p = \frac{1}{2}$ and $q = \frac{9}{22}$;

e) $\frac{3}{8}w + \frac{5}{6}y - \frac{3}{4}z$ for $w = \frac{1}{3}, y = \frac{8}{15}$, and $z = \frac{2}{3}$.

8. The cost, C, in dollars, of installing a steel-panel fence is
$$C = 7l + 15p + 80,$$
where l is the length of the fence, in metres, and p is the number of posts required. Find the total cost when

a) $l = 120$ m and $p = 41$; b) $l = 32$ m and $p = 12$;

c) the fence is 65 m long and requires 25 posts;

d) 85 posts are required for a 250 m fence.

9. The formula for the volume, V, of a cylinder is $V = \pi r^2 h$, where r is its radius and h its height. Find the volume of a cylinder whose radius is 4 cm and whose height is 15 cm. ($\pi \doteq 3.14$)

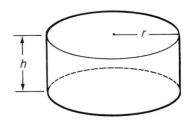

10. The area, A, of an ellipse is given by the formula $A = \pi ab$. Calculate the area when $a = 31$ cm and $b = 19$ cm.

11. The intelligence quotient (IQ) is a measure of a student's intellectual ability. The formula is $IQ = \frac{100\ m}{p}$, where m is the mental age and p the physical age. Calculate the IQ of a student who is

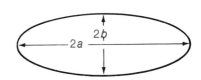

a) 14 years old and has a mental age of 16;

b) 15 years old and has a mental age of 13.

1 - 4   Like Terms

Terms that have exactly the same variable are called **like terms**.

x, $7x$, $0.8x$, and $24x$ are all like terms.

m^2, $8m^2$, $\frac{1}{2}m^2$, and $0.1m^2$ are all like terms.

$8pq$, $101pq$, and $10.5pq$ are all like terms.

x and x^2 are different variables, so $3x$ and $4x^2$ are not like terms. No two of the terms $7x$, $7x^2$, $7x^3$, $7m$, m^2, xz, or $9xy$ are like terms.

Like terms can be combined into a single term:

Since $3x + 5x$ means

$$x + x + x \quad + \quad x + x + x + x + x, \text{ or } 8x,$$

then $3x + 5x = 8x$.

> To combine like terms, add or subtract their coefficients.

Example 1. Simplify:

    a) $8x + 3x$;   b) $12a - a$

    c) $7z + z + 2z$;   d) $12y + 6 + 3y$.

Solution. a) $8x + 3x = 11x$ b) $12a - a = 11a$

    c) $7z + z + 2z = 10z$ d) $12y + 6 + 3y$

$$= 15y + 6$$

In (d), $15y$ and 6 cannot be combined because they are not like terms.

"Simplify" means combining like terms. An expression may have more than one kind of like term. The like terms should be grouped according to their kind and then the terms of each group combined separately.

Example 2. Simplify: $26x + 11y - 12x + 3x - 5y$.

Solution. $26x + 11y - 12x + 3x - 5y$

    $= 26x - 12x + 3x \quad + \quad 11y - 5y$

    $= 17x + 6y$

$17x$ and $6y$ are not like terms; they cannot be combined.

Exercises 1 - 4

A 1. Simplify:

a) $4a + 7a$ b) $19m - 6m$ c) $42x + 29x$

d) $14p - 5p$ e) $21g - 16g$ f) $12b + 37b$

g) $36p - 29p + 14p$ h) $6r + 47r - r$

i) $13w + w - 9w$ j) $18c + 49c - 26c + 3c$

2. Simplify:

a) $7x + 5x + 8y - 3y$ b) $18m - 7m + 6p + 11p$

c) $9a + 23b - 4a - 11b$ d) $52x + 31y - 31x - 2y$

e) $44u + 17v - 41u + 4v$ f) $7j + 13k - 5j - k$

g) $4s + 5t + 19t + 37s$ h) $28x + 15y - 19x - 11y$

i) $6a + 9b + 7c + 3a - 5b - c$

j) $14x + 17y - 5x + 11z - 6y - 2z$

3. Simplify where possible:

a) $4m + 5 - 3m$ b) $2c + d + 3c - d$

c) $5a + 3b + 5a$ d) $3x + 2y$

e) $8u + 3v + 11v + 7$ f) $5m + 4$

g) $7x + y - 2x$ h) $4a + 5b + 3c$

i) $15x + 3y - 9x + z - 3x$ j) $10p + 5 + 8q - 3p - 2$

4. Simplify:

a) $23a + 42b - 17a + 18b$

b) $12x + 10y - 6x - 6y + x$

c) $45m + 15n + 7 - 5m - 5n$

d) $32c + 10 - 15c + 4d - 3$

e) $23a + 7a + 13 - 2a$

f) $16x + 17y + x - y$

g) $2a + 3b - a + 4$

h) $48p + 16q + 3r - 18p - 3r$

i) $67x + 15 - 52x + y - x$

B 5. Evaluate:

a) $4a + 7a$, for $a = 3$ b) $19m - 6m$, for $m = 2.5$

c) $42x + 29x$, for $x = 7$ d) $14p - 5p$, for $p = \frac{2}{3}$

e) $23b + 17b$, for $b = 5$ f) $64k - 44k$, for $k = \frac{3}{4}$

g) $4x + 7x + 11x$, for $x = 4$

h) $16y + 29y - 15y$, for $y = \frac{1}{2}$

i) $84c - 59c + 5c$, for $c = 12$

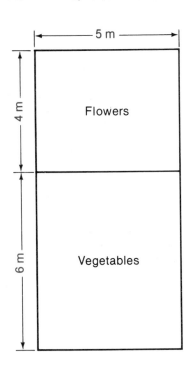

1-5 From Products to Sums and Differences

Mr. Ying grows flowers and vegetables in a rectangular garden with the dimensions shown. The total area of Mr. Ying's garden, in square metres, can be calculated in two ways:

1. Total area: width × length
$$= 5(4 + 6)$$

2. Total area:
   area with flowers + area with vegetables
$$= 5(4) + 5(6)$$

Since the area must be the same by either method:
$$5(4 + 6) = 5(4) + 5(6)$$

This is an example of the distributive law. It shows a product expanded into a sum.

If variables are used instead of numbers, a general statement for the distributive law is obtained.

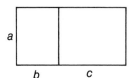

$$a(b + c) = ab + ac$$

Is $5(9 - 4) = 5(9) - 5(4)$?

$5(9 - 4) = 5 \times 5$, or 25; and $5(9) - 5(4) = 45 - 20$, or 25.

We see that we can also expand a product into a difference:

$$a(b - c) = ab - ac$$

Example 1. Use the distributive law to expand these products:

a) $5(9 + 4)$ b) $3(8 - 6)$

c) $6(x - 4)$ d) $3(4b + 8)$

Solution.

a) $5(9 + 4)$
   $= 5(9) + 5(4)$
   $= 45 + 20$
   $= 65$

b) $3(8 - 6)$
   $= 3(8) - 3(6)$
   $= 24 - 18$
   $= 6$

c) $6(x - 4)$
   $= 6x - 6(4)$
   $= 6x - 24$

d) $3(4b + 8)$
   $= 3(4b) + 3(8)$
   $= 12b + 24$

The distributive law can be extended to more than two terms.

Example 2. Expand: $4(2x + y - 3)$

Solution. $4(2x + y - 3) = 4(2x) + 4y - 4(3)$
$$= 8x + 4y - 12$$

Some expressions must be expanded before they can be simplified.

Example 3. Simplify: $12(3p + q) + 8(q + 2p)$

Solution.
$$12(3p + q) + 8(q + 2p) = 36p + 12q + 8q + 16p$$
$$= 36p + 16p + 12q + 8q$$
$$= 52p + 20q$$

Exercises 1-5

A 1. Expand:
   a) $3(m - 8)$ b) $18(x + 5)$ c) $11(p + 7)$
   d) $23(a - 9)$ e) $7(2p + 6)$ f) $4(a - b + 15)$
   g) $12(3m - 9n)$ h) $8(7a + b - 1)$ i) $6(2s + 11t - 5)$

   2. Simplify:
   a) $5(m + 3) + 63$ b) $18(2x + 4) - 27$
   c) $14 + 3(6x + 7)$ d) $96 + 7(3a - 12)$
   e) $17(3x + 5) - 2$ f) $7a + 3(2a - 9 - b)$
   g) $15e + 5(12 + e - 4f)$ h) $6t + 9(3t - 4) - 12t$
   i) $2(5x - 7) - 3x - x$ j) $8w + 6(3w + 5) - 19$
   k) $3(c + 4) + 2(2c - 3)$ l) $12t + 3(5 - 2t) - 7$
   m) $5m + 2 + 3(4m + 1) - 2m$
   n) $4(2c + 5d) + 2(3c - 7d)$
   o) $5(a + 3) + 2(a - 5) + (a - 1)$
   p) $4(2a + 5b + 3) + 3(a - 6b - 1)$
   q) $10(2k + 1) + 5(3k - 1) - k$

   3. Simplify:
   a) $3(2x + 5y) + 7(4x - 2y)$ b) $5(7x + 2y) + 3(x - 2y)$
   c) $5(3m + 6n) + 8(9m - 2n)$ d) $5(4a + 16b) + 2(17a - 29b)$
   e) $11(8k + 4l) + 3(2k - 13l)$ f) $3(p + 2q) + 7(2p + q)$
   g) $11(3r + 2s) + 7(2p + q)$ h) $6(4u + 7v) + 9(u - v)$
   i) $10(3a + 2b + c) + 5(a - b - c)$
   j) $8(12x + 5y + 4) + 3(2x - 4y + 2)$

B 4. Simplify:
   a) $0.5(2.8x + 1.2y) + 1.5(1.4x + 0.2y)$
   b) $1.4(3.5x + 6y) + 2.8(0.5x - 2.5y)$
   c) $2.6(1.5x + 5y) + 5.2(4.5x - 2y)$
   d) $3.8(4.5y + 2.5x) + 7.5(1.8x - 2.2y)$

1 - 6   Writing Sums and Differences as Products

The distributive law: $a(b + c) = ab + ac$

can be written in the form: $\mathbf{ab + ac = a(b + c)}$.

That is, it can be used to write certain sums and differences as products.

Example 1. Simplify: $27(63) + 27(37)$

Solution. Since both terms have the multiplier 27 in common, we can use the distributive law as follows:
$$27(63) + 27(37) = 27(63 + 37)$$
$$= 27(100)$$
$$= 2700$$

Example 2. Express $5x + 5y - 5z$ as a product.

Solution. Since each term has the coefficient 5, we can write:
$$5x + 5y - 5z = 5(x + y - z)$$

Example 3. Express $3a + 5ab$ as a product.

Solution. The variable, a, is common to both terms. Therefore, we may use the distributive law, and write:
$$3a + 5ab = a(3 + 5b)$$

Exercises   1 - 6

A   1. Express as products and simplify:

a) $5(31) + 5(19)$ b) $6(18) + 6(12)$

c) $8(11) + 8(9)$ d) $7(67) - 7(47)$

e) $4(53) - 4(23)$ f) $9(91) - 9(41)$

g) $11(0.5) + 11(1.5)$ h) $15(2.25) - 15(0.25)$

i) $50(0.6) - 50(0.5)$ j) $44(2.75) - 44(2.5)$

2. Express as products:

a) $3a + 3b$ b) $7x - 7y$ c) $2c + 2d$

d) $4m + 4$ e) $5p - 5$ f) $6q + 6$

g) $ab + ac$ h) $xy - xz$ i) $pq + p$

j) $5x + 3xy$ k) $7u + 12uv$ l) $9ac + 7bc$

m) $7st - 6t$ n) $abc + bcd$ o) $4w - 4wx$

3. Express as products:

a) $3bc + 3bd$ b) $4ay - 4az$

c) $10mn - 10n$ d) $4xy + 5y$

e) $6mn - 6m$ f) $7ab + b$

g) $3a + 3b + 3c$ h) $7u - 7v + 7$

i) $5p - 5q - 5r$ j) $2mx - 2nx + 2kx$

k) $5ac + 5ad - 5a$ l) $3xy - 3xz + 3x$

B   4. In one week, Sarah worked the following hours after school:
Monday—4 h,   Tuesday—3 h,   Wednesday—3.5 h,
Thursday—2.5 h,   Friday—5 h.
If she receives $2.25 per hour, how much did she earn that week?

5. Every night for a week, Loretta phoned Robert in Miami. Each call costs $0.82 for the first minute and $0.72 for every minute after that. If the lengths of her calls were 15 min, 17 min, 9 min, 21 min, 19 min, 15 min, and 7 min, what was the cost of her week's calls?

C   6. Two circles, whose circumferences are x and y, are touching. What is the distance between their centres?

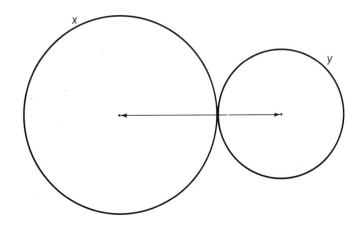

7. Express as products:

a) $6x + 2y$ b) $10a - 15b$

c) $12m + 20n$ d) $6x + 9y - 15$

e) $30a - 24b + 6$ f) $6m - 8n - 2$

g) $6ab + 15ac$ h) $15xy - 12xz$

i) $8am - 12m$ j) $6ax - 4bx + 10cx$

k) $14ax + 35ay - 7a$ l) $30rs - 20rt + 10r$

1-7   Order of Operations

Tim and Theresa each simplified this expression:

$$3 \times 5 + 4 \times 2.$$

Tim thought:	Theresa thought:
$3 \times 5 = 15$,	$3 \times 5 = 15$
$15 + 4 = 19$,	$4 \times 2 = 8$
$19 \times 2 = 38$	$15 + 8 = 23$

Only one answer can be right. Theresa's answer is the correct one.

By agreement, in any sequence of operations multiplication and division are performed first followed by addition and subtraction.

Example 1. Simplify: $9 + 3 \times 10 - 12 \div 4$

Solution. $9 + 3 \times 10 - 12 \div 4$

$$= 9 + 30 - 3$$
$$= 36$$

Frequently, parts of expressions are contained in parentheses, ( ), sometimes called round brackets. Operations inside the parentheses are performed first. If the parentheses are contained within square brackets, [ ], usually just called brackets, the operations within parentheses are performed first followed by the other operations within the brackets.

Example 2. Simplify: $3 + [5 + 2(10 - 8 \div 4) - 1] \div 5$

Solution. $3 + [5 + 2(10 - 2) - 1] \div 5$

$$= 3 + [5 + 2 \times 8 - 1] \div 5$$
$$= 3 + [5 + 16 - 1] \div 5$$
$$= 3 + 20 \div 5$$
$$= 3 + 4$$
$$= 7$$

A fraction bar is also a grouping symbol. It indicates that the numerator and denominator must each be simplified before carrying out the remaining operations.

Example 3. Simplify: $\dfrac{3}{4} \times \dfrac{7-4}{2+2} - \dfrac{3}{4 \times 2}$

Solution. $\dfrac{3}{4} \times \dfrac{7-4}{2+2} - \dfrac{3}{4 \times 2}$

$= \dfrac{3}{4} \times \dfrac{3}{4} - \dfrac{3}{8}$

$= \dfrac{9}{16} - \dfrac{3}{8}$

$= \dfrac{9}{16} - \dfrac{6}{16}$

$= \dfrac{3}{16}$

Algebraic expressions obey the same rules as numerical expressions.

Example 4. Simplify: $7\left[23x - \dfrac{8x - 2x}{3} - 3(5x - x)\right]$

Solution. $7\left[23x - \dfrac{8x - 2x}{3} - 3(5x - x)\right]$

$= 7\left[23x - \dfrac{6x}{3} - 3(4x)\right]$

$= 7\left[23x - 2x - 12x\right]$

$= 7\left[9x\right]$

$= 63x$

Exercises 1-7

A   1. Simplify:

    a) i) $20 - 8 + 2$ ii) $20 - (8 + 2)$

    b) i) $3 + 5 \times 7$ ii) $(3 + 5) \times 7$

    c) i) $6 \times 9 - 4$ ii) $6 \times (9 - 4)$

    d) i) $12 \div 3 + 1$ ii) $12 \div (3 + 1)$

    e) i) $27 - 16 - 4$ ii) $27 - (16 - 4)$

B   2. Simplify:

    a) $20 - 2 \times 3 + 5$ b) $(20 - 2) \times 3 + 5$

    c) $20 - 2 \times (3 + 5)$ d) $(20 - 2) \times (3 + 5)$

    e) $20 - (2 \times 3 + 5)$ f) $(20 - 2 \times 3) + 5$

   3. Use parentheses with the expression $3 + 5 \times 4 - 2$ so that it simplifies to: a) 16;   b) 21;   c) 30.

4. Use parentheses with the following expressions so that they simplify to the answers given:

   a) $1 + 3 \times 5 + 7$ answer: 27

   b) $4 + 4 + 4 \times 4$ answer: 48

   c) $2 \times 4 + 6 + 8$ answer: 28

   d) $5 + 5 \times 5 + 5$ answer: 100

   e) $48 \div 8 - 2 \times 3$ answer: 24

5. Simplify:

   a) $4 + 5 \times 3 - 6$

   b) $17 + 15 \div 3(4 + 1)$

   c) $(5 + 3) \times 6 - 12 \div (5 - 1)$

   d) $8 + 5(7 - 3) \div 2$

   e) $9 + \dfrac{4 + 16}{5 \times 10} - 3$

   f) $60 - \dfrac{9(12 \div 2 + 6)}{5 \times 4 + 4} - 2 \times 10$

   g) $15 \div 3 + 4(9 + 7) - 21$

   h) $(1 - \frac{1}{2})(1 - \frac{1}{3})(1 - \frac{1}{4})(1 - \frac{1}{5})$

   i) $24 - (8 + 2) \div 2 + 10(2 + 1)$

   j) $39 - 2[3 + 2 \times (10 - 3)] - 5$

   k) $7 \times 5 - 5[7 - 6 \div (2 + 1)] + 3 \times 4$

   l) $\dfrac{20 \div 5 + 3}{6 \div 2 + 6} + \dfrac{(6 - 4) \div 2}{8 \div (4 \times 2)}$

6. Using all the digits in the year 1984, the operations $+$, $-$, $\times$, $\div$, and parentheses, if required, form expressions for all the natural numbers from 2 to 12. Example: $(9 + 8 - 1) \div 4 = 4$

7. Evaluate:

   a) $9a - 4$ and $9(a - 4)$ for $a = 7$;

   b) $3m + 21$ and $3(m + 21)$ for $m = 13$;

   c) $6p - 9$ and $6(p - 9)$ for $p = 27$;

   d) $5x - 4$ and $5(x - 4)$ for $x = 12$;

   e) $11s + 3$ and $11(s + 3)$ for $s = 14$.

8. Evaluate:

   a) $7(m + n) - 4$ for $m = 8, n = 5$;

   b) $3(p - q) + 19$ for $p = 26, q = 14$;

   c) $9(a - b) - 10$ for $a = 17, b = 8$;

   d) $6(r + s) + 37$ for $r = 29, s = 16$;

   e) $13(u - v) - 43$ for $u = 9, v = 3$.

9. Simplify:

a) $12x - (5x + 3x)$ b) $81y - (17y - 9y)$

c) $(24a - 9a) - (17a - 7a)$ d) $(8m - 2m) - (42m - 39m)$

e) $(63d - 29d) - (16d + 11d)$ f) $\frac{1}{2}(5z - 3z) + \frac{1}{4}(6z - 2z)$

10. Simplify:

a) $3[14x - (5x + 3x)]$ b) $8[29a - (14a - 5a)]$

c) $2[42y - (25y - 14y) - 4(3y + y)]$

d) $6[5x + 3(12x - 7x) - 2(12x - 4x)]$

11. Simplify:

a) $14a + 7b - (9a + 3a) - 2b$

b) $16m - (3m + 4m) + 19n - 6n$

c) $45s + 27t - 29s - (15t + 6t)$

d) $18x + 38y - (7x - 4x) - (22y - 8y)$

e) $53p - (19p + 6p) + 26q - (11q + 9q)$

12. Simplify:

a) $4x + \dfrac{15x - 6x}{3}$ b) $5a + \dfrac{12a + 8a}{5} - 3a$

c) $\dfrac{2m + 14m}{4} - \dfrac{16m + 4m}{10}$ d) $\dfrac{4(5m - 2m)}{2} + \dfrac{3(7m - m)}{9}$

e) $\dfrac{5(9s - 5s)}{2} - \dfrac{2(6s + 3s)}{6}$ f) $\dfrac{12(14y - 2y)}{144} - \dfrac{7(2y + 5y)}{49}$

1-8 Equations

A mathematical sentence which uses an "equals" sign, =, to relate two expressions is called an **equation**. An equation contains at least one variable. Examples of equations are:

$$56 - n = 19 \quad \text{and} \quad 7d = 28.$$

For any equation like those above, there is exactly one value of the variable that will make both sides the same. Finding that value is called **solving** the equation.

Example 1. Solve: a) $56 - n = 19$;   b) $7d = 28$

Solution. a) $56 - n = 19$
56 less *some number* is 19.
Then $n = 37$

b) $7d = 28$
7 times *some number* is 28.
Then $d = 4$

The equations in *Example 1* were solved by *inspection.* Knowledge of the basic number facts and the operations of arithmetic quickly suggested the solutions. Only very simple equations can be solved by inspection. In less simple cases, *systematic trial* can be used.

Example 2. Solve: $54 - 7y = 26$

Solution. $54 - 7y = 26$

If $y = 5$,
$54 - 7y = 54 - 7(5)$
$= 54 - 35$
$= 19$
19 is too small.

If $y = 2$,
$54 - 7y = 54 - 7(2)$
$= 54 - 14$
$= 40$
40 is too large.

Since 26 lies between 19 and 40, the solution lies between 5 and 2.
If $y = 4$, $\quad 54 - 7y = 54 - 7(4)$
$= 54 - 28$
$= 26$
The solution is $y = 4$.

When solving a problem by a formula, values must be given for all but one of the variables in the formula. The resulting equation can sometimes be solved by inspection, and often by systematic trial.

Example 3. The area, A, of a trapezoid is given by the formula $A = \frac{1}{2}h(a + b)$, where a and b are the lengths of the parallel sides and h is the distance between them. If A is 198 cm², h is 12 cm, and a is 15 cm, calculate the value of b.

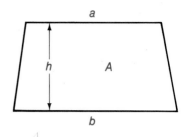

Solution. Substitute the given values in the formula:

$$198 = \frac{1}{2}(12)(15 + b)$$
$$= 6(15 + b)$$

If b = 15 cm,   If b = 20 cm,
$6(15 + b)$ $6(15 + b)$
$= 6(15 + 15)$ $= 6(15 + 20)$
$= 6(30)$ $= 6(35)$
$= 180$ $= 210$

180 is too small.   210 is too large.
Since 198 lies between 180 and 210, the solution lies between 15 and 20.
If b = 18 cm, $6(15 + b) = 6(15 + 18)$
$$= 6(33)$$
$$= 198$$

The value of b is 18 cm.

Exercises 1 - 8

A   1. Solve:

     a) $x + 17 = 32$ b) $29 - x = 12$ c) $x - 7 = 27$
     d) $x + 26 = 61$ e) $43 + x = 79$ f) $11 - x = 11$

   2. Solve:

     a) $4z = 24$ b) $7s = 63$ c) $12q = 132$
     d) $\frac{1}{3}y = 12$ e) $8v = 4$ f) $0.5w = 25$

   3. Solve:

     a) $m - 3\frac{1}{2} = 5$ b) $8\frac{2}{3} + x = 12$ c) $t + 2\frac{3}{8} = 5\frac{3}{8}$
     d) $12\frac{7}{10} - q = 7$ e) $w - 3\frac{2}{3} = 5$ f) $1.3 + z = 3.3$

   4. Solve:

     a) $3 + 2n = 11$ b) $7 + 3m = 13$ c) $7x - 5 = 30$
     d) $4c - 1 = 23$ e) $24 - 3y = 15$ f) $9k - 27 = 36$

5. Solve.
   a) $(a + 2)(3 + 4) = 49$ b) $(b - 11)(5 + 1) = 60$
   c) $(x - 3)(5 + 1) = 24$ d) $(y - 2)(6 + 2) = 24$
   e) $(2x + 1)(3 + 2) = 25$ f) $(5w + 5)(5 + 5) = 100$

6. Solve:
   a) $2x + 7 = 17$ b) $28 - 5m = 18$ c) $6a - 4 = 20$
   d) $9 + 3y = 57$ e) $8s - 7 = 153$ f) $11t + 19 = 140$

7. Solve:
   a) $3(x - 4) = 15$ b) $7(m + 2) = 42$ c) $4(a - 7) = 48$
   d) $8(2p + 3) = 56$ e) $12(y - 9) = 96$ f) $11(11y - 1) = 110$

8. Solve:
   a) $6(m + 2) = 42$ b) $8(a - 4) = 48$ c) $3(x + 7) = 27$
   d) $7(y + 3) = 84$ e) $5(2s - 1) = 35$ f) $4(5t + 4) = 76$

9. Solve:
   a) $2 + 3(b - 1) = 14$ b) $5(c + 7) - 4 = 51$
   c) $8(t - 3) - 11 = 5$ d) $15 + 4(c + 1) = 39$
   e) $9(2s - 1) - 5 = 40$ f) $16 + 3(3d - 9) = 43$

10. Some children receive an allowance, A, in dollars, according to the formula $A = 0.25t - 1$, where t is the child's age.
    a) Calculate the allowance for a child who is
       i) 10 years old;   ii) 15 years old.
    b) How old is a child who receives
       i) $2.00?   ii) $3.00?

B   11. A rectangle is 20 cm long. Determine its width if
    a) the area is 300 cm²;   b) the perimeter is 66 cm.

12. The number of hours of sleep, n, that a person under 21 years of age needs is given by the formula $n = 17 - \frac{1}{2}a$, where a is the person's age in years.
    a) How much sleep does a 10-year-old need?
    b) How old is a person who needs 10 h sleep?

13. A trapezoid has an area of 256 cm². Use the formula in *Example 3* to find
    a) the distance between the parallel sides if their lengths are 9 cm and 23 cm;
    b) the length of the other parallel side if the length of one side is 10 cm and the distance between them is 32 cm.

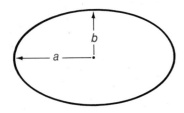

14. The area, A, of an ellipse is given by the formula $A = \pi ab$. If $A = 785$ mm², $a = 25$ mm, and $\pi = 3.14$, find b.

15. To determine how far away you are from the centre of a storm, count the number of seconds, *t*, between a flash of lightning and the sound of thunder. Substitute your value for *t* in the formula $d = \frac{8t}{25}$ in order to find *d*, the distance in kilometres.

a) How far away is a storm when
   i) *t* = 5 s?   ii) *t* = 10 s?   iii) *t* = 3.5 s?

b) Find the value for *t* when the storm is
   i) 8 km away;   ii) 6 km away.

16. The volume, *V*, of a cone is given by the formula $V = \frac{1}{3}\pi r^2 h$, where *r* is the radius of the base and *h* is the height. Find the approximate height of a cone that has a volume of 1769 cm³ and a base with a radius of 13 cm. ($\pi \doteq 3.14$)

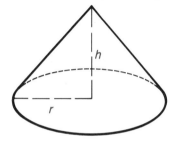

C 17. Typing speed, *S*, in words per minute, is calculated by the formula $S = \frac{w - 10e}{5}$, where *w* is the number of words typed in 5 min and *e* the number of errors made in the same period.

a) How many words must be typed in 5 min if, when 5 errors are made, the typing speed is 40 words/min?

b) How many errors are made for the typing speed to be 30 words/min when 180 words are typed in 5 min?

18. Weekly mathematics tests have 15 questions. Each question has either one, two, or three answers.

a) If, on one test, 9 questions have one answer, 4 questions have two answers, and 2 questions have three answers, how many answers are there altogether?

b) If a test has a total of 29 answers, 6 questions having one answer and 5 questions having three answers, how many questions have two answers?

c) One test has 8 questions with only one answer and there are a total of 24 answers. How many questions have two answers and how many have three?

Review Exercises

1. Jim's car used 6.0 L of gasoline on a trip of 75 km. How much gasoline will it use on a trip of 125 km?

2. A rectangular kitchen sink measures 35 cm by 42 cm and is 16 cm deep. If the drain empties it at the rate of 24 L/min, how long will it take to empty when it is three-quarters full?

3. A hilltop is 20.8 km from an airport runway. How far apart will they appear on a map drawn to a scale of 1 cm : 16 km?

4. The formula for the area, A, of a circle is $A = \pi r^2$, where r is the radius and $\pi \doteq 3.14$. What is the area of
   a) the face of a watch with radius 15 mm?
   b) the face of a clock with radius 10.5 cm?

5. The annual interest rate, r, paid on an instalment loan which is to be repaid in 16 monthly payments is given by $r = \dfrac{24c}{15A} \times 100\%$, where c is the total interest charged and A is the amount borrowed. Find r if $c = \$125$ and $A = \$1000$.

6. Evaluate:
   a) $9 \times 3y$ for $y = 1, 4, 9, 16$
   b) $8s - 7$ for $s = 2, \ 5.4, \ 3\frac{1}{4}$
   c) $5p + 8q$ for $p = 13, q = 6$
   d) $6s - t$ for $s = 1.7, t = 2.9$
   e) $14u + v - 9w$ for $u = 6, v = 29, w = 12$
   f) $2\frac{1}{4}x - 3y$ for $x = \frac{2}{3}, y = \frac{3}{8}$
   g) $7y + 9y - 4y$ for $y = 2\frac{1}{4}$
   h) $21s - 6s + 4t - t$ for $s = 0.4, t = 1.2$
   i) $3.2x + 4.1y - z$ for $x = 0.3, y = 1.1, z = 2.4$
   j) $\frac{3}{8}a - \frac{2}{5}b + \frac{5}{2}c$ for $a = \frac{5}{3}, b = \frac{3}{8}, c = \frac{4}{25}$

7. Simplify:
   a) $23x - 11x$ b) $4m + 13m$ c) $15x - 9x + 3x$
   d) $12a + 4b - 5a + 3b$ e) $9x + 4y - x - 2y$
   f) $14c + d - 11c + 3d$ g) $2(3x + 5y) - 4x$
   h) $2(5a + 7) + 17$ i) $5(2m + 7n) + 3(m - 4n)$
   j) $3(2x + 4y) + (3x - y)$ k) $7(r + 3s) + 2(2r - 9s)$

8. Express as products:
   a) $14x + 3xy$ b) $3uv - 5v$ c) $7ab - 14a$
   d) $2abc - 6bcd$ e) $13c + 52d$ f) $68 - 17a$
   g) $\frac{1}{4}x - \frac{3}{4}y$ h) $2\pi r + \pi rb$ i) $abc + 2bcd - 4abd$

9. Lois buys five items costing \$16.50, \$23.85, \$7.65, \$11.95, and \$2.35. If the sales tax on each item is 6%, how much tax does she pay?

10. Simplify:
    a) $3 + 4 \times 6 - 2$ b) $8(4 + 6 \div 2) - 5$
    c) $6 \times 9 - 3 \times 5$ d) $6 + 8 \div 2 + 5 \times 3$
    e) $18 - (2 + 7) \div 3 - 6$
    f) $6 + 2(4 + 8) - 5$
    g) $7 + \frac{5 - 3}{4 + 6} - 3$
    h) $\frac{6 + 4}{3 \times 5} - \frac{2 \times 3}{7 + 5}$
    i) $4 - \frac{6 - 3}{2 \times 4} + \frac{2}{3} \times \frac{3}{4}$
    j) $20 \div 5 + 3(7 - 2)$
    k) $6 + [2 + 4(7 - 5) - 6] - 9$
    l) $16 \div 2 \times 5 - 3 \times 13$

11. Simplify:
    a) $7[7a - (3a + 2a)]$ b) $9[11b - (8b - b)]$
    c) $11[36x - (27x - 15x) - 3(4x + x)]$
    d) $4[(1 - 4z) + 5(3 + 2z) + 7(z + 2)]$

12. Simplify:
    a) $x + \frac{3x + 9x}{6}$ b) $19w - \frac{13w + 7w}{4} - 13w$
    c) $\frac{17m - 2m}{5} - \frac{8m - 2m}{3}$ d) $\frac{3(4a - 2a)}{2} + \frac{7(3a + 7a)}{10}$
    e) $3\left[y - \frac{y - 9y}{8} + 3y\right]$ f) $11\left[\frac{11x - 2x}{3} - \frac{4(2x + 5x)}{14}\right]$

13. Solve:
    a) $3x + 7 = 19$ b) $5 + 2x = 17$ c) $2\frac{1}{2} + x = 7\frac{1}{2}$
    d) $(x + 2)(4 + 5) = 45$ e) $(m - 3)(2 + 4) = 36$
    f) $4(3z + 2) = 32$ g) $2(x - 3) = 26$
    h) $5(t - 2) - 7 = 38$ i) $6(2x - 8) + 1 = 25$

14. A rectangle is 35 cm long. Determine its width if its perimeter is
    a) 80 cm;   b) 100 cm;   c) 115 cm;   d) 138 cm.

Decide What Information Is Needed.

PROBLEM SOLVING STRATEGY

Too Much Information

Decide what information is needed to solve the problem. Extract that information from the given data and ignore the rest.

Example 1. A bus operates between an airport and the city centre 25 km away. It makes 10 round trips per day carrying an average of 42 passengers per trip. The fare each way is $2.50. What are the receipts from one day's operation?

Solution. Information needed:
  i) cost per trip per passenger
  ii) average number of passengers
  iii) number of trips

Data not needed: the length of a trip
One day's receipts: (i) × (ii) × (iii)
$2.50 × 42 × (10)(2) = $2100
The total receipts for one day are $2100.

Too Little Information

Sometimes missing data can be found in reference books.

Example 2. A news item stated that Canadians owed a total of $35 760 000 000 on consumer goods they had bought. What does this debt amount to per person?

Solution. The information missing is the population of Canada. This can be found from such reference books as *Quick Canadian Facts*, or from Statistics Canada. In mid 1979 the population was 23 600 000. Thus, the debt per person is
$\frac{\$35\ 760\ 000\ 000}{23\ 600\ 000}$, or $1515.
The debt per person is approximately $1500.

Sometimes when data is missing, a useful approximation can be made by making reasonable assumptions and estimates.

Example 3. How many times will your heart beat in your lifetime?

Solution. Two items of information are missing:
   i) how long you will live;
   ii) the average rate of your heartbeat.
The average human lifespan is usually taken as 70 years.
Heartbeat (relaxing): 72 beats/min;
                    (exercising): 200 beats/min
An average of 85 beats/min is a reasonable estimate.

70 years = 70 × 365 d
         = 70 × 365 × 24 h
         = 70 × 365 × 24 × 60 min
         = 70 × 365 × 24 × 60 × 85 heartbeats
         = 3 127 320 000 heartbeats

Your heart will beat about 3 billion times in your lifetime.

Exercises

1. If you save $5.00 per week, how long will it take you to save enough to buy a 10-speed bicycle?

2. Janet used 27.5 L of gasoline in driving 225 km. The cost of the gasoline was $8.50. What distance was she driving per litre of gasoline?

3. The diameters of Mars, Venus, and Mercury, as percents of Earth's diameter, are:
   Mars 53%,   Venus 96%,   Jupiter 1120%.
   Determine the diameters of Mars, Venus, and Jupiter.

Cable car

4. A cable car carrying tourists 2.5 km up a mountain makes 25 round trips each day. The car carries an average of 15 passengers per trip for which they each pay a round-trip fare of $3.00.
   a) What do the fares total each week?
   b) How far does the cable car travel in one week?

5. Keith is practising his tennis strokes by hitting the ball against a wall. How long will it take him to make 10 000 strokes?

6. The world's largest swimming pool is in Casablanca, Morocco. It is 480 m long by 75 m wide.
   a) About how many cubic metres of water does the pool contain?
   b) About how long would it take to fill the pool?

Mathematics Around Us

Niagara Falls is Moving!

The flow of the Niagara River is about 5700 m³/s. This great volume of water causes erosion at Niagara Falls so that the Falls moved upstream about 264 m between 1700 and 1900. Hydro-electric power plants, requiring diversion of some of the water around the Falls, were opened in 1905, 1922, 1954, and 1960, so that nowadays as much as 75% of the water may go round, not over, the Falls. This has halved the rate of erosion.

Questions

1. About how far did Niagara move upstream in 1800?

2. About how far does Niagara Falls move upstream each year now?

3. a) How far has it moved since you were born?

   b) How far will it move in your lifetime?

4. How long would it take to move the length of your classroom?

5. Use the map to answer the following:

   a) When will Niagara Falls reach the Three Sister Islands?

   b) When was Niagara Falls at the location of the Rainbow Bridge?

6. What important assumption did you make in answering the above questions?

2 Integers

Ste. Agathe
Altitude: 2350 m

Valleville
Altitude: 350 m

It is 8°C in Valleville.

The temperature decreases 6.5°C for every 1000 m increase in altitude.

If there is precipitation at the village of Ste. Agathe, will it be in the form of rain or snow?

(See *Example 6* in Section 2-3.)

2-1 Using Integers

Numbers such as 1, 2, 3, ..., sometimes written +1, +2, +3, ..., are called *positive* whole numbers. Numbers such as −1, −2, −3, ... are called *negative* whole numbers. The positive whole numbers, the negative whole numbers, and zero make up the set of **integers**.

$$I = \{..., -3, -2, -1, 0, 1, 2, 3, ...\}$$

There are many opposites in everyday life—up-down, left-right, debt-credit—to name a few. When they involve numbers, integers are used to describe them.

City	Temperature °C	
	Night Low	Yesterday High
Charlottetown	1	3
Chicago	− 5	0
Edmonton	−17	−15
Fredericton	− 4	− 4
Halifax	1	1
Miami	15	22
New York	− 1	2
Quebec City	−14	−12
Regina	−13	−12
St. John's	− 1	7
Toronto	− 5	0
Victoria	0	7
Winnipeg	−17	− 3

On a thermometer, 0° represents the freezing point of water on the Celsius scale. The opposite of *above freezing* is *below freezing*.

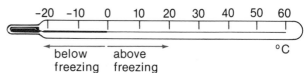

As the diagram of the thermometer shows, positive integers indicate above freezing, negative integers below freezing. From the table we see that, of the thirteen cities listed, only Charlottetown, Halifax, and Miami were above freezing during the night. Edmonton, Fredericton, Quebec City, Regina, and Winnipeg stayed below freezing during the day.

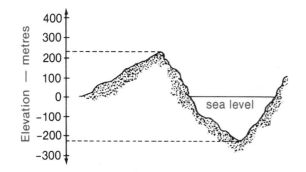

In land elevations, positive integers indicate *above sea level*, negative integers *below sea level*.

The sectional drawing shows a high point of +225 and a low point of −225.

+225 means 225 m above sea level;
−225 means 225 m below sea level.

+225 and −225 are called **opposites**.

Integers can be represented on the number line.

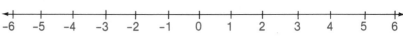

Any integer on the number line is *greater than* all integers to its *left* and *less than* all integers to its *right*.

We say that -2 is greater than -5, and write: $-2 > -5$.

We also say -5 is less than -2, and write: $-5 < -2$.

Example 1. Compare: -6 and 2; -5 and -1; 4 and -3; 0 and -5.

Solution.

Think	Say	Write
-6 is to the left of 2.	-6 is less than 2.	$-6 < 2$
-5 is to the left of -1.	-5 is less than -1.	$-5 < -1$
4 is to the right of -3.	4 is greater than -3.	$4 > -3$
0 is to the right of -5.	0 is greater than -5.	$0 > -5$

Example 2. Arrange the integers -5, 6, -3, -1, 0, 4, -2 in order

a) from least to greatest;

b) from greatest to least.

Solution. Circle the given integers on the number line.

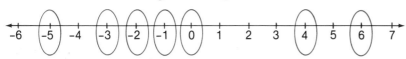

a) The given integers are ordered from least to greatest when the circled integers are read from left to right.

b) The given integers are ordered from greatest to least when the circled integers are read from right to left.

Exercises 2 -1

A   1. Write using integers:

a) a gain of $9   b) a loss of $21

c) 80° above freezing   d) 20° below freezing

e) a profit of $50   f) a loss of $75

g) a debt of $81   h) a depth of 12 000 m

i) 15 s to blast-off   j) an altitude of 3000 m

2. If $+100$ represents a gain in altitude of 100 m, what would the following integers represent?

a) -300 b) $+25$ c) -100 d) 2

3   If −5 represents a debt of $5, what would the following integers represent?

a) −12   b) +7   c) 15   d) −53

4. State the opposite of

a) a gain of $10;   b) an altitude loss of 500 m;

c) a 3 kg loss of mass;   d) a temperature of −14°C;

e) an elevation of 100 m above sea level;

f) +18;   g) −11;

h) the opposite of −7;   i) the opposite of 5.

5. State the greater or greatest integer:

a) −3, 2   b) 5, −6   c) −4, −1   d) −3, 0

e) 2, −5, 3   f) −9, −1, −4

g) −11, 10, 2, −14, 5   h) −8, −9, 0, −2, −6

6. State the lesser or least integer:

a) 1, −2   b) −6, −3   c) −1, 4   d) 0, −2

e) 2, −9, −3   f) −5, 1, −1

g) −4, 7, −10, 1, −5, 3   h) 3, −2, 15, −18, 7, 0

7. Arrange in order from least to greatest:

a) 3, −1, 5, −4   b) −2, 8, −10, 5

c) −1, 4, −8, −2, 5, 0   d) 9, −3, 2, −8, −1, 6

8. Arrange in order from greatest to least:

a) −2, 7, 1, −4   b) −3, 0, 2, −1

c) 5, −8, −2, 8, 0, 3   d) −10, 10, −7, 5, −8, −1

B   9. Which integer is

a) 3 less than 1?   b) 2 more than −1?

c) 6 more than −4?   d) 8 less than 5?

e) 7 more than the opposite of 3?

f) 5 less than the opposite of −1?

g) 10 more than the opposite of 5?

10. Which integer is

a) 6 less than 2?   b) 4 more than −9?

c) 3 more than 0?   d) 5 less than −2?

e) 2 more than the opposite of −3?

f) 1 less than the opposite of −1?

g) 9 less than the opposite of 2?

2 - 2   Addition of Integers

The number line is very useful in showing the addition of two
or more integers. Study these examples.

Example 1. Simplify: a) $(+1) + (+3)$;   b) $(-1) + (-3)$;
                           c) $(+1) + (-3)$;   d) $(-1) + (+3)$.

Solution.   a) $(+1) + (+3)$
              Start at $+1$.   Move 3 to the right.

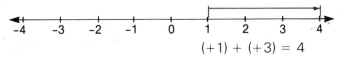

$(+1) + (+3) = 4$

b) $(-1) + (-3)$
    Start at -1.   Move 3 to the left.

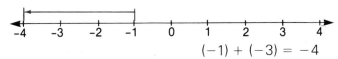

$(-1) + (-3) = -4$

c) $(+1) + (-3)$
    Start at $+1$.   Move 3 to the left.

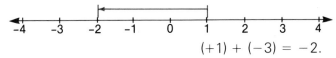

$(+1) + (-3) = -2.$

d) $(-1) + (+3)$
    Start at -1.   Move 3 to the right.

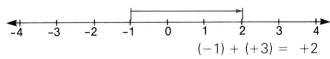

$(-1) + (+3) = +2$

> Addition of integers can be shown by moves on the number
> line. Start at the first integer. Then, move to the *right* for
> *positive* integers and move to the *left* for *negative* integers.

Example 2. Simplify: $(+5) + (-7) + (+3) + (-4)$.

Solution. Start at $+5$.

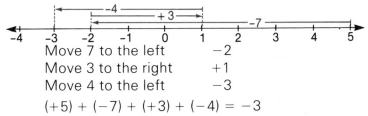

Move 7 to the left -2
Move 3 to the right $+1$
Move 4 to the left -3

$(+5) + (-7) + (+3) + (-4) = -3$

When adding several integers, it is often helpful to add the integers with the same sign first and then use the number line to obtain the final sum.

Example 3. Simplify: $(-25) + (+16) + (-11) + (-28) + (+34)$

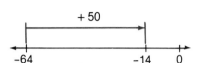

Solution. $(-25) + (+16) + (-11) + (-28) + (+34)$
$= (-25) + (-11) + (-28)$ $+$ $(+16) + (+34)$
$= (-64) + (+50)$
$= -14$

Example 4. An elevator is at the 14th floor. It goes down 8 floors, then down 5 more floors, then up 4 floors, then down 1 floor. At which floor is the elevator now?

Solution. $(+14) + (-8) + (-5) + (+4) + (-1)$
$= (+14) + (+4)$ $+$ $(-8) + (-5) + (-1)$
$= (+18) + (-14)$
$= +4$
The elevator is now at the 4th floor.

Example 5. If $a = -3, b = +7, c = -6$, find the value of $a + b + c$.

Solution. $a + b + c = (-3) + (+7) + (-6)$
$= (-9) + (+7)$
$= -2$

The number line is a useful device for showing addition of integers. However, after a little practice you should be able to add integers without its help.

Exercises   2 - 2

A   1. Simplify:

a) $(-6) + (+2)$ b) $(+8) + (-5)$ c) $(+5) + (-8)$

d) $(-3) + (+8)$ e) $(+3) + (+2)$ f) $(-6) + (-4)$

g) $(+5) + (+3)$ h) $(+7) + (-5)$ i) $(-4) + (+6)$

j) $(+12) + (-3)$ k) $(-5) + (+11)$ l) $(+9) + (-9)$

m) $(+60) + (-25)$ n) $(-65) + (-15)$ o) $(+40) + (-90)$

p) $(-45) + (+70)$ q) $(-37) + (+77)$ r) $(+33) + (-63)$

s) $(-325) + (+245)$ t) $(-780) + (-190)$

u) $(+310) + (-570)$ v) $(-313) + (+209)$

w) $(-1019) + (+3028)$ x) $(+3377) + (-4376)$

2. Simplify:
   a) $(-3) + (-4) + (+8)$ b) $(+2) + (-5) + (-7)$
   c) $(-2) + (-5) + (-9)$ d) $(+7) + (-8) + (+2)$
   e) $(-13) + (+10) + (+27)$ f) $(+29) + (-101) + (+71)$
   g) $(+38) + (+29) + (-57)$ h) $(-17) + (-64) + (+81)$

3. Simplify:
   a) $(-8) + (-7) + (+14) + (+1)$
   b) $(-13) + (+2) + (+19) + (-7)$
   c) $(+32) + (+43) + (-29) + (-11)$
   d) $(-71) + (-92) + (+143) + (-5)$
   e) $(-103) + (+99) + (-111) + (+109)$
   f) $(-211) + (+199) + (+12) + (+100)$

4. Simplify:
   a) $5 + (-3) + 7$ b) $(-5) + (-4) + 6$
   c) $2 + (-9) + 4$ d) $(-8) + 1 + (-2)$
   e) $(-13) + 27 + (-11)$ f) $37 + (-21) + (-52)$
   g) $18 + 39 + (-71)$ h) $(-87) + 78 + (-13)$
   i) $91 + (-27) + 19 + (-72)$ j) $(-32) + 22 + (-42) + 18$

5. Simplify:
   a) $(+4) + (-4)$ b) $(-7) + (+7)$ c) $(-36) + (+36)$
   d) $(+81) + (-81)$ e) $(-23) + (+23)$ f) $(-57) + (+57)$
   g) What can you conclude about the sum of an integer and its opposite?

6. Find the value of $x + y + z$ when
   a) $x = -1,\ y = -2,\ z = 3$;   b) $x = -10,\ y = -12,\ z = 6$;
   c) $x = 1,\ y = 11,\ z = -10$;   d) $x = -5,\ y = 9,\ z = -7$;
   e) $x = -23,\ y = 31,\ z = -17$;   f) $x = 139,\ y = -98,\ z = -79$

B 7. Write an addition statement for each of the following:
   a) A football team is on the 15 m line. It gains 17 m, and on the next play it is penalized 5 m.
   b) Deposit $85 in an account. Write cheques for $29 and $37. Deposit a further $52, and write a cheque for $66.
   c) A man's mass was 80 kg. He went on a diet and lost 7 kg. While on vacation he gained 5 kg.

C 8. Solve:
   a) $7 + y = 3$ b) $2 + x = -5$ c) $-6 + k = -2$
   d) $1 + m = 4$ e) $-7 = -7 + m$ f) $-11 = -5 + x$

2 - 3 Subtraction of Integers

Subtraction is used to find a difference or a change. The change is always the final value less the first value.

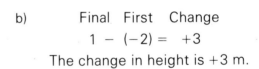

| Final value | − | First value | = | Change |

Example 1. What is the change of height when

 a) a hawk at 20 m rises to a height of 25 m?

 b) a dolphin at a depth of 2 m leaps to a height of 1 m above the water surface?

 c) a diver 5 m above a pool dives to a depth of 2 m?

 d) a trout at a depth of 1 m swims to a depth of 3 m?

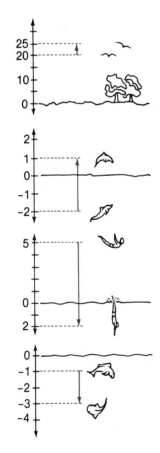

Solution. a)   Final   First   Change
$$25 - 20 = 5$$
The change in height is +5 m.

b)   Final   First   Change
$$1 - (-2) = +3$$
The change in height is +3 m.

c)   Final   First   Change
$$-2 - (+5) = -7$$
The change in height is −7 m.

d)   Final   First   Change
$$-3 - (-1) = -2$$
The change in height is −2 m.

For any subtraction problem there is a corresponding addition problem with the same answer. This leads us to a rule for subtraction.

Example 2. Simplify: a) i) $(+2) - (-3)$;   ii) $(+2) + (+3)$;
                                      b) i) $(-3) - (-3)$;   ii) $(-3) + (+3)$;
                                      c) i) $(-2) - (+4)$;   ii) $(-2) + (-4)$;

Solution. a) i) $(+2) - (-3)$ ii) $(+2) + (+3)$

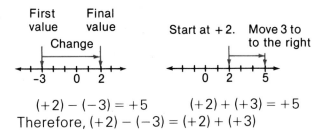

$$(+2) - (-3) = +5 \qquad (+2) + (+3) = +5$$
Therefore, $(+2) - (-3) = (+2) + (+3)$

b) i) $(-3) - (-3) = 0$ ii) $(-3) + (+3) = 0$
Therefore, $(-3) - (-3) = (-3) + (+3)$

c) i) $(-2) - (+4) = -6$ ii) $(-2) + (-4) = -6$
Therefore, $(-2) - (+4) = (-2) + (-4)$

The solutions to *Example 2* show that adding the opposite
of an integer gives the same result as subtracting the integer.

> To subtract an integer, add its opposite.

Example 3. Simplify: a) $(+6) - (-2)$;   b) $(-7) - (+3)$.

Solution. a) $(+6) - (-2)$ b) $(-7) - (+3)$
                 $= (+6) + (+2)$ $= (-7) + (-3)$
                 $= 8$ $= -10$

Sometimes subtraction and addition occur in the same
exercise.

Example 4. Simplify: a) $(+7) - (-2) + (-5) - (+3) + (+1)$
                          b) $5 - 9 + 2 - 8 - 3 + 6$

Solution. a) $(+7) - (-2) + (-5) - (+3) + (+1)$
                 $= (+7) + (+2) + (-5) + (-3) + (+1)$
                 $= (+10) + (-8)$
                 $= 2$
            b) $5 - 9 + 2 - 8 - 3 + 6$
                 $= 5 + 2 + 6 - 9 - 8 - 3$
                 $= 13 - 20$
                 $= -7$

Example 5. If $x = -2$, $y = +5$, $z = -6$, find the value of $x - y - z$.

Solution. $x - y - z = (-2) - (+5) - (-6)$
$= (-2) + (-5) + (+6)$
$= -1$

We can now answer the question at the beginning of the chapter.

Example 6. It is 8°C in Valleville (altitude 350 m). The temperature decreases 6.5°C for every 1000 m increase in altitude. If there is precipitation in Ste. Agathe (altitude 2350 m), will it be rain or snow?

Solution. The difference in altitude is 2000 m. Therefore, the temperature in Ste. Agathe is 2 × 6.5°C, or 13°C, lower than the temperature in Valleville.
Temperature in Ste. Agathe:
8°C − 13°C = −5°C
Any precipitation in Ste. Agathe will probably be snow.

Exercises 2 - 3

A 1. The temperature was −6°C. It is now 4°C. How much did the temperature change?

2. A balloon was 600 m above the ground. It is now 250 m above the ground. What is its change in altitude?

3. What is the temperature change
   a) from −12°C to +8°C?   b) from −17°C to −5°C?
   c) from +27°C to −27°C?   d) from −6°C to + 18°C?

4. What is the altitude change
   a) from 3170 m to 525 m?   b) from −265 m to 425 m?
   c) from −350 m to −580 m?   d) from −900 m to −250 m?

5. Simplify:
   a) $(+4) - (+6)$ b) $(+7) - (+2)$ c) $(-8) - (+4)$
   d) $(+3) - (+1)$ e) $(+6) - (-1)$ f) $(-4) - (-3)$
   g) $(-6) - (+3)$ h) $(-2) - (+5)$ i) $(+2) - (+1)$
   j) $(+1) - (-5)$ k) $(-4) - (-2)$ l) $(+2) - (-4)$
   m) $0 - (-2)$ n) $0 - (+3)$ o) $(-5) - 0$

6. Simplify:
   a) $(+45) - (-15)$ b) $(-23) - (-13)$ c) $(-14) - (+66)$
   d) $(-145) - (-35)$ e) $(+68) - (+98)$ f) $(-72) - (-42)$
   g) $(+75) - (-15)$ h) $(-187) - (-42)$ i) $(-27) - (+43)$
   j) $(+81) - (+93)$ k) $(-101) - (+12)$ l) $(+456) - (+567)$
   m) $(-505) - (-65)$ n) $(+505) - (+75)$ o) $(-987) - (+513)$

7. Simplify:
   a) $(-9) - (+2) + (-3) - (+5)$ b) $(+8) + (4) - (+6) - (-3)$
   c) $(+8) - (+3) - (-4) - (-7)$ d) $(-6) - (-3) - (-7) + (-8)$
   e) $(-10) + (+6) - (+5) - (+7)$ f) $(+1) - (-6) - (+3) - (-4)$

8. Simplify:
   a) $5 - 2 - 8 + 3 - 1$ b) $-4 + 6 + 2 - 7 - 3$
   c) $-1 - 5 + 9 - 2 + 3$ d) $7 - 2 - 6 + 4 - 8 + 2$
   e) $-3 - 9 + 1 - 5 + 7 - 4$ f) $17 - 14 - 2 + 13 - 9 - 10$

9. Find the value of $x - y$ if
   a) $x = -3$ and $y = -4$;   b) $x = 7$ and $y = -3$;
   c) $x = -5$ and $y = -2$;   d) $x = 18$ and $y = -6$.

10. If $a = 5, b = -2, c = -3$, find the values of the following:
    a) $a - b$ b) $b - a$ c) $a - b - c$
    d) $a + b + c$ e) $a - (b - c)$ f) $c - b - a$

B 11. Simplify:
    a) $(-4 + 6) - (3 - 7)$ b) $(8 - 5) - (-4 + 6)$
    c) $(-8 + 3) - (-2 - 5)$ d) $(-9 + 4) - (-3 - 7)$
    e) $(7 - 4) - (8 - 3)$ f) $(-6 + 2) - (7 - 9)$

12. The lowest temperature ever recorded in Canada was $-63°C$ in the Yukon in 1947. The highest was $45°C$ in Saskatchewan in 1937. What is the difference between these two temperatures?

13. The greatest temperature change in North America in a single day was from $+7°C$ to $-49°C$ in Montana. What is the difference between these temperatures?

Yukon

Saskatchewan

Canada

C 14. Solve:
    a) $x - (+5) = +3$ b) $m - (+6) = -4$
    c) $+7 - (+3) = t$ d) $-8 - (-5) = y$
    e) $+11 - w = +3$ f) $x - (+3) = -6$
    g) $-23 - x = +23$ h) $+17 - (+19) = y$

15. The time difference between Toronto and Vancouver is 3 h.

   a) If an airplane leaves Vancouver for Toronto at 08:00 and the flying time is 4 h 10 min, what time does it arrive in Toronto?

   b) On the return flight, if the airplane leaves Toronto at 07:30 and the flying time to Vancouver is 4 h 50 min, what time does it arrive in Vancouver?

16. The time of day changes 1 h for every 15° difference in longitude. This means that when it is midday in London, England (0° longitude), it is midnight at the date line (180° longitude); a new day is just starting there. It is 12 h ahead of London, 17 h ahead of Ottawa. Times are usually compared with the time at London.

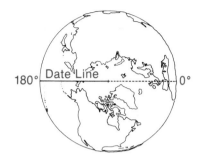

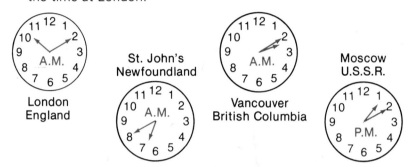

London England — St. John's Newfoundland — Vancouver British Columbia — Moscow U.S.S.R.

Standard Time Difference, in Hours, Between London (England) and Other Cities					
Athens	+ 2	Halifax	−4	Rome	+ 1
Bangkok	+ 7	Jakarta	+7	Peking	+ 8
Bogota	− 5	Jerusalem	+2	Santiago	− 4
Brasilia	− 3	Mexico City	−6	Washington	− 5
Canberra	+10	Moscow	+3	Wellington	+12
Dublin	0	Ottawa	−5	Vancouver	− 8

Halifax −4 means that Halifax is 4 h behind London. Moscow +3 means that Moscow is 3 h ahead of London. Therefore Moscow is 3 h − (−4 h), or 7 h ahead of Halifax. Wellington +12, Canberra +10 means that Wellington is 2 h ahead of Canberra.

a) When it is 08:00 in London, what time is it in
   i) Ottawa?   ii) Mexico City?   iii) Jerusalem?
   iv) Canberra?   v) Rome?   vi) Athens?

b) When it is 22:00 in Jakarta, what time is it in
   i) Bangkok?   ii) Peking?   iii) Wellington?
   iv) Bogota?   v) Halifax?   vi) Santiago?

c) When it is 21:00 in Ottawa, what time is it in
   i) Vancouver?   ii) Dublin?   iii) Washington?
   iv) Peking?   v) Brasilia?   vi) Santiago?

17. Team Canada is playing the Moscow Selects and the game is being televised live via satellite.

   a) If the game is in Moscow at 8 p.m., refer to the table in Exercise 16 and state what time a viewer would be watching it in
      i) Halifax;   ii) Vancouver;   iii) Ottawa.

   b) If the game is in Toronto (same time zone as Ottawa) at 8 p.m., state what time someone in Moscow would be watching it.

18. Scenes of an earthquake in Chile are sent via television satellite from Santiago at 16:00. Refer to the table of Exercise 16 and say what time the transmission is received in these places.

   a) London   b) Ottawa   c) Bogota
   d) Mexico City   e) Moscow   f) Vancouver

19. In August 1978, three Americans made the first crossing of the Atlantic Ocean by balloon. The graph shows the altitude of the balloon along the flight path.

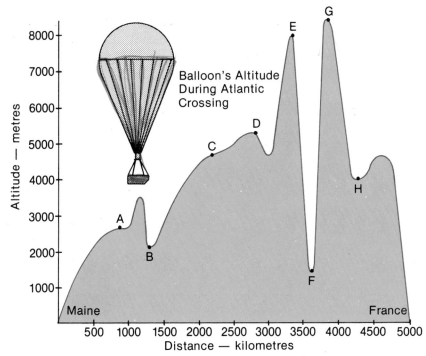

Balloon's Altitude During Atlantic Crossing

State the approximate change in altitude:
   a) from *A* to *B*;   b) from *B* to *C*;   c) from *A* to *C*;
   d) from *E* to *F*;   e) from *F* to *G*;   f) from *E* to *G*;
   g) from *D* to *H*;   h) from *B* to *F*;   i) from *D* to *F*.

Mathematics Around Us

The Importance of the St. Lawrence Seaway

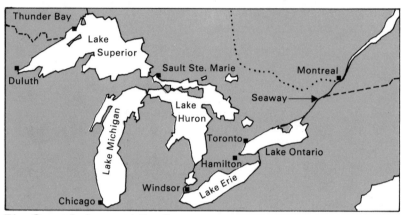

The Great Lakes and the Seaway

Lock 3	Ecluse 3
Sailing distance to	Distances nautiques jusquà

Port Colborne 34 km L. Erie	Port Weller L. Ontario 9 km

Toronto 56 km

961 km Sault Ste. Marie

1447 km Chicago	London 5763 km

Sept Iles 1357 km

1402 km Thunder Bay

Montreal 475 km

426 km Windsor	Hamilton 62 km

A Signpost for Ships

The St. Lawrence Seaway, opened in 1959, is the world's longest artificial seaway, stretching 304 km from Montreal harbour to Lake Ontario. Its canal depth of 8.25 m (as against 4.25 m of the old seaway system) and seven locks enable ocean-going ships to get through to and from the Great Lakes. By using the Welland canal and the canal and locks at Sault Ste. Marie as well, they can sail as far inland as Thunder Bay, Ontario, and Duluth, Minnesota. Within seven years of the Seaway's opening, the amount of cargo traffic had increased fourfold.

Questions

1. What is the change in elevation when going
   a) from Lake Ontario to Lake Erie?
   b) from Lake Erie to Lake Huron?
   c) from Thunder Bay to Sept Iles?
   d) from Thunder Bay to Montreal?

2. According to the sign, how far is it
   a) from Windsor to Thunder Bay?
   b) from Hamilton to Montreal?
   c) from Thunder Bay to Sept Iles?
   d) from Chicago to London?

3. Where is the sign posted?

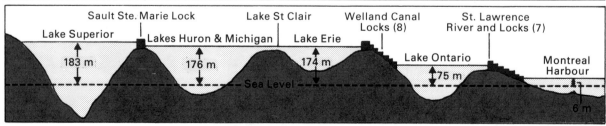

Levels and Depths of Lakes and Seaway

2 - 4   Multiplication of Integers

In order to multiply integers, we must define what we mean by each of the following:

$$(+3)(+5) \quad (+3)(-5) \quad (-3)(+5) \quad (-3)(-5)$$

$$\begin{aligned}(+3)(+5) &= (3)(5)\\ &= 5 + 5 + 5\\ &= +15\end{aligned}$$

$$\begin{aligned}(+3)(-5) &= (3)(-5)\\ &= (-5) + (-5) + (-5)\\ &= -15\end{aligned}$$

$$\begin{aligned}(-3)(+5) &= (5)(-3)\\ &= (-3) + (-3) + (-3) + (-3) + (-3)\\ &= -15\end{aligned}$$

$$(-3)(-5) = ?$$

To find an answer for $(-3)(-5)$ we recall the distributive law for positive integers:

$$a(b + c) = ab + ac.$$

If the distributive law is also to apply to negative integers, then, for example:

$$-3[(+5) + (-5)] = (-3)(+5) + (-3)(-5).$$

That is,
$$0 = (-3)(+5) + (-3)(-5)$$
$$0 = -15 + (-3)(-5)$$
But
$$0 = -15 + (+15)$$
Therefore, $(-3)(-5) = +15$

The product of two integers with like signs is positive. The product of two integers with unlike signs is negative.

×	Positive	Negative
Positive	Positive	Negative
Negative	Negative	Positive

Example 1. Simplify: a) $(-2)(+6)$;

b) $(-7)(-8)$;

c) $(+16)(-14)$.

Solution. a) $(-2)(+6) = -12$

b) $(-7)(-8) = +56$

c) $(+16)(-14) = -224$

Example 2. Simplify: $(-3)(+2) - (-6)(-2)$.

Solution. $(-3)(+2) - (-6)(-2) = (-6) - (+12)$
$$= (-6) + (-12)$$
$$= -18$$

Example 3. If $a = -2$, $b = -3$, and $c = +5$, evaluate $3abc - a$.

Solution. $3abc - a = (+3)(-2)(-3)(+5) - (-2)$
$$= (-6)(-15) + (+2)$$
$$= (+90) + (+2)$$
$$= 92$$

Example 4. Evaluate: $5a - 3b - 2a - 7b - 6a + 3b$, when $a = -4$ and $b = -1$.

Solution. To simplify the arithmetic, collect like terms first, use the distributive law, then substitute.

$5a - 3b - 2a - 7b - 6a + 3b$
$= 5a - 2a - 6a - 3b - 7b + 3b$
$= (5 - 2 - 6)a + (-3 - 7 + 3)b$
$= -3a - 7b$
$= (-3)(-4) + (-7)(-1)$
$= +12 + 7$
$= 19$

Example 5. Evaluate $(x - 6)(x + 5)$ for:

a) $x = -3$;   b) $x = -7$

Solution. a) $(x - 6)(x + 5) = (-3 - 6)(-3 + 5)$
$$= (-9)(+2)$$
$$= -18$$

b) $(x - 6)(x + 5) = (-7 - 6)(-7 + 5)$
$$= (-13)(-2)$$
$$= +26$$

Exercises 2 - 4

A 1. Simplify:

   a) $(-5)(+6)$ b) $(+7)(-8)$ c) $(-7)(-9)$ d) $(+6)(+9)$

   e) $(-12)(+5)$ f) $(-3)(-13)$ g) $(+8)(-9)$ h) $(-5)(+5)$

   i) $(-5)(-5)$ j) $(+7)(-6)$ k) $(+3)(-4)$ l) $(-4)(-8)$

2. Simplify:

   a) $(+16)(-5)$ b) $(-18)(+3)$ c) $(-14)(-4)$ d) $(-17)(-9)$

   e) $(-11)(+28)$ f) $(+19)(-11)$ g) $(+36)(-72)$

   h) $(+47)(-16)$ i) $(-69)(-89)$ j) $(-74)(-18)$

3. Find the products:

   a) $(-2)(+5)(-7)$ b) $(-3)(-4)(-2)$ c) $(+6)(-5)(+4)$

   d) $(-1)(+3)(-3)$ e) $(+2)(-3)(-3)$ f) $(-2)(-2)(-3)$

   g) $(+5)(-1)(-1)(-1)$ h) $(-2)(-2)(-2)$

   i) $(-12)(+15)(-6)$ j) $(-1)(-2)(-3)(-4)$

4. Simplify:

   a) $(-2)(+3) + (-6)(-2)$ b) $(-4)(-3) + (-1)(-2)$

   c) $(-2)(-6) - (+5)(-2)$ d) $(-3)(+7) - (-1)(-5)$

   e) $(-2)(+8) - (-3)(-3)$ f) $(+4)(-7) + (-8)(+6)$

   g) $(-3)(-9) + (-2)(+7)$ h) $(-7)(-9) - (-6)(-7)$

   i) $(-2)(-2)(+1) + (-3)(-3)(-2)$

   j) $(-5)(-2)(-2) - (+2)(-1)(-1)$

5. Evaluate the following when $a = -2$ and $b = +3$:

   a) $3a + 5b$ b) $4a - 2b$ c) $(a - b)(a - b)$

   d) $(a - b)(a + b)$ e) $a(a - b)$ f) $b(a - b)$

6. Evaluate the following when $m = 7$ and $n = -6$:

   a) $5m + 3n - 2$ b) $-2m - 5n + 3$ c) $3(m + n) - n$

   d) $n(m - 9)$ e) $n(n - m + 3)$ f) $4m + 4n + 4$

7. Evaluate the following when $a = -3$ and $b = +1$:

   a) $5a - 3b - 2a + 4b$ b) $6a - 2b + 3a + b$

   c) $-4a + 5b - 2b + a - b$ d) $-2a + 6b + a - 2b - 3a$

   e) $7a - 2b - a - b$ f) $5a + 8b - 3a - 10b + 5b$

8. Evaluate $(x - 4)(x + 1)$ for these values of x:

   a) 2 b) -2 c) 4 d) -1

9. Evaluate $(n - 9)(n + 3)$ for these values of n:

   a) 5   b) -7 c) 11   d) -2

B 10. What must be true of two integers if their product is

   a) positive?   b) negative?   c) zero?

11. By comparing the answers in Exercise 3, state a general rule for:

   a) the product of an even number of negative numbers;

   b) the product of an odd number of negative numbers.

12. For the years 1975 to 2000, the approximate population, P, in thousands, of a city is given by the formula:
$$P = (y - 1981)(1995 - y) + 500,$$
   where y is the year. Find the population in the years:

   a) 1975   b) 1980   c) 1983   d) 1985

   e) 1988   f) 1990   g) 1995   h) 2000

13. For the years 1977 to 1989, a company's approximate profits, P, in millions of dollars, is given by the formula:
$$P = (y - 1970)(y - 1985) + 60,$$
   where y is the year. Find the profits in these years:

   a) 1977   b) 1978   c) 1979   d) 1980

   e) 1982   f) 1984   g) 1986   h) 1988

C 14. Find the missing factor:

   a) $(+5)(\ \ ) = -20$ b) $(-2)(\ \ ) = 16$

   c) $(\ \ )(+7) = -56$ d) $(\ \ )(-6) = 54$

   e) $(-8)(\ \ ) = -72$ f) $(\ \ )(-12) = -96$

   g) $-32 = (+8)(\ \ )$ h) $48 = (-16)(\ \ )$

   i) $-39 = (\ \ )(-13)$ j) $-63 = (\ \ )(-9)$

15. Find the missing factor:

   a) $(-2)(+3)(\ \ ) = -24$ b) $(+4)(\ \ )(+2) = -32$

   c) $(\ \ )(-1)(-6) = 42$ d) $(-3)(\ \ )(+5) = -45$

   e) $96 = (-2)(+6)(\ \ )$

   f) $-36 = (-6)(-6)(\ \ )$

   g) $56 = (\ \ )(-2)(-2)$

   h) $-81 = (-9)(\ \ )(+3)$

   i) $(-10)(\ \ )(-8) = -240$

   j) $-360 = (-2)(+6)(-10)(\ \ )$

2 - 5   Division of Integers

Division is the inverse of multiplication.

We know $28 \div 7 = 4$ because $7 \times 4 = 28.$

The same is true when negative integers are involved.

We know $(-20) \div (+5) = -4$

because $(+5) \times (-4) = -20.$

Example 1.   Simplify: a) $(+6) \div (-3)$;   b) $(-24) \div (-6)$.

Solution. a) $(+6) \div (-3)$

Since $(-3)(-2) = +6$

then, $(+6) \div (-3) = -2$

b) $(-24) \div (-6)$

Since $(-6)(+4) = -24$

then, $(-24) \div (-6) = +4$

Example 2.   Simplify: a) $(+15) \div (+3)$;   b) $(-20) \div (-2)$;

c) $(+24) \div (-8)$;   d) $(-33) \div (+3)$.

Solution. a) $(+15) \div (+3) = +5$;

b) $(-20) \div (-2) = +10$;

c) $(+24) \div (-8) = -3$;

d) $(-33) \div (+3) = -11$.

The above examples suggest the following rules:

> The quotient of two integers with like signs is positive.
> The quotient of two integers with unlike signs is negative.

Example 3.   Simplify: a) $\frac{+63}{-9}$;   b) $\frac{-42}{-7}$.

Solution. a) $\frac{+63}{-9} = -7$ b) $\frac{-42}{-7} = +6$

Example 4.   Simplify: a) $\frac{(-8)(-9)}{(+3)(-4)}$;   b) $\frac{-24}{+4} - \frac{+10}{-2}$.

Solution.   a) Simplify numerator and denominator before
dividing.

$\frac{(-8)(-9)}{(+3)(-4)} = \frac{+72}{-12}$

$= -6$

b) Perform the divisions before subtracting.

$$\frac{-24}{+4} - \frac{+10}{-2} = -6 - (-5)$$
$$= -6 + (+5)$$
$$= -1$$

Example 5. Evaluate $\frac{4a - b}{b - a}$ when $a = -2$ and $b = -3$.

Solution. Simplify numerator and denominator before dividing.

$$\frac{4a - b}{b - a} = \frac{(4)(-2) - (-3)}{(-3) - (-2)}$$
$$= \frac{-8 + (+3)}{(-3) + (+2)}$$
$$= \frac{-5}{-1}$$
$$= 5$$

Exercises 2 - 5

A 1. Simplify:

a) $(-48) \div (+4)$ b) $(-36) \div (-4)$ c) $32 \div (-8)$

d) $-18 \div 3$ e) $(-60) \div (-12)$ f) $(-40) \div (-5)$

g) $(-54) \div (+9)$ h) $(+84) \div (-7)$ i) $(-91) \div (-7)$

2. Simplify:

a) $\frac{-36}{+4}$ b) $\frac{+46}{-2}$ c) $\frac{-18}{-9}$ d) $\frac{-85}{+5}$

e) $\frac{-49}{-7}$ f) $\frac{+81}{-9}$ g) $\frac{-76}{-19}$ h) $\frac{-121}{+11}$

3. Simplify:

a) $\frac{(-4)(+10)}{-8}$ b) $\frac{(+6)(-15)}{-5}$ c) $\frac{(-10)(+12)}{(+5)(-3)}$

d) $\frac{(-15)(-20)}{(-10)(+3)}$ e) $\frac{(-50)(+9)}{(+15)(+6)}$ f) $\frac{(+14)(-16)}{(-8)(-7)}$

g) $\frac{(-27)(+18)}{(+6)(-9)}$ h) $\frac{(-5)(+9)(-24)}{(-3)(+4)}$ i) $\frac{(-6)(-8)}{(-2)(-1)(-3)}$

4. Simplify:

   a) $\dfrac{-30}{+5} + \dfrac{+15}{-3}$ b) $\dfrac{-20}{+10} + \dfrac{+8}{-2}$ c) $\dfrac{-9}{-3} - \dfrac{+12}{+4}$

   d) $\dfrac{+14}{-2} - \dfrac{-16}{+8}$ e) $\dfrac{-36}{+4} + \dfrac{-56}{-8}$ f) $\dfrac{-42}{+7} - \dfrac{+54}{-6}$

   g) $\dfrac{-63}{-7} - \dfrac{-56}{-8}$ h) $\dfrac{-81}{-9} + \dfrac{-72}{-8}$ i) $\dfrac{+35}{+7} + \dfrac{+48}{-6}$

B 5. Find the missing number:

   a) $(+40) \div (\ \ ) = -10$ b) $(\ \ ) \div (-5) = -7$

   c) $\dfrac{(\ \ )}{(-3)} = 6$ d) $\dfrac{(-20)}{(\ \ )} = -4$ e) $-3 = \dfrac{(\ \ )}{(-2)}$

   f) $(-65) \div (\ \ ) = -13$ g) $(\ \ ) \div (-9) = 3$

   h) $2 = (-28) \div (\ \ )$ i) $-11 = (\ \ ) \div (-3)$

6. Find the missing number:

   a) $(+3132) \div (\ \ ) = -87$ b) $(+1972) \div (\ \ ) = -29$

   c) $(\ \ ) \div (-47) = +63$ d) $(-4676) \div (\ \ ) = -167$

   e) $-33 = (\ \ ) \div (-44)$ f) $48 = (-1008) \div (\ \ )$

7. What must be true of two integers if their quotient is

   a) positive?   b) negative?   c) zero?

8. If $a = -6, b = 3$, and $c = -12$, find the value of

   a) $\dfrac{a}{b} + \dfrac{c}{b}$ b) $\dfrac{c}{a} - \dfrac{a}{b}$ c) $\dfrac{c - a}{b}$ d) $\dfrac{2b}{a} + \dfrac{c}{2a}$

9. Find the value of $y - \dfrac{2x}{y}$ for these values of x and y:

   a) $x = 18$ b) $x = -30$ c) $x = -44$ d) $x = -21$
   $\ \ \ \ y = -9$ $\ \ \ \ y = -5$ $\ \ \ \ y = 4$ $\ \ \ \ y = -6$

C 10. If $x > 0, y > 0$, and $z < 0$, decide which of the following expressions are always positive or always negative.

   a) $\dfrac{x}{y}$ b) $\dfrac{x + y}{z}$ c) $\dfrac{xy}{z}$ d) $\dfrac{y - z}{x}$

   e) $\dfrac{z - x}{y}$ f) $\dfrac{z}{x + y}$ g) $\dfrac{z - x}{z}$ h) $\dfrac{y - z}{x + y}$

11. If $p > 0, q < 0$, and $r < 0$, decide which of the following expressions are always positive or always negative.

   a) $\dfrac{2p}{q}$ b) $-p + q$ c) $\dfrac{q + r}{p}$

   d) $\dfrac{q}{r-}$ e) $\dfrac{r - p}{p}$ f) $\dfrac{rq}{p - q}$

2 - 6   Order of Operations With Integers

Now that we have reviewed the four basic operations with integers, let us review the order in which they must be performed.

- Operations within grouping symbols are performed first, starting with the innermost and working outward.

- Multiplication and division are performed in order from left to right.

- Lastly, addition and subtraction are performed in order from left to right.

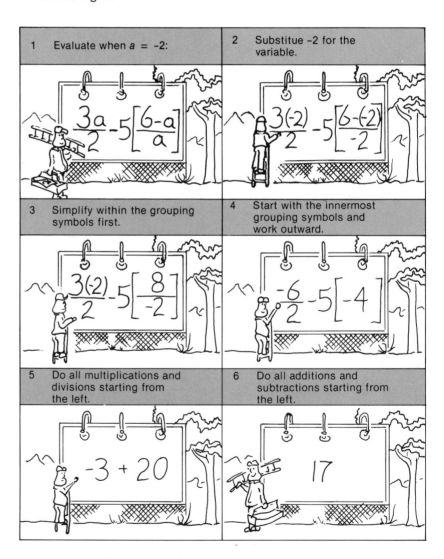

1. Evaluate when $a = -2$:

$$\frac{3a}{2} - 5\left[\frac{6-a}{a}\right]$$

2. Substitue –2 for the variable.

$$\frac{3(-2)}{2} - 5\left[\frac{6-(-2)}{-2}\right]$$

3. Simplify within the grouping symbols first.

$$\frac{3(-2)}{2} - 5\left[\frac{8}{-2}\right]$$

4. Start with the innermost grouping symbols and work outward.

$$\frac{-6}{2} - 5[-4]$$

5. Do all multiplications and divisions starting from the left.

$$-3 + 20$$

6. Do all additions and subtractions starting from the left.

$$17$$

Example 1. Simplify: $-3[2 - 5(2 - 8)(-1 + 3)]$.

Solution.
$$-3[2 - 5(2 - 8)(-1 + 3)]$$
$$= -3[2 - 5(-6)(2)]$$
$$= -3[2 + 30(2)]$$
$$= -3[2 + 60]$$
$$= (-3)(62)$$
$$= -186$$

Example 2. Simplify: $\dfrac{4n + 7(3n + 5)}{(2 + n)(2 - n)}$ when $n = -3$.

Solution.
$$\frac{4n + 7(3n + 5)}{(2 + n)(2 - n)} = \frac{4(-3) + 7(-9 + 5)}{(2 - 3)(2 + 3)}$$
$$= \frac{4(-3) + 7(-4)}{(-1)(5)}$$
$$= \frac{-12 - 28}{-5}$$
$$= \frac{-40}{-5}$$
$$= 8$$

Sometimes, terms inside grouping symbols cannot be combined. In such cases, the distributive law is used.

Example 3. Simplify: a) $-2(5 - 2x)$;   b) $3x - 5(4 - 2x)$

Solution. a) $-2(5 - 2x) = -2(5) - (-2)(2x)$
$$= -10 - (-4x)$$
$$= -10 + 4x$$

b) $3x - 5(4 - 2x) = 3x - 5(4) - 5(-2x)$
$$= 3x - 20 + 10x$$
$$= 13x - 20$$

Exercises 2 - 6

A 1. Simplify:

a) $3(-2 + 6) - 5(4 - 1)$ b) $(-5)(-4) + (-6)(3)$

c) $-2(-4 + 3) + 3(-1 - 5)$ d) $(-3)(-1)(5) - (-2)(-4)(-1)$

e) $5(2 - 6)(2 - 6)$ f) $7(7 - 2) - 5(-3 - 8) + 19$

2. Simplify:

a) $\dfrac{-15}{3} - \dfrac{-10}{5}$ b) $\dfrac{(-30)(4)}{(6)(-2)}$ c) $\dfrac{-7 + 3(-1 + 4)}{-2}$

d) $\dfrac{4(-5 + 3) - 2(-1 + 5)}{-6 + 2}$ e) $\dfrac{35 - 81}{27 - 4} - \dfrac{(-4)(3 - 10)}{8 - 15}$

3. Simplify:

a) $(12 + 8) \div (2 - 6)$

b) $(-3 + 4)(8 - 10) - (7 - 9)(4 - 1)$

c) $(6 - 2 + 3)(-7 + 5 - 1)$

d) $(4 - 9)(2 + 3) + (8 - 2)(-3 + 2)$

e) $\dfrac{(-5 + 2)(-4 - 6)}{3 - 9}$

f) $\dfrac{5(-3 - 4) - (-6)(13 - 6)}{(-1)(11 - 4)}$

4. Find the value of each of these expressions for $a = -2, b = 5$, and $c = -3$:

a) $a - b + c - b + a - c$ b) $3b - 2c - a$

c) $(a + b)(c - a)$ d) $(2a - b)(b + 2c)$

e) $(a + c) \div b$ f) $(b - 5c) \div (b + c - 9a)$

B 5. Simplify:

a) $1 - (x - 3)$ b) $6a + b - (b + 2a)$

c) $-(x - y)$ d) $-4(5 - z)$

e) $3y - 4(y - 7) + 18$ f) $-5 - (-6 - x) + (x - 6)$

6. Simplify:

a) $3x - 4(4 - 3x)$ b) $3(2x - 5) - 2(x + 6)$

c) $4(a - 2b) - 3(2a + b)$ d) $(p + 2q) - 5(q - p)$

e) $5n - (m + 2n) + 3(m - 2n)$

f) $4(x - y) + 5(y - 2x) - 3y$

C 7. Evaluate these expressions for $x = -2$:

a) $\dfrac{(x - 2)(x - 3)}{20}$ b) $(\dfrac{3x - 2}{2})(\dfrac{5x - 2}{3})$

c) $\dfrac{1 + 3(3x - 1)}{10}$ d) $\dfrac{4x - 8}{x + 4} - \dfrac{8 - 4x}{4 + x}$

8. Evaluate these expressions for $p = -1$ and $q = 4$:

a) $q - p(p - 1)(q - 1)$ b) $(p + q)(p - q)(q - p)$

c) $\dfrac{q - 2p}{3} + p(\dfrac{2p - 3q}{7})$ d) $\dfrac{q - p + 3(3p + q)}{p(8p + q)}$

9. Simplify:

a) $5x - [3(x + y) - 4(2y - x)] + 3y$

b) $[2(a - 3b) + (b - 2a)] - 3(a + b)$

c) $2[3(a - b) - 4(b - a)]$

d) $3x - 2[2(y - x) - 3(x + 2y)] - 8y$

Review Exercises

1. Arrange in order from least to greatest:
   a) $-5, 3, -4, -7, -9, 6$ b) $-6, 6, -5, -9, 9, -7$

2. Simplify:
   a) $(-5) + (+6)$ b) $(-5) + (-3)$ c) $(-6) + (+9)$
   d) $(-1) + (+9)$ e) $(-7) + (+11)$ f) $(+12) + (-13)$
   g) $(-4) + (+3) + (-5)$ h) $(+2) + (-7) + (-4)$
   i) $(-43) + (-17) + (+5)$ j) $(-39) + (+10) + (+31) + (-7)$

3. The greatest temperature variation in a single day in Alberta's chinook belt was from $+17°C$ to $-28°C$. What was the change in temperature?

*Coat of Arms
Alberta*

4. Solve:
   a) $(+5) + x = 0$ b) $(-13) + y = 0$ c) $(+7) + y = (-1)$
   d) $w + (-5) = (-2)$ e) $(+4) + x = (+3)$ f) $v + (-3) = (+2)$
   g) $r + (-6) = (-4)$ h) $g + (+5) = (-4)$ i) $x - (+4) = (+2)$
   j) $n + (+3) = (-5)$ k) $(-5) - p = (+2)$ l) $r - (-9) = (+1)$

5. Simplify:
   a) $(+7) - (+4)$ b) $(+8) - (+2)$ c) $(+9) - (+2)$
   d) $(+5) - (-2)$ e) $(+7) - (-2)$ f) $(+8) - (-5)$
   g) $(+6) - (-3)$ h) $(+5) - (-5)$ i) $0 - (-5)$

6. Evaluate $2a - b + 3c$ for
   a) $a = +3, b = -2, c = -1;$ b) $a = -2, b = -1, c = -3.$

7. Simplify:
   a) $(-8) - (+3) + (-5) - (+7)$ b) $(-14) - (+12) - (+3)$
   c) $(-103) - (+27) - (-100)$ d) $(+283) - (-20) + (-60)$
   e) $(+70) - (+90) - (-100)$ f) $(-100) + (-70) - (+20)$
   g) $(-30) - (-72) + (-43)$ h) $(+981) - (-19) - (+891)$
   i) $(+45) + (-100) - (-10) - (-40)$
   j) $(-230) - (-300) - (-50) - (-40) + (+230)$

8. Simplify:
   a) $(-8 + 5) - (17 - 9)$ b) $(-30 - 20) - (-20 - 30)$
   c) $(93 - 84) - (-67 + 89)$ d) $(-9 - 17) - (11 - 27)$
   e) $(3 - 7) + (5 - 9) - (2 - 7)$
   f) $(284 - 180) - (-3 + 99) + (109 - 47)$
   g) $(-1000 - 2000) + (-1000 - 1000) - (100 - 100)$

9. If $k = -1$, $l = -3$, and $m = +2$, find the value of
   a) $(k + l) + (k - l)$;
   b) $(k - l) - m - l$;
   c) $-2k + (3l - m) + 2l$;
   d) $(m - k) - 3l + 2k$;
   e) $(-3m + l) - k + m$;
   f) $l + (2k - m) - 2$;
   g) $5 - k + l - 3m$;
   h) $m - k + l + 2m + k + l$.

10. Simplify:
    a) $(-7)(+8)$ b) $(-6)(-9)$ c) $(+5)(-7)$ d) $(+3)(-9)$
    e) $(-12)(-6)$ f) $(-10)(-8)$ g) $(-14)(+3)$ h) $(-4)(+15)$
    i) $(-11)(-10)$ j) $(+12)(-12)$ k) $(-16)(-12)$ l) $(-150)(+3)$

11. Simplify:
    a) $(-6)(+3)(-4)$ b) $(-8)(-2)(+7)$ c) $(+6)(+5)(-3)$
    d) $(-8)(-2)(-3)$ e) $(+5)(-3)(-4)$ f) $(-7)(-2)(-5)$

12. Simplify:
    a) $(-2)(-3)(-4)(-5)$ b) $(+8)(-2)(+6)(-3)$
    c) $(-5)(+8)(-2)(-10)$ d) $(+6)(+7)(-2)(0)$

13. Simplify:
    a) $(-16) \div (+4)$ b) $(-18) \div (-9)$ c) $(-48) \div (-16)$
    d) $(+81) \div (-3)$ e) $(-64) \div (-8)$ f) $(-54) \div (+18)$
    g) $(-108) \div (+36)$ h) $(+121) \div (-11)$ i) $(-144) \div (-36)$

14. Simplify:
    a) $\dfrac{(-8)(-12)}{(-24)(-4)}$ b) $\dfrac{(-5)(+39)}{(-13)(-3)}$ c) $\dfrac{(+121)(-7)}{(+77)(-11)}$

    d) $\dfrac{(-42)(+6)}{(+14)(-9)}$ e) $\dfrac{(+65)(-15)}{(+25)(+3)}$ f) $\dfrac{(+85)(+70)}{(-50)(-17)}$

    g) $\dfrac{-40}{+5} + \dfrac{+18}{-6}$ h) $\dfrac{-42}{+6} - \dfrac{-63}{+7}$ i) $\dfrac{+42}{-3} + \dfrac{-42}{+7}$

    j) $\dfrac{-49}{+7} - \dfrac{+26}{-13}$ k) $\dfrac{-96}{-16} - \dfrac{+132}{+12}$ l) $\dfrac{+85}{+17} + \dfrac{-95}{+19}$

15. If $a = -64$, $b = 8$, and $c = -4$, evaluate:
    a) $\dfrac{a}{b} + \dfrac{b}{c}$ b) $\dfrac{a}{c} - \dfrac{6c}{b}$ c) $\dfrac{a}{bc} + \dfrac{ac}{4b}$

16. A pilot is flying at an altitude of 5000 m where the temperature is $-21°C$. The nearby airport where he intends to land is at an altitude of 1000 m and the control tower reports precipitation. If the temperature increases 6.5°C for every 1000 m decrease in altitude, will the precipitation be rain or snow?

3 Rational Numbers

The salesmen answer questions concerning this car's fuel economy by using the equation: $y = \frac{-36}{5}x + \frac{29}{2}$, where y is the rate of fuel consumption in litres per 100 km and x is the fraction of driving done on the highway. What is the rate of fuel consumption when

a) three-quarters of the driving is on the highway?

b) two-thirds of the driving is in the city?

(See *Example 1* in Section 3-4.)

3 - 1 Positive and Negative Fractions

Just as each integer has its opposite, so also does each fraction. A positive number has an opposite negative number, and a negative number has an opposite positive number. This can be seen when the numbers are represented on the number line.

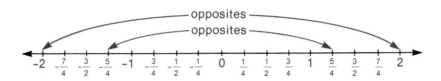

The opposite of the integer 2 is the integer -2, and vice versa. The opposite of the fraction $\frac{5}{4}$ is the fraction $-\frac{5}{4}$ and vice versa.

Remember that: $-2 = -\frac{8}{4}$, or $\frac{-8}{4}$, or $\frac{8}{-4}$,

and $-\frac{3}{4} = \frac{-3}{4}$, or $\frac{3}{-4}$.

Any number that can be written in the form $\frac{m}{n}$, where m and n are integers and $n \neq 0$, is called a **rational number**.

Example 1. Which of the following are rational numbers?

$$\frac{3}{2}, \frac{-8}{24}, 3\frac{1}{4}, 5, \frac{-6}{-7}, 0, 2.5, -11.27, -\frac{60}{-12}$$

Solution. All are rational numbers because each is in, or can be written in, the form of a quotient with the denominator not equal to zero.

$\frac{3}{2}, \frac{-8}{24}, \frac{-6}{-7}$, and $-\frac{60}{-12}$ are already in quotient form. The others can be written in this form.

$$3\frac{1}{4} = \frac{13}{4}, \qquad 5 = \frac{5}{1}, \qquad 0 = \frac{0}{1}$$

$$2.5 = \frac{25}{10}, \quad \text{and} \quad -11.27 = -\frac{1127}{100}.$$

Equivalent rational numbers can be written for each of the numbers in *Example 1.* For example:

$$5 = \frac{10}{2} = \frac{25}{5} = \frac{-30}{-6} = \cdots$$

and $\quad \frac{-6}{-7} = \frac{-12}{-14} = \frac{-18}{-21} = \frac{24}{28} = \cdots$

The quotient of two integers with like signs is positive.

That is, we can raise a rational number to higher terms by multiplying the numerator and the denominator by the same number. Similarly, we can reduce a rational number to lower terms by dividing the numerator and the denominator by the same number. For example:

$$\frac{-8}{24} = -\frac{8}{24} = -\frac{8 \div 8}{24 \div 8} = -\frac{1}{3}$$

and $-\frac{60}{-12} = \frac{60}{12} = \frac{60 \div 12}{12 \div 12} = 5.$

The quotient of two integers with unlike signs is negative.

Example 2. Are any of these rational numbers equivalent?

$$\frac{18}{24}, \quad \frac{-3}{2}, \quad \frac{-6}{8}, \quad \frac{16}{-12}, \quad \frac{-9}{-12}, \quad \frac{-20}{-15}$$

Solution. Reduce each fraction to its lowest terms.

$$\frac{18}{24} = \frac{3}{4}, \quad \frac{-3}{2} = -\frac{3}{2}, \quad \frac{-6}{8} = -\frac{3}{4}$$

$$\frac{16}{-12} = -\frac{4}{3}, \quad \frac{-9}{-12} = \frac{3}{4}, \quad \frac{-20}{-15} = \frac{4}{3}$$

$\frac{18}{24}$ and $\frac{-9}{-12}$ are equivalent rational numbers since they both reduce to $\frac{3}{4}$.

Example 3. Arrange the following numbers in order from least to greatest.

$$\frac{3}{4}, \quad \frac{-2}{5}, \quad \frac{-7}{-10}, \quad -\frac{-5}{12}, \quad -\frac{4}{-15}, \quad -\frac{-9}{-20}$$

Solution. $\frac{3}{4} = \frac{45}{60},$ $\frac{-2}{5} = -\frac{24}{60},$

$$\frac{-7}{-10} = \frac{7}{10} = \frac{42}{60} \qquad -\frac{-5}{12} = \frac{5}{12} = \frac{25}{60},$$

$$-\frac{4}{-15} = \frac{4}{15} = \frac{16}{60}, \qquad -\frac{-9}{-20} = -\frac{9}{20} = -\frac{27}{60}$$

Ordered from least to greatest:

$$-\frac{27}{60}, \quad -\frac{24}{60}, \quad \frac{16}{60}, \quad \frac{25}{60}, \quad \frac{42}{60}, \quad \frac{45}{60}$$

$$-\frac{-9}{-20}, \quad \frac{-2}{5}, \quad -\frac{4}{-15}, \quad -\frac{-5}{12}, \quad \frac{-7}{-10}, \quad \frac{3}{4}$$

Example 4. Express 3.2, -0.25, and -4.5 as quotients in lowest terms.

Solution. $3.2 = \frac{32}{10}$ $-0.25 = -\frac{25}{100}$ $-4.5 = -\frac{45}{10}$

$$= \frac{16}{5} \qquad\qquad = -\frac{1}{4} \qquad\qquad = -\frac{9}{2}$$

Exercises 3-1

A   1. Write the temperature shown on each thermometer.

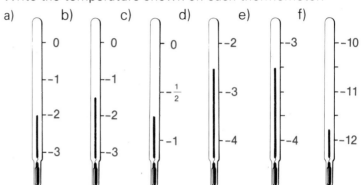

2. Write the rational numbers for the points indicated.

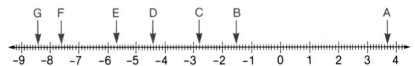

3. Reduce to lowest terms:

   a) $\dfrac{5}{-10}$ b) $\dfrac{10}{-15}$ c) $\dfrac{-12}{-30}$ d) $-\dfrac{6}{15}$ e) $-\dfrac{-6}{11}$

   f) $-\dfrac{-6}{18}$ g) $-\dfrac{4}{-14}$ h) $-\dfrac{-14}{-25}$ i) $-\dfrac{-15}{-35}$ j) $-\dfrac{-24}{-72}$

B   4. Show the positions of the rational numbers A to H on a number
   line. Then list them from least to greatest.

   $A \quad \dfrac{5}{-10}$ $B \quad \dfrac{-3}{2}$ $C \quad \dfrac{-7}{-28}$ $D \quad \dfrac{-5}{20}$

   $E \quad -\dfrac{12}{16}$ $F \quad -\dfrac{9}{-6}$ $G \quad \dfrac{10}{-8}$ $H \quad -\dfrac{-18}{-9}$

5. Express each of the following with 72 as the denominator. Then
   list them from greatest to least:

   $-\dfrac{3}{8}, \qquad \dfrac{-1}{3}, \qquad \dfrac{1}{-4}, \qquad -\dfrac{-2}{9}, \qquad \dfrac{-7}{-18}, \qquad -\dfrac{-13}{-36},$

6. List these numbers from least to greatest:

   $\dfrac{4}{-8}, \qquad \dfrac{-3}{-15}, \qquad \dfrac{-16}{40}, \qquad -\dfrac{19}{-60}, \qquad \dfrac{14}{30}, \qquad \dfrac{13}{20},$

7. Which five of these rational numbers are equivalent?

   $\dfrac{-2}{-3} \qquad \dfrac{4}{-6} \qquad \dfrac{-3}{4} \qquad \dfrac{-12}{-20} \qquad \dfrac{15}{-20}$

   $\dfrac{12}{-16} \qquad \dfrac{6}{-10} \qquad -\dfrac{-6}{-8} \qquad -\dfrac{-8}{-12} \qquad \dfrac{9}{-12}$

3 - 2   Multiplication and Division of Rational Numbers in the Form $\frac{m}{n}$

The rules for multiplication and division of rational numbers in the form $\frac{m}{n}$ are the same as those for multiplication and division of common fractions. The signs obey the same rules as when operating with integers.

Example 1. Simplify: $(-\frac{2}{3})(-\frac{9}{11})$.

Solution. $(-\frac{2}{3})(-\frac{9}{11})$

$$= \frac{2}{\overset{}{\underset{1}{\cancel{3}}}} \times \frac{\overset{3}{\cancel{9}}}{11}$$

$$= \frac{6}{11}$$

Numerator and denominator are divided by the common factor 3.

Example 2. Simplify: $6\frac{1}{4} \div (-\frac{5}{8})$.

Solution. $6\frac{1}{4} \div (-\frac{5}{8})$

$$= \frac{25}{4} \times (-\frac{8}{5})$$

$$= \frac{\overset{5}{\cancel{25}}}{\underset{1}{\cancel{4}}} \times (-\frac{\overset{2}{\cancel{8}}}{\underset{1}{\cancel{5}}})$$

$$= -10$$

Recall that dividing by a fraction is the same as multiplying by its reciprocal.

Example 3. Simplify: $-2\frac{1}{4} \times (-\frac{10}{-21}) \div (-6\frac{3}{7})$

Solution. $-2\frac{1}{4} \times (-\frac{10}{-21}) \div (-6\frac{3}{7})$

$$= (-\frac{9}{4}) \times (\frac{10}{21}) \div (-\frac{45}{7})$$

$$= (-\frac{\overset{1}{\cancel{9}}}{\underset{2}{\cancel{4}}}) \times (\frac{\overset{1}{\cancel{10}}}{\underset{3}{\cancel{21}}}) \times (-\frac{\overset{1}{\cancel{7}}}{\underset{1}{\underset{\cancel{8}}{\cancel{45}}}})$$

$$= \frac{1}{6}$$

Example 4. If $p = -3$, $q = -5$, and $r = 4$, evaluate:

a) $\dfrac{4p - 2q}{r}$;

b) $\dfrac{p + q}{p - q} \times (-\dfrac{p}{4r})$.

Solution. a) $\dfrac{4p - 2q}{r} = \dfrac{4(-3) - 2(-5)}{4}$

$$= \dfrac{-12 + 10}{4}$$

$$= \dfrac{-2}{4}, \text{ or } -\dfrac{1}{2}$$

b) $\dfrac{p + q}{p - q} \times \left(-\dfrac{p}{4r}\right) = \dfrac{(-3) + (-5)}{(-3) - (-5)} \times \left[-\dfrac{(-3)}{4(4)}\right]$

$$= \dfrac{(-8)}{2} \times \dfrac{3}{16}$$

$$= \dfrac{(-\cancel{8})}{2} \times \dfrac{3}{\cancel{16}_2}$$

$$= -\dfrac{3}{4}$$

Example 5. If $x = -\dfrac{2}{3}$ and $y = \dfrac{-1}{4}$, evaluate $\dfrac{-6}{7}xy$.

Solution. $\dfrac{-6}{7}xy = \left(\dfrac{-6}{7}\right)\left(-\dfrac{2}{3}\right)\left(\dfrac{-1}{4}\right)$

$$= -\left(\dfrac{\cancel{6}}{7}\right)\left(\dfrac{\cancel{2}}{\cancel{3}}\right)\left(\dfrac{1}{\cancel{4}}\right)$$

$$= -\dfrac{1}{7}$$

Example 6. Substitute $-\dfrac{2}{3}y$ for x in the expression $-\dfrac{9}{4}x$, and simplify.

Solution. $-\dfrac{9}{4}x = -\dfrac{9}{4}\left(-\dfrac{2}{3}y\right)$

$$= \left(-\dfrac{\cancel{9}^3}{\cancel{4}_2}\right)\left(-\dfrac{\cancel{2}}{\cancel{3}_1}y\right)$$

$$= \dfrac{3}{2}y$$

Exercises 3 - 2

A 1. Simplify:

a) $\left(\dfrac{1}{2}\right)\left(\dfrac{-3}{5}\right)$

b) $\left(\dfrac{-2}{3}\right)\left(\dfrac{6}{-7}\right)$

c) $\left(\dfrac{-1}{4}\right)\left(\dfrac{-2}{-3}\right)$

d) $\left(\dfrac{-3}{-8}\right)\left(\dfrac{12}{-21}\right)$

e) $\left(\dfrac{15}{-28}\right)\left(\dfrac{-21}{45}\right)$

f) $-\left(\dfrac{-5}{12}\right)\left(\dfrac{36}{-25}\right)$

g) $\left(-2\dfrac{1}{3}\right)\left(\dfrac{-6}{5}\right)$

h) $2\dfrac{2}{3} \times \left(-2\dfrac{1}{4}\right)$

i) $12\dfrac{1}{2} \times 1\dfrac{3}{5}$

2. Simplify:

a) $\left(-\frac{2}{3}\right) \div \left(\frac{5}{7}\right)$ b) $\left(\frac{-7}{10}\right) \div \left(\frac{4}{-9}\right)$ c) $\left(\frac{5}{-8}\right) \div \left(\frac{-3}{-4}\right)$

d) $\left(\frac{-12}{-25}\right) \div \left(\frac{8}{-15}\right)$ e) $\left(\frac{-8}{21}\right) \div \left(\frac{-4}{35}\right)$ f) $\left(-\frac{10}{33}\right) \div \frac{25}{44}$

g) $6\frac{1}{4} \div \left(-2\frac{1}{2}\right)$ h) $15\frac{1}{8} \div 5\frac{1}{2}$ i) $\left(-3\frac{1}{3}\right) \div \left(-4\frac{1}{6}\right)$

3. Simplify:

a) $\left(\frac{4}{-9}\right) \times \left(\frac{-21}{-32}\right) \times \left(\frac{-3}{14}\right)$ b) $\left(\frac{-10}{27}\right) \times \left(\frac{-8}{20}\right) \times \left(\frac{-45}{-28}\right)$

c) $\left(\frac{-6}{-25}\right) \div \left(\frac{-2}{-21}\right) \div \left(\frac{14}{-45}\right)$ d) $\left(\frac{12}{-39}\right) \div \left(\frac{-10}{-9}\right) \div \left(\frac{18}{-5}\right)$

e) $\left(\frac{15}{-32}\right) \times \left(\frac{-4}{5}\right) \div \left(-\frac{9}{16}\right)$ f) $\left(\frac{-12}{28}\right) \div \left(\frac{-8}{-15}\right) \times \left(\frac{-14}{-25}\right)$

g) $2\frac{1}{2} \div \left(-3\frac{1}{3}\right) \times 2\frac{2}{3}$ h) $\left(-3\frac{3}{4}\right) \times 1\frac{3}{5} \div \left(-1\frac{1}{5}\right)$

B 4. If $x = -2$, $y = -3$, and $z = -5$, evaluate:

a) $\frac{x + y}{z}$ b) $\frac{z - y}{x}$ c) $\frac{x - z}{y - x}$

d) $\frac{3x}{z} \times \frac{x + y}{y}$ e) $\frac{x}{y} \times \frac{3x}{z - y}$ f) $\frac{z}{y - z} \div \frac{y + x}{x}$

g) $\frac{2x}{y} \times \frac{z - x}{y}$ h) $\frac{3z}{x - z} \times \frac{y + z}{x}$ i) $\frac{5x}{z - 2y} \div \frac{2z}{y - 2x}$

5. If $x = -\frac{4}{5}$, $y = \frac{3}{-4}$, and $z = \frac{1}{2}$, evaluate:

a) $3x$ b) $\frac{-5}{7}y$ c) $\frac{9}{4}x$

d) xy e) $-yz$ f) xyz

g) $-\frac{2}{3}xz$ h) $\frac{5}{3}xy - \frac{8}{3}yx$ i) $5xz + 4y$

6. Substitute (i) $-\frac{2}{15}y$, (ii) $-\left(\frac{4}{-25}\right)y$, (iii) $\frac{3}{-10}y$ for x in the following expressions, and simplify: a) $\frac{5}{8}x$ b) $-\frac{5}{2}x$

C 7. Solve for x:

a) $\frac{1}{2}\left(\frac{x}{3}\right) = \frac{5}{6}$ b) $\left(\frac{-2}{3}\right)\frac{x}{5} = -\frac{8}{15}$ c) $\frac{5}{9}\left(\frac{x}{3}\right) = -\frac{10}{27}$

d) $\frac{x}{4}\left(\frac{-2}{3}\right) = \frac{-1}{2}$ e) $\left(\frac{-5}{12}\right)\frac{x}{2} = \frac{5}{4}$ f) $\left(\frac{3}{-10}\right)\left(\frac{x}{-9}\right) = -\frac{1}{6}$

g) $\frac{3}{5}x \div \frac{5}{4} = -\frac{24}{25}$ h) $\frac{x}{4}\left(\frac{-2}{5}\right) = \frac{-5}{10}$

3 - 3 Addition and Subtraction of Rational Numbers in the Form $\frac{m}{n}$

The rules for addition and subtraction of rational numbers in the form $\frac{m}{n}$ are the same as those for addition and subtraction of common fractions. The signs obey the same rules as when operating with integers. You will find the work easier if you first change any rational numbers with negative denominators to their equivalents with positive denominators.

Example 1. Simplify: a) $\frac{3}{4} + (\frac{2}{-3})$; b) $(\frac{-3}{4}) - (\frac{2}{-3})$.

Solution. a)
$$\frac{3}{4} + (\frac{2}{-3})$$
$$= \frac{3}{4} + (\frac{-2}{3})$$
$$= \frac{9}{12} + (\frac{-8}{12})$$
$$= \frac{1}{12}$$

b)
$$(\frac{-3}{4}) - (\frac{2}{-3})$$
$$= (\frac{-3}{4}) - (\frac{-2}{3})$$
$$= (\frac{-9}{12}) - (\frac{-8}{12})$$
$$= -\frac{1}{12}$$

Example 2. Simplify: $3(-\frac{5}{6}) + 5(-\frac{9}{8}) - 2(-\frac{3}{4})$

Solution.
$$3(-\frac{5}{6}) + 5(-\frac{9}{8}) - 2(-\frac{3}{4})$$
$$= -\frac{5}{2} - \frac{45}{8} + \frac{3}{2}$$
$$= -\frac{20}{8} - \frac{45}{8} + \frac{12}{8}$$
$$= -\frac{53}{8}$$

Example 3. If $p = \frac{-2}{3}$, $q = \frac{3}{10}$, and $r = -\frac{1}{2}$, evaluate:

a) $p + 2q - 3r$; b) $\frac{7}{8}p - \frac{5}{4}q - \frac{1}{3}r$.

Solution. a)
$$p + 2q - 3r = (\frac{-2}{3}) + 2(\frac{3}{10}) - 3(-\frac{1}{2})$$
$$= -\frac{2}{3} + \frac{3}{5} + \frac{3}{2}$$
$$= -\frac{20}{30} + \frac{18}{30} + \frac{45}{30}$$
$$= \frac{43}{30}$$

b) $\frac{7}{8}p - \frac{5}{4}q - \frac{1}{3}r = \frac{7}{8}(\frac{-2}{3}) - \frac{5}{4}(\frac{3}{10}) - \frac{1}{3}(-\frac{1}{2})$

$$= -\frac{7}{12} - \frac{3}{8} + \frac{1}{6}$$

$$= -\frac{14}{24} - \frac{9}{24} + \frac{4}{24}$$

$$= -\frac{19}{24}$$

Exercises 3 - 3

A   1. Simplify:

  a) $\frac{3}{4} + \frac{2}{3}$

  b) $\frac{5}{7} - \frac{2}{5}$

  c) $\frac{3}{8} - \frac{5}{6}$

  d) $\frac{-5}{12} + (\frac{-3}{8})$

  e) $\frac{2}{-9} + \frac{5}{6}$

  f) $-\frac{4}{5} - \frac{2}{3}$

  g) $\frac{3}{-4} - (\frac{-2}{5})$

  h) $\frac{9}{10} - (\frac{3}{-4})$

  i) $-(\frac{-3}{8}) - (\frac{5}{-4})$

  2. Simplify:

  a) $\frac{7}{3} + \frac{21}{4}$

  b) $\frac{47}{8} - \frac{8}{3}$

  c) $\frac{13}{2} - \frac{49}{5}$

  d) $\frac{17}{5} - \frac{35}{4}$

  e) $-\frac{14}{3} + \frac{12}{5}$

  f) $\frac{9}{7} - \frac{9}{5}$

  g) $\frac{17}{3} - \frac{11}{4}$

  h) $-\frac{13}{5} + \frac{11}{6}$

  i) $\frac{43}{3} - \frac{47}{7}$

  3. Simplify:

  a) $\frac{-2}{3} + (\frac{1}{-4}) - (\frac{-5}{6})$

  b) $\frac{3}{2} - (\frac{3}{-8}) - \frac{3}{4}$

  c) $-\frac{5}{8} + (\frac{-1}{-6}) - (\frac{2}{-3})$

  d) $\frac{3}{-10} - \frac{3}{4} - (\frac{-5}{8})$

  e) $2\frac{1}{4} + 5\frac{2}{3} - 4\frac{5}{6}$

  f) $4\frac{2}{5} - 6\frac{7}{10} - 2\frac{1}{3}$

  g) $-7\frac{5}{6} + 2\frac{1}{6} - 3\frac{2}{3}$

  h) $8\frac{1}{4} - 5\frac{3}{7} - 12\frac{1}{2}$

  i) $6\frac{1}{2} + (\frac{-2}{3}) - 1\frac{3}{4} + (\frac{4}{-3})$

  j) $-2\frac{2}{9} - 7\frac{1}{3} + (\frac{-1}{6}) + 5\frac{11}{18}$

B   4. If $x = -2$ and $y = 5$, evaluate:

a) $\dfrac{x}{y}$

b) $-\dfrac{y}{x}$

c) $\dfrac{2x}{-y}$

d) $\dfrac{x}{3} - \dfrac{y}{2}$

e) $\dfrac{3}{4} + \left(\dfrac{2}{-y}\right)$

f) $\dfrac{3}{y} - \dfrac{x}{8}$

g) $\dfrac{x}{3} + \dfrac{3}{x}$

h) $\dfrac{x + y}{y} - \dfrac{y}{x}$

i) $\dfrac{x - y}{4} + \dfrac{x}{y}$

5. Evaluate $x + y - z$ when

a) $x = \dfrac{2}{3},\ y = -\dfrac{1}{5},\ z = -\dfrac{1}{3}$;

b) $x = \dfrac{5}{-8},\ y = \dfrac{3}{4},\ z = \dfrac{2}{3}$;

c) $x = \dfrac{5}{-6},\ y = -\dfrac{1}{4},\ z = \dfrac{-3}{-8}$;

d) $x = \dfrac{8}{7},\ y = -\dfrac{3}{4},\ z = \dfrac{3}{2}$

6. If $r = -\dfrac{1}{2},\ s = \dfrac{2}{3},$ and $t = -\dfrac{1}{4},$ evaluate:

a) $r + s + t$

b) $r - s + t$

c) $-r - s - t$

d) $2r + 3s - 4t$

e) $r - 4s + 3t$

f) $3r - 2s - 5t$

g) $\dfrac{3}{5}r - \dfrac{3}{8}s + \dfrac{6}{5}t$

h) $\dfrac{-3}{4}r + \dfrac{5}{2}s - \dfrac{1}{3}t$

i) $\dfrac{5r}{3} - \dfrac{5s}{6} + \dfrac{t}{2}$

j) $\dfrac{2r}{3} - \dfrac{s}{4} - \left(\dfrac{2t}{-5}\right)$

C   7. Solve for x:

a) $\dfrac{2}{5} + \dfrac{x}{5} = \dfrac{6}{5}$

b) $\dfrac{2}{5} + \dfrac{x}{5} = \dfrac{-6}{5}$

c) $\dfrac{3}{-7} - \dfrac{x}{7} = -\dfrac{5}{7}$

d) $\dfrac{3}{-7} - \dfrac{x}{7} = \dfrac{5}{7}$

e) $\dfrac{7}{8} - \dfrac{4}{x} = \dfrac{11}{8}$

f) $\dfrac{5}{9} - \dfrac{8}{x} = \dfrac{1}{9}$

8. If $x > 0$, $y < 0$, and $z < 0$, which of the following expressions are always positive?

a) $\dfrac{x}{y}$

b) $\dfrac{xy}{z}$

c) $\dfrac{x}{yz}$

d) $\dfrac{y}{xz}$

e) $\dfrac{x}{y + z}$

f) $\dfrac{x - y}{z}$

g) $\dfrac{x}{x - y}$

h) $\dfrac{x - y}{x - z}$

i) $\dfrac{x}{y} + \dfrac{x}{z}$

j) $\dfrac{y}{z} - \dfrac{x}{y}$

9. Shares of the stocks of publicly owned companies are bought and sold at the Stock Exchange. Newspapers list each day's transactions in the manner shown.

Stock	Sales	High	Low	Close	Change
Scot Paper	413	$ $9\frac{3}{4}$	$9\frac{3}{4}$	$9\frac{3}{4}$	$-\frac{3}{4}$
Scot York	14 850	$ $7\frac{1}{2}$	7	$7\frac{1}{2}$	$+\frac{1}{8}$
Scotts A	1 000	$ $8\frac{1}{2}$	$8\frac{1}{2}$	$8\frac{1}{2}$	
Seagram	3 815	$$35\frac{1}{2}$	$35\frac{1}{4}$	$35\frac{1}{4}$	$-\frac{3}{4}$
Seco Cem	100	$ $9\frac{1}{2}$	$9\frac{1}{2}$	$9\frac{1}{2}$	$-\frac{1}{4}$
Selkirk A	1 200	$18	17	18	$+2\frac{1}{4}$
Shaw Pipe	10 900	$$12\frac{1}{2}$	$12\frac{1}{4}$	$12\frac{3}{8}$	$-\frac{1}{8}$
Shell Can	9 119	$$17\frac{1}{8}$	17	$17\frac{1}{8}$	$+\frac{1}{8}$

The clipping shows that on one day 10 900 shares of the stock of Shaw Pipe were traded. The highest price paid for a share was $12\frac{1}{2}$ and the lowest price was $12\frac{1}{4}$. At the end of the day's trading the price being paid for a share was $12\frac{3}{8}$. This was down $\frac{1}{8}$ on the previous day's closing price.

a) What happened to Scot York stock that day?

b) How much would 500 shares of Seagram's cost at the low price for the day?

c) How many shares of Seco Cem could have been bought for $1900?

d) Calculate the previous day's closing price for each stock.

10. An editor allows a value for each letter and space when composing and laying out the draft of a book. The values are:

$\frac{1}{2}$ for i, I, punctuation, letter l, numeral 1

$1\frac{1}{2}$ for m, w, mathematical signs

1 for all other digits and lower case letters

2 for spaces and all but two of the capital letters

$2\frac{1}{2}$ for M, W

Find the count value of each of these lines:

a) The answer given was $109\frac{1}{2}$.

b) Mass is measured in kilograms. Weight is in newtons.

c) Simplify 3.14 ÷ 4 × 7.2 correct to 2 decimal places.

3 - 4 Evaluating Expressions With Rational Numbers

Many of the formulas used in science, business, and industry involve rational numbers. Consider the problem at the beginning of this chapter.

Example 1. The rate of fuel consumption of a certain model of car is given by the formula

$$y = \frac{-36}{5}x + \frac{29}{2},$$

where x is the fraction of driving on the highway, and y is the rate of fuel consumption in litres per 100 km. What will be the rate of fuel consumption when

a) three-quarters of the driving is on the highway?

b) two-thirds of the driving is in the city?

Solution. a) Substituting $\frac{3}{4}$ for x in the formula, we get:

$$y = \frac{-\overset{9}{\cancel{36}}}{5}\left(\frac{3}{\underset{1}{\cancel{4}}}\right) + \frac{29}{2}$$

$$= \frac{-27}{5} + \frac{29}{2}$$

$$= -5.4 + 14.5, \text{ or } 9.1$$

The rate of fuel consumption will be approximately 9.1 L/100 km.

b) If two-thirds of the driving is in the city, one-third must be on the highway. Substituting $\frac{1}{3}$ for x in the formula, we get:

$$y = \frac{-\overset{12}{\cancel{36}}}{5}\left(\frac{1}{\underset{1}{\cancel{3}}}\right) + \frac{29}{2}$$

$$= -2.4 + 14.5, \text{ or } 12.1$$

The rate of fuel consumption will be approximately 12.1 L/100 km.

Example 2. An old reference book gives temperatures in Fahrenheit degrees. If C and F are the temperatures in Celsius and Fahrenheit degrees respectively, the formula relating them is

$$C = \frac{5}{9}(F - 32).$$

What Celsius temperature corresponds to
a) a room temperature of 77°F?
b) an oven temperature of 350°F?
c) a weather temperature of −40°F?

Solution. Substituting the given temperatures for F, we get:

a) $C = \frac{5}{9}(77 - 32)$

$= \frac{5}{9}(45)$

$= 25$

b) $C = \frac{5}{9}(350 - 32)$

$= \frac{5}{9}(318)$

$\doteq 176.7$

c) $C = \frac{5}{9}(-40 - 32)$

$= \frac{5}{9}(-72)$

$= -40$

The corresponding temperatures are:
a) 25°C;   b) 176.7°C   c) −40°C.

Example 3. A car-rental firm uses the formula
$C = \frac{39}{2}D + \frac{2}{25}(d - 50)$ to compute the cost, C,
in dollars, to a customer who uses the car for D
days to travel d km $(d > 50)$. How much will the
firm charge a customer who uses the car for three
days and travels 735 km?

Solution. $C = \frac{39}{2}D + \frac{2}{25}(d - 50)$

Substitute 3 for D and 735 for d in the formula.

$C = \frac{39}{2}(3) + \frac{2}{25}(735 - 50)$

$= \frac{117}{2} + \frac{2}{25}(685)$

$= 58.5 + 54.8$, or 113.3

The customer's bill will be for $113.30.

Exercises   3 - 4

A 1. Simplify when x is the value given:

a) $\frac{2}{3}x + \frac{3}{4}$, $x = \frac{-3}{4}$

b) $4(\frac{2}{5}x - \frac{2}{3})$, $x = \frac{1}{2}$

c) $x[\frac{-3}{8}x + (\frac{1}{-4})]$, $x = -\frac{1}{3}$

d) $(1 - \frac{1}{5}x)(3 - 2x)$, $x = -\frac{1}{2}$

2. A car-rental firm uses the formula $C = \frac{37}{2}D + \frac{3}{32}(d - 75)$ to compute the cost, C, in dollars, to a customer who uses one of their cars for D days to travel d km ($d > 75$). How much will the firm charge a customer who uses the car for five days and travels 955 km?

B 3. The cost, C, in dollars per hour, of operating a certain type of aircraft is given by the formula:

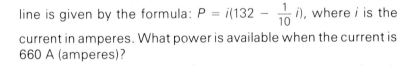

$$C = 900 + \frac{m}{200} + \frac{20\,000\,000}{m}$$

where m is the cruise altitude in metres. Find the hourly cost of operating the aircraft at 8000 m;   at 10 000 m.

4. The power, P, in kilowatts, delivered by a high-voltage power line is given by the formula: $P = i(132 - \frac{1}{10}i)$, where i is the current in amperes. What power is available when the current is 660 A (amperes)?

pitch

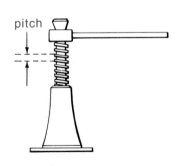

5. The efficiency, E, of a jack in machine-shop technology is given by the formula:

$$E = \frac{h(1 - \frac{1}{2}h)}{h + \frac{1}{2}},$$

where h is determined by the pitch of the thread. What is the efficiency of a jack whose value of h is $\frac{3}{5}$? $\frac{2}{3}$?

6. The focal length, f, of a concave spherical mirror is related to the object and image distances, p and q respectively, by the formula $\frac{1}{f} = \frac{1}{p} + \frac{1}{q}$.

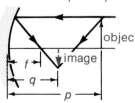

object

image

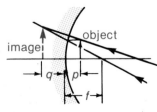

object

image

Calculate the values of f for these values of p and q:

a) $p = 3$, $q = 2$

b) $p = \frac{3}{4}$, $q = \frac{3}{5}$

c) $p = \frac{9}{20}$, $q = -\frac{3}{4}$

d) $p = 1.6$, $q = -2.4$

7. $A = P(1 + \frac{9t}{100})$ is the formula giving the amount of money, A, that an investment, P, is worth after being invested for t years at 9% per annum simple interest. Calculate A for the following investments:

a) $900 for 3 years

b) $12 000 for $2\frac{1}{2}$ years

c) $8000 for $1\frac{2}{3}$ years

d) $60 for 16 months

8. A market analysis shows that the monthly revenue, R, in dollars, from the sale of a certain model of TV set is according to the formula $R = (650 - 25x)(200 + 10x)$, where $x(x < 26)$ is the number of $25 reductions on the regular price of $650. Calculate the monthly revenue if the price of the set is

a) reduced by $100;

b) increased to $700.

9. A car brakes and decelerates uniformly. The distance, d, in metres, that it travels in t s is given by the formula
$$d = ut - 3.5t^2,$$
where u is its speed, in metres per second, before the brakes are applied. How far does it travel when

a) $u = 25$ m/s and $t = 5$ s?   b) $u = 35$ m/s and $t = 5.5$ s?

C 10. A car-rental firm uses the formula $C = \frac{37}{2}D + \frac{3}{32}(d - 75)$ to compute the cost, C, in dollars, to customers who use one of their cars for D days to travel d km $(d > 75)$. If a customer's bill was $126 for three days of use, how far did the customer drive?

11. Each year, Sheila uses the annual interest on the money in her savings account to buy Christmas presents. In this account, savings, S, grow to an amount, A, in t years according to the formula: $A = S(1 + 0.08t)$. What sum must be in her savings account for her Christmas money to amount to

a) $100?

b) a total of $480 in 3 years?

12. Use the formula in *Example 2* to find the Fahrenheit equivalent of

a) 100°C;   b) 215°C;   c) 5°C;   d) −20°C.

PROBLEM SOLVING STRATEGY

Use Simpler Numbers.

When first reading through a problem, it may not be apparent which operations are necessary to solve it. In such cases, using simpler numbers or rounded numbers may make the choice of operations more obvious.

Example 1. 42.5 L of a liquid have a mass of 48.2 kg. What is the mass of 32.5 L? of 17.5 L?

With Simpler Numbers, the problem might read:

4 L have a mass of 5 kg, what is the mass of 3 L?

If 4 L have a mass of 5 kg,

then 1 L has a mass of $\frac{5}{4}$ kg,

and 3 L have a mass of $\frac{5}{4} \times 3$ kg.

To solve the original problem then, we divide the number of kilograms by the number of litres to find the mass per litre, and then multiply by the number of litres.

Solution. Mass of 32.5 L: $\frac{48.2}{42.5} \times 32.5$ kg $\doteq$ 36.86 kg

Mass of 17.5 L: $\frac{48.2}{42.5} \times 17.5$ kg $\doteq$ 19.85 kg

Exercises

Use simpler numbers to determine which operations are needed to solve each problem. Then select one of the alternatives given.

1. A man jogs at 3.8 m/s. How many seconds will it take him to jog 225 m?

   a) 225×3.8 b) $3.8 \div 225$ c) $225 \div 3.8$

2. A cyclist travels 28.5 km in 1.75 h. At this speed, how many kilometres would she travel in 2.75 h?

   a) $\frac{1.75}{28.5} \times 2.75$ b) $\frac{2.75}{1.75} \times 28.5$ c) $\frac{1.75}{2.75} \times 28.5$

3. 9 men working for 7 h earn a total of $362.25. What would 8 men, working at the same rate for 8 h, earn, in dollars?

   a) $\frac{362.25}{7} \times \frac{9}{8} \times 8$ b) $\frac{362.25}{9} \times \frac{7}{8} \times 8$

   c) $\frac{362.25}{8} \times \frac{9}{8} \times 7$ d) $362.25 \times \frac{8 \times 8}{9 \times 7}$

Rounding numbers makes them easier to work with mentally. Not only can the necessary operations be determined, but the estimated answer serves as a safeguard against incorrect calculations.

Example 2. The average speed of a bus is 87.5 km/h. How long will it take to travel 1280 km?

With Rounded Numbers, the problem might read:
If the average speed is 100 km/h, how long will it take to travel 1300 km?

If the bus travels 100 km in 1 h,

then it will travel 1300 km in $\frac{1300}{100}$ h, or 13 h.

To solve the original problem, then, we divide the distance by the speed. The answer must be fairly close to 13 h.

Solution. Time to travel 1280 km: $\frac{1280}{87.5}$ h $\doteq$ 14.6 h

Had we obtained 1.46 or 146 from our calculations, our estimate would tell us that our answer was incorrect.

4. Show what rounded numbers you would use and estimate the cost of each of the following:
   a) 7 kg of apples at $1.95/kg
   b) 19 kg of meat at $3.05/kg
   c) 3 dozen cabbages at 95¢ each
   d) 24 grapefruit at 3 for 98¢
   e) 11 kg of butter at $3.75/kg
   f) 13 L of milk at 91¢/L
   g) 2 dozen bars of chocolate at 59¢ a bar
   h) 48 cans of apple juice at 75¢ a can
   i) Thirty-five 17¢ stamps
   j) 37 L of gasoline at 32.5¢/L

5. Jill's room is 5.3 m long, 3.7 m wide, and 2.85 m high. The door is 90 cm by 2.15 m. The window is 2.5 m by 1.5 m. If 1 L of paint will cover an area of 20 m² and costs $4.25, how much will it cost to give the walls two coats of paint? Give an approximate answer only but show the rounded figures.

3 - 5 Expressing Rational Numbers in Decimal Form

Rational numbers can be expressed in decimal form. Sometimes the decimal form of a rational number is a **terminating decimal**:

$$\frac{1}{4} = 0.25 \qquad \frac{1}{8} = 0.125 \qquad \frac{1}{25} = 0.04$$

That is, after a certain decimal place, all the rest of the digits are zeros. ($\frac{1}{4} = 0.250\,000\ldots$). Sometimes the decimal form is a **repeating decimal**:

$$\frac{1}{12} = 0.083\,333\,333\ldots, \text{ written } 0.08\overline{3}, \text{ or } 0.08\dot{3}$$

$$\frac{26}{11} = 2.363\,636\,363\ldots, \text{ written } 2.\overline{36}, \text{ or } 2.\dot{3}\dot{6}$$

$$\frac{22}{7} = 3.142\,857\,142\ldots, \text{ written } 3.\overline{142\,857}, \text{ or } 3.\dot{1}4285\dot{7}$$

That is, the decimal does not terminate but a digit or a sequence of digits repeats.

The following examples further illustrate the fact that a rational number can always be expressed as either a repeating or a terminating decimal.

Example 1. Express as a decimal: a) $\frac{3}{8}$; b) $\frac{35}{11}$.

Solution. a) Divide 3 by 8:

$$
\begin{array}{r}
0.375 \\
8\overline{)3.000} \\
\underline{2\,4} \\
60 \\
\underline{56} \\
40 \\
\underline{40} \\
00
\end{array}
$$

zero remainder ⟶ 00

$\frac{3}{8} = 0.375$, a terminating decimal

b) Divide 35 by 11:

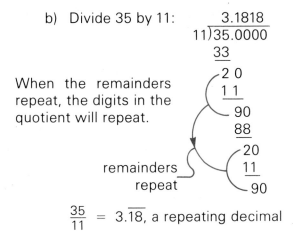

$$\begin{array}{r} 3.1818 \\ 11\overline{)35.0000} \\ 33 \\ \hline 2\,0 \\ 1\,1 \\ \hline 90 \\ 88 \\ \hline 20 \\ 11 \\ \hline 90 \end{array}$$

When the remainders repeat, the digits in the quotient will repeat.

remainders repeat

$\frac{35}{11} = 3.\overline{18}$, a repeating decimal

Example 2. Express $\frac{100}{7}$ as a repeating decimal.

Solution.

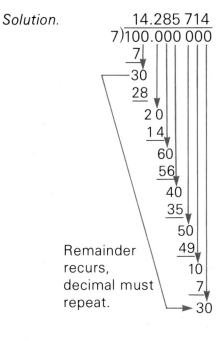

$$\begin{array}{r} 14.285\,714 \\ 7\overline{)100.000\,000} \\ 7 \\ \hline 30 \\ 28 \\ \hline 2\,0 \\ 1\,4 \\ \hline 60 \\ 56 \\ \hline 40 \\ 35 \\ \hline 50 \\ 49 \\ \hline 10 \\ 7 \\ \hline 30 \end{array}$$

As we divide 7 into 100, we bring down zeros. So long as we get different remainders at each stage, the quotient does not repeat. As soon as a remainder recurs, the quotient repeats. Since there are only 7 possible remainders (0, 1, 2, ... 6), we know that $100 \div 7$ must repeat on or before the seventh digit.

Remainder recurs, decimal must repeat.

We write: $\frac{100}{7} = 14.\overline{285\,714}$

Example 3. In football, the pass-completion average of a quarterback is found by dividing the number of passes completed by the number attempted.

  a) Calculate the lifetime passing averages of the following outstanding quarterbacks;

  b) List their averages in order from greatest to least.

Name	Attempts	Completions
Otto Graham	2626	1464
Len Dawson	3366	1905
Fran Tarkenton	4449	2459
Bart Starr	3149	1808
Johnny Unitas	5186	2830

Solution. a)

Name	Pass-completion Average
Otto Graham	$1464 \div 2626 = 0.557\,501\,9\ldots$
Len Dawson	$1905 \div 3366 = 0.565\,953\,6\ldots$
Fran Tarkenton	$2459 \div 4449 = 0.552\,708\,4\ldots$
Bart Starr	$1808 \div 3149 = 0.574\,150\,5\ldots$
Johnny Unitas	$2830 \div 5186 = 0.545\,699\,9\ldots$

  b) $0.574\,150\,5\ldots, \quad 0.565\,953\,6\ldots, \quad 0.557\,501\,9\ldots,$
      $0.552\,708\,4\ldots, \quad 0.545\,699\,9\ldots.$

We can summarize the sets of numbers we have used up to now as follows:

  i) The **natural**, or counting, **numbers**: $N = \{1, 2, 3, 4, \ldots\}$

  ii) The **integers**: $I = \{\ldots, -3, -2, -1, 0, 1, 2, 3, \ldots\}$

  iii) The **rational numbers**, Q. These are all the numbers that can be written in the form $\frac{m}{n}$, where m and n are integers and $n \neq 0$.

There is one more set of numbers that should be noted at this time:

  iv) The **real numbers**, R. These are all the rationals *and* those numbers having decimal representations that are *non-terminating* and *non-repeating*, such as:

$$\pi \doteq 3.141\,592\,653\,589\,793\,238\,462\,643\,383\,279\,50\ldots.$$

$$e \doteq 2.718\,281\,828\,459\,045\,235\,360\,287\,471\,352\,66\ldots.$$

$$\sqrt{3} \doteq 1.732\,050\,807\,568\,877\,293\,527\,446\,341\,505\,87\ldots.$$

Exercises   3 - 5

A   1. Write these numbers to eight decimal places:

   a) $3.\overline{23}$ b) $42.\overline{307}$ c) $-81.4\dot{6}$

   d) $690.0\overline{45}$ e) $-2.6\dot{5}1\dot{3}$ f) $2.\dot{6}51\dot{3}$

   g) $0.0\overline{69}$ h) $-3.5\dot{1}\dot{4}$ i) $-0.007\dot{4}$

2. Write these repeating decimals in short form:

   a) 6.3333...   b) 0.171 717 17...   c) 42.135 135...

   d) 0.036 363 6...   e) -38.348 348...   f) -46.233 33...

   g) -0.717 171...   h) 813.813 813...   i) 15.315 151 5...

3. Write as terminating or repeating decimals:

   a) $\dfrac{3}{5}$ b) $\dfrac{2}{-3}$ c) $\dfrac{4}{9}$ d) $-\dfrac{3}{8}$

   e) $\dfrac{7}{21}$ f) $\dfrac{-3}{22}$ g) $\dfrac{7}{15}$ h) $-\dfrac{1}{6}$

   i) $\dfrac{5}{16}$ j) $\dfrac{-17}{27}$ k) $\dfrac{11}{12}$ l) $\dfrac{11}{13}$

B   4. In the hockey play-offs, a goalkeeper let in 11 goals in 7 games. The goalkeeper for the opposing team let in only 8 goals in the same number of games. Calculate the "goals against" average for each goalkeeper correct to two decimal places.

5. In baseball, a player's batting average is found by dividing the number of hits by the number of official times at bat, and rounding to three decimal places.

   a) Calculate the batting average of these famous players:

Batter	Year	Times at Bat	Number of Hits	Batting Average
Hugh Duffy	1894	539	236	
Ty Cobb	1911	591	248	
Babe Ruth	1924	529	200	
Lou Gehrig	1927	584	218	
Ted Williams	1941	456	185	
Rod Carew	1972	535	170	
Billy Williams	1972	594	191	
Dave Cash	1975	699	213	

   b) List the batters in the order of their batting averages from greatest to least.

6. The following numbers are real. For those ending with ...,
assume whatever pattern you see continues.

$$\frac{1}{12}, \qquad -1.8, \qquad 0.616\,611\,611\,161\ldots, \qquad 0, \qquad 2\frac{3}{4} \qquad 0.\overline{3}$$

$$7, \qquad -13.85\overline{7\,62}, \qquad -17\frac{7}{17}, \qquad 6.432\,432\ldots, \qquad 0.625, \qquad \frac{0}{21}$$

Which of the above numbers are:

a) natural numbers?   b) integers? c) rational numbers?

7. Arrange these fractions from greatest to least:

$$\frac{6}{7} \qquad \frac{5}{8} \qquad \frac{9}{11} \qquad \frac{10}{13} \qquad \frac{13}{15}$$

C 8. a) Simplify $\frac{2}{3} + \frac{5}{6}$, and then write in decimal form.

b) Write $\frac{2}{3}$ and $\frac{5}{6}$ in decimal form and then find their sum.
How does your answer compare with that for part (a)?

c) Repeat the procedure of parts (a) and (b) for the following:

  i) $\frac{3}{4} + \frac{2}{5}$ ii) $\frac{5}{8} - \frac{1}{4}$ iii) $\frac{1}{6} - \frac{5}{9}$

  iv) $\frac{2}{9} - \frac{5}{11}$ v) $\frac{7}{16} + \frac{5}{12}$ vi) $\frac{29}{37} - \frac{11}{37}$

9. Use a calculator to express these fractions in decimal form.
What do you notice?

  a) $\frac{5}{173}$ b) $\frac{50}{173}$ c) $\frac{500}{173}$ d) $\frac{5000}{173}$

10. Use your conclusion from Exercise 9 to express the following in
decimal form as accurately as possible.

  a) $\frac{1}{810}$ b) $\frac{6}{5293}$ c) $\frac{6.9}{9572.6}$ d) $\frac{2.3 \times 6.4}{168.7 \times 24.9}$

11. a) Use a calculator to verify that $\frac{1}{7} = 0.\overline{142\,857}$ and that

$\frac{2}{7} = 0.\overline{285\,714}$. Notice that the repeating digits for

the two fractions are the same. They can be arranged in a
circle, as shown.

b) Find the decimal representations of $\frac{3}{7}, \frac{4}{7}, \frac{5}{7}, \frac{6}{7},$ and decide

if they also fit the circle of digits. If they do, we can say that
the digits of the decimal representations of this set of fractions form a *cyclic pattern*.

c) Investigate the cyclic patterns for:

  i) $\frac{1}{13}, \frac{2}{13}, \frac{3}{13}, \ldots;$ ii) $\frac{1}{14}, \frac{2}{14}, \frac{3}{14}, \ldots;$ iii) $\frac{1}{21}, \frac{2}{21}, \frac{3}{21}, \ldots$

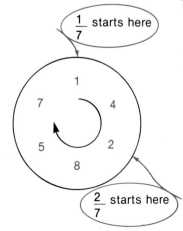

Review Exercises

1. Express as quotients in lowest terms:
   a) -2.5 b) 0.125 c) -1.8

2. Which of these rational numbers are equivalent?

   $$\frac{-6}{9}, \quad \frac{16}{-25}, \quad \frac{21}{28}, \quad -\frac{14}{21}, \quad \frac{-30}{-45}, \quad \frac{10}{-15}, \quad \frac{-9}{12}$$

3. Arrange the following from least to greatest:

   $$-\frac{15}{16}, \quad \frac{-43}{48}, \quad \frac{11}{12}, \quad \frac{5}{6}, \quad -\frac{-31}{-32}, \quad \frac{2}{3}$$

4. a) List four rational numbers between 0 and 1.
   b) List four rational numbers between -1 and -2.

5. Simplify:
   a) $-\frac{2}{3} \times \frac{7}{8}$ b) $(\frac{5}{-8})(\frac{-9}{12})$ c) $(\frac{3}{-5})(\frac{-7}{-8})$

   d) $(\frac{13}{15})(\frac{30}{-39})$ e) $(-\frac{16}{25})(-\frac{-20}{-24})$ f) $(-1\frac{7}{8})(-2\frac{3}{5})$

   g) $(-\frac{6}{5})(\frac{12}{-15})(\frac{-25}{36})$ h) $(2\frac{2}{3})(-1\frac{1}{5})(\frac{-4}{7})$ i) $(\frac{-5}{7})(-1\frac{1}{2})(\frac{-14}{-15})$

6. Simplify:
   a) $\frac{3}{7} \div (\frac{-9}{14})$ b) $-3\frac{1}{4} \div (\frac{2}{-3})$ c) $\frac{7}{-8} \div (-2\frac{1}{4})$

   d) $10\frac{1}{4} \div (-2\frac{1}{2})$ e) $\frac{-24}{-35} \div (\frac{16}{-21})$ f) $-5\frac{1}{2} \div (-2\frac{1}{3})$

   g) $\frac{-3}{5} \div (\frac{-5}{-12}) \div (\frac{-9}{10})$ h) $3\frac{3}{5} \div (-3) \div 1\frac{1}{2}$

7. Simplify:
   a) $\frac{3}{5} + \frac{4}{7}$ b) $\frac{5}{12} + \frac{3}{8}$ c) $\frac{2}{9} + \frac{7}{12}$

   d) $\frac{3}{11} + (\frac{-5}{11})$ e) $\frac{13}{-24} + (\frac{-7}{24})$ f) $\frac{-2}{3} + (\frac{-4}{9})$

   g) $\frac{4}{-5} + \frac{14}{15}$ h) $\frac{-3}{-7} + (\frac{-2}{5})$ i) $\frac{-4}{9} + (\frac{17}{-21})$

   j) $-\frac{5}{12} + (\frac{7}{-9})$ k) $-2\frac{2}{15} + 3\frac{1}{6}$ l) $4\frac{2}{3} + (-7\frac{3}{4})$

8. Express in decimal form to three places:
   a) $\frac{3}{8}$ b) $\frac{-4}{7}$ c) $\frac{7}{-12}$ d) $2\frac{2}{9}$

   e) $-3\frac{5}{16}$ f) $\frac{2}{3}$ g) $-\frac{5}{11}$ h) $\frac{-3}{-13}$

9. Simplify:

   a) $\dfrac{7}{9} - \dfrac{1}{6}$ b) $\dfrac{5}{6} - \dfrac{3}{10}$ c) $\dfrac{7}{8} - \dfrac{5}{12}$

   d) $\dfrac{-5}{8} - \dfrac{3}{8}$ e) $\dfrac{17}{-20} - (\dfrac{-12}{20})$ f) $\dfrac{-7}{8} - (\dfrac{-1}{4})$

   g) $\dfrac{9}{11} - (\dfrac{-3}{5})$ h) $\dfrac{-3}{-4} - (\dfrac{-2}{3})$ i) $5\dfrac{17}{24} - (-2\dfrac{3}{8})$

   j) $-8\dfrac{2}{7} - (-3\dfrac{5}{7})$ k) $-12\dfrac{2}{9} - 4\dfrac{5}{6}$ l) $-4\dfrac{2}{3} - (-7\dfrac{3}{4})$

10. Simplify:

   a) $2\dfrac{1}{2} - 3\dfrac{2}{3} + 1\dfrac{1}{4}$ b) $2\dfrac{1}{2} - 1\dfrac{1}{4} \div \dfrac{5}{4}$

   c) $\dfrac{-6}{5} + (\dfrac{10}{-2})(\dfrac{-3}{5})$ d) $[\dfrac{3}{-4} - (\dfrac{-3}{4})] \div 2$

   e) $-6(\dfrac{4}{5} - \dfrac{1}{2})$ f) $(\dfrac{3}{5})(-\dfrac{1}{2})(\dfrac{-6}{3}) + \dfrac{1}{5}$

   g) $(\dfrac{3}{4})(\dfrac{1}{-2}) + (\dfrac{5}{6})(\dfrac{-1}{3})$

   h) $\dfrac{3}{8} \times \dfrac{2}{3} - (\dfrac{1}{2})(\dfrac{-5}{6}) + (\dfrac{3}{5})(\dfrac{3}{-4})$

   i) $[2\dfrac{1}{2} \div (\dfrac{-4}{5})] - (\dfrac{3}{-4})(\dfrac{-8}{9})$

   j) $[(\dfrac{-3}{5}) \div (\dfrac{4}{-5})] - (\dfrac{1}{2})(\dfrac{2}{3}) \div (-\dfrac{1}{6})$

   k) $[-6\dfrac{1}{2} \times (\dfrac{-4}{5})][(-\dfrac{2}{3}) \div 2\dfrac{1}{2} + 3\dfrac{3}{5}]$

   l) $[(\dfrac{1}{-2}) \div (\dfrac{-2}{3})] + 4[-3 + (\dfrac{1}{2})(\dfrac{-5}{6})]$

11. If $x = -3$, $y = 4$, $z = -2$, and $w = -5$, evaluate:

   a) $\dfrac{7x}{y}$ b) $\dfrac{xz}{w}$ c) $-2x - (\dfrac{y}{-w})$

   d) $\dfrac{5}{3}(\dfrac{x}{y}) - \dfrac{1}{2}(\dfrac{z}{w})$ e) $\dfrac{3y - wz}{-x}$ f) $\dfrac{-x}{z} + \dfrac{y}{w}$

   g) $(\dfrac{-2}{5})(\dfrac{w}{x}) - \dfrac{3}{2}(\dfrac{y}{w})$ h) $\dfrac{yw}{-x} - \dfrac{xz}{y}$ i) $\dfrac{-3}{4}xz + \dfrac{1}{2}y - \dfrac{z}{2}$

12. A car rental firm uses the formula: $C = \dfrac{41}{2}D + \dfrac{1}{8}(d - 80)$ to compute the cost, C, in dollars, to a customer who uses one of their cars for D days to travel d km ($d > 80$). How much will a customer pay for using a car for 4 days and driving 800 km?

13. The following numbers are real. For those ending with ...,
    assume whatever pattern you see continues.

$$\frac{1}{8}, \quad 13.25, \quad 0, \quad -13, \quad \frac{0}{1}, \quad 0.707\,007\,000\,7\ldots\ldots,$$

$$-11\frac{3}{8}, \quad 1, \quad 7.692\,307\,692\ldots, \quad 5.8\overline{5}, \quad 5\frac{1}{2}, \quad 5.\overline{127\,16}$$

Which of the above numbers are

a) natural numbers? b) integers? c) rational numbers?

14. The focal length, f, of a concave spherical mirror is related to the
    object and image distances, p and q respectively, by the formula
    $\frac{1}{f} = \frac{1}{p} + \frac{1}{q}$. Find f for these values of p and q:

a) $p = 2, \quad q = 8$ b) $p = 7.5, \quad q = 15$

c) $p = 2.4, \quad q = -4.8$ d) $p = 12.5, \quad q = -62.5$

Adding and Subtracting Rational Numbers in Fraction Form

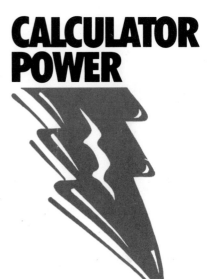

CALCULATOR POWER

Simple hand calculators will multiply and divide fractions if the numbers are keyed in as they appear. This is not the case when it comes to adding and subtracting fractions. If we key in:

$$2 \;\boxdot\; 5 \;\boxplus\; 3 \;\boxdot\; 4 \;\boxminus\;$$

in order to simplify $\frac{2}{5} + \frac{3}{4}$ we get 0.85. This is incorrect because the last operation causes $\frac{2}{5}$ as well as 3 to be divided by 4. However, by altering the sequence of operations, the correct result can be obtained.

$$\frac{a}{b} + \frac{c}{d} \quad \text{can be written} \quad \frac{ad + bc}{bd}$$

$$= (\frac{ad + bc}{b})\frac{1}{d}$$

$$= (\frac{ad}{b} + c)\frac{1}{d}$$

The sequence: $a \;\boxtimes\; d \;\boxdot\; b \;\boxplus\; c \;\boxdot\; d \;\boxminus\;$

gives the correct result.

Example 1. Simplify: $\frac{2}{5} + \frac{3}{4}$

Sequence: 2 ⊠ 4 ⊡ 5 ⊞ 3 ⊡ 4 ▱

Display: 1.15

The common fraction can be obtained by multiplying the displayed result by the product of the denominators, 20, to obtain the numerator:

$$\frac{2}{5} + \frac{3}{4} = 1.15, \text{ or } \frac{1.15 \times 20}{20}. \text{ That is, } \frac{23}{20}.$$

Example 2. Simplify: $-\frac{5}{6} + \frac{3}{8}$

Sequence: 0 ⊟ 5 ⊠ 8 ⊡ 6 ⊞ 3 ⊡ 8 ▱

Display: -0.4583333

The product of the denominators is 48. Multiply the display result by 48, and round to the nearest whole number to find the numerator:

$$-\frac{5}{6} + \frac{3}{8} = -\frac{22}{48}, \text{ or } -\frac{11}{24}$$

What change in the sequence is needed to simplify: $\frac{2}{5} - \frac{3}{4}$?

Exercises

Use your calculator to simplify:

1. $\frac{3}{5} + \frac{1}{4}$

2. $\frac{7}{8} + \frac{5}{6}$

3. $\frac{2}{3} + \frac{3}{8}$

4. $\frac{5}{8} + \frac{7}{12}$

5. $\frac{-4}{9} + \frac{3}{4}$

6. $\frac{5}{6} - \frac{1}{8} + \frac{2}{7}$

Cumulative Review   (Chapters 1 - 3)

1. Evaluate:

   a) $3m - 2n + 7p$ for $m = \dfrac{4}{3}$, $n = -3$, $p = -\dfrac{5}{2}$

   b) $4a - 3b + 5c$ for $a = -\dfrac{3}{4}$, $b = -\dfrac{5}{3}$, $c = \dfrac{2}{5}$

   c) $\dfrac{5}{6}x - \dfrac{3}{8}y - \dfrac{7}{9}z$ for $x = -18$, $y = 24$, $z = -27$

2. Simplify:

   a) $7t - 5t + 2t$ b) $-12a + 7a - 4a$

   c) $7ab + 3ab - ab$ d) $14b - 9b - 3b$

   e) $4(3x - 7) - 2(7x - 5)$ f) $-5(2a - 3) - 4(a + 2)$

3. Simplify:

   a) $4 + 5 \times 3 - 6$ b) $12 \div 6 + 3 - 20 \div 5$

   c) $9 - (4 + 8) \div 2 - 5$ d) $3 \times 4 + 2 \times (6 \div 3)$

   e) $2 \times 3 - 4 + 5 \times 6$ f) $2 \times (3 - 4) + 5 \times 6$

4. Simplify:

   a) $(-7) - (-3) + (+5)$ b) $(-17) + (-12) + (-4)$

   c) $(-30) - (-90) - (-60)$ d) $(+84) - (-76) - (+14)$

   e) $(-4 - 9) - (-14 - 7)$ f) $(7 - 9) - (-12 + 6)$

   g) $-(8 + 13 - 7) + (11 - 5)$ h) $(-17 + 8) - (13 - 4 + 26)$

5. Simplify:

   a) $(-5)(+4)(-3)$ b) $(-7)(-3)(-6)$

   c) $(-4)(-2)(+3)(-5)$ d) $(+8)(-3)(-5)(+2)$

   e) $(-5)(+8)(-11)(+2)$ f) $(+9)(+8)(+5)(-5)$

6. Simplify:

   a) $\dfrac{(+7)\,(-5)}{(-28)\,(-15)}$ b) $\dfrac{(+169)\,(-63)}{(-9)\,(-13)}$

   c) $\dfrac{+39}{-3} + \dfrac{-48}{+6}$ d) $\dfrac{-40}{+8} - \dfrac{-64}{+16}$

   e) $\dfrac{(+5)\,(-7)}{-91} + \dfrac{(-6)\,(+4)}{39}$ f) $\dfrac{-51}{(+17)\,(-3)} - \dfrac{95}{(-5)\,(-19)}$

7. If $x = -7$, $y = +14$, $z = -21$, evaluate:

   a) $\dfrac{x}{y} - \dfrac{y}{z}$ b) $\dfrac{3x}{z} + \dfrac{2z}{y}$

8. Simplify:

   a) $\left(\frac{2}{-3}\right)\left(\frac{-5}{-6}\right)$

   b) $\left(-\frac{7}{8}\right)\left(\frac{11}{-14}\right)\left(\frac{-16}{22}\right)$

   c) $\frac{3}{5} \div \left(\frac{-9}{7}\right)$

   d) $8\frac{1}{3} \div (-2) \div \left(-1\frac{1}{4}\right)$

   e) $\frac{5}{4} \times \left(\frac{-32}{35}\right) \div \frac{-7}{8}$

   f) $1\frac{19}{20} \times \left(-\frac{2}{13}\right) \div \frac{24}{25}$

9. How much fuel should a car use on a trip of 350 km if its Ministry of Transport rating is 8 L/100 km?

10. Simplify:

    a) $-\frac{3}{5} + \left(\frac{4}{-3}\right)\left(-\frac{6}{5}\right)$

    b) $\left(\frac{2}{5}\right)\left(-\frac{3}{2}\right)\left(\frac{-4}{3}\right) + \frac{1}{6}$

    c) $\left(\frac{3}{4}\right)\left(-\frac{5}{2}\right) + \left(\frac{1}{6}\right)\left(\frac{-5}{4}\right)$

    d) $\left[\left(\frac{-6}{5}\right) \div \left(\frac{4}{-5}\right)\right] - \left(\frac{5}{2}\right)\left(\frac{1}{3}\right) \div \left(\frac{-5}{6}\right)$

11. If $a = 3$, $b = -4$, $c = -2$, $d = -6$, evaluate:

    a) $1\frac{1}{3}\left(\frac{c}{b}\right) - \frac{3}{2}\left(\frac{a}{d}\right)$

    b) $\frac{3b - cd}{-a}$

    c) $\frac{bd}{-a} - \frac{ac}{b}$

    d) $-\frac{3}{4}ac + \frac{1}{2}b - \frac{cd}{3}$

12. A circular patio has a diameter of 5.5 m.
    a) Calculate the area of the patio. Use $\pi \doteq 3.14$.
    b) If it cost $4.80/m² to build the patio, calculate the total cost to the nearest dollar.

13. The rate of fuel consumption of a certain model of car is given by the formula: $y = \frac{-41}{5}x + \frac{31}{2}$, where x is the fraction of driving on the highway and y is the rate of fuel consumption in litres per 100 km. What will be the rate of fuel consumption when
    a) two-fifths of the driving is on the highway?
    b) one-third of the driving is in the city?

14. The following numbers are real:
    $$0.625, \quad \frac{1}{15}, \quad 7, \quad -7, \quad 4.8\overline{1}, \quad 3\frac{1}{3}, \quad 0, \quad \frac{0}{4}$$
    $$13.121\ 121\ 112\ 111\ 121\ ..., \quad -4.2, \quad 5.\overline{416}, \quad \pi$$
    Which of the above numbers are
    a) natural numbers?   b) integers?   c) rational numbers?

A car's fuel economy is given by the formula: $y = \frac{-36}{5}x + \frac{29}{2}$, where y is the rate of fuel consumption in litres per 100 km and x is the fraction of driving done on the highway. What fraction of the driving is on the highway if the car is averaging 10.9 L/100 km? (See *Example 3* in Section 4-4.)

4 - 1   Isolating the Variable

In Chapter 1 we said that an equation is a mathematical sentence which uses an equals sign to relate two expressions. Simple equations, such as $x - 4 = 10$, were solved by inspection as follows:

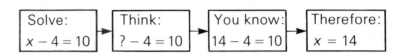

Solve:	Think:	You know:	Therefore:
$x - 4 = 10$	$? - 4 = 10$	$14 - 4 = 10$	$x = 14$

To solve harder equations, it is helpful to think of a level balance. The masses in each pan can be changed, but as long as the total on each side is the same, the balance remains level. The same rule applies to equations.

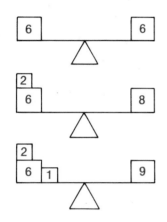

> Whatever change is made to one side of an equation must be made to the other side.

Any change is made in order to reduce the equation to its solution: $x = a$, where a is a number called the **root** of the equation. In this form, the variable is said to be isolated.

Example 1. Solve for x: $x + 4 = 10$.

Solution. $x + 4$ balances 10.
Subtract 4 from both sides to isolate x.
$$x + 4 - 4 = 10 - 4$$
$$x = 6$$

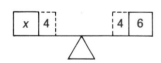

Example 2. Solve for x: $x - 4 = 10$.

Solution. $x - 4$ balances 10.
Add 4 to both sides to isolate x.
$$x - 4 + 4 = 10 + 4$$
$$x = 14$$

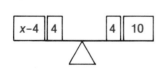

To solve an equation in the form	To solve an equation in the form
$x + b = c,$	$x - b = c,$
where b and c are numbers, isolate the variable by subtracting b from both sides.	where b and c are numbers, isolate the variable by adding b to both sides.

Example 3.   Solve for x: $\frac{x}{3} = 10$

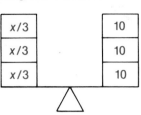

Solution. $\frac{x}{3}$ balances 10.

Multiply both sides by 3 to isolate x.

$$\frac{x}{3} \times 3 = 10 \times 3$$

$$x = 30$$

Example 4.   Solve for x: $2x = 10$

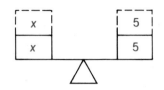

Solution. $2x$ balances 10.

Divide both sides by 2 to isolate x.

$$2x \div 2 = 10 \div 2$$

$$x = 5$$

To solve an equation in the form $$\frac{x}{b} = c,$$ where b and c are numbers, $b \neq 0$, isolate the variable by multiplying both sides by b.	To solve an equation in the form $$bx = c,$$ where b and c are numbers, $b \neq 0$, isolate the variable by dividing both sides by b.

Exercises 4 - 1

A 1. Solve:

a) $x + 5 = 11$ b) $z - 3 = 10$ c) $4 + y = -9$

d) $-8 = 2 + x$ e) $3 = y - 5$ f) $5x = 30$

g) $\frac{1}{3}z = -4$ h) $48 = -6y$ i) $-12 = \frac{x}{2}$

B 2. Solve:

a) $y + 7 = 16$ b) $w - 23 = 61$ c) $20 = 10 + z$

d) $-2 = x - 14$ e) $13 + x = 13$ f) $-11 = n + 21$

g) $3.7 + p = 5.2$ h) $x + 1.35 = 0.85$ i) $z - \frac{1}{4} = 1\frac{3}{4}$

3. Solve:

a) $\frac{w}{4} = 3$ b) $13x = 169$ c) $\frac{q}{15} = 5$

d) $144 = 36p$ e) $\frac{1}{29}m = 5$ f) $56 = -7r$

g) $\frac{1}{13}x = \frac{2}{13}$ h) $2.5y = -10$ i) $8.5 = 1.7x$

4 - 2 Solutions Requiring Two or More Steps

Many equations require more than one step to isolate the variable. To solve them, isolate the term containing the variable first.

Example 1. Solve: a) $-5w + 9 = 21$; b) $7.3 = 6y - 2$.

Solution. a) $-5w + 9 = 21$ b) $7.3 = 6y - 2$

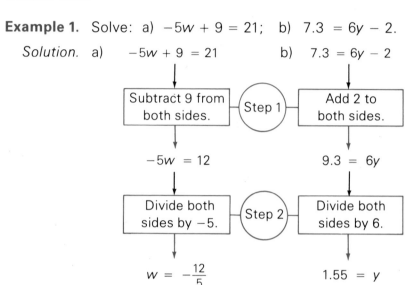

Subtract 9 from both sides.	Step 1	Add 2 to both sides.

$-5w = 12$ $9.3 = 6y$

Divide both sides by -5.	Step 2	Divide both sides by 6.

$$w = -\frac{12}{5}$$ $$1.55 = y$$

Expanding by the distributive law often helps to simplify an equation.

Example 2. Solve: $3(x - 2) = 8$.

Solution. $3(x - 2) = 8$

Expand by the distributive law.

$3x - 6 = 8$

Add 6 to both sides.

$3x = 14$

Divide both sides by 3.

$$x = \frac{14}{3}$$

Example 3. The cost, C, in dollars, of taking a class of n students on a weekend excursion to Ottawa is given by the formula: $C = 180 + 35n$. If the cost was $1685, how many students went?

Solution. Substitute 1685 for C:

$$1685 = 180 + 35n$$
$$1685 - 180 = 35n$$
$$1505 = 35n$$
$$n = \frac{1505}{35}, \text{ or } 43$$

43 students went on the excursion.

Exercises 4 - 2

A 1. Solve:

a) $2w - 5 = 11$ b) $5n - 8 = 12$ c) $8 - 2u = 12$

d) $0 = 7p - 35$ e) $-9p - 81 = 0$ f) $10 = -3x - 5$

g) $-11z - 2 = 20$ h) $3 - 2y = -7$ i) $-9 = 8b - 1$

j) $17 = 5q + 2$ k) $8 - 3z = -1$ l) $-13 = 4p - 1$

2. Solve:

a) $8t + 7 = -10$ b) $9p - 2 = 6$ c) $-5r + 6 = 8$

d) $4x + \frac{3}{4} = 1\frac{3}{4}$ e) $3x - \frac{1}{4} = 2$ f) $-1 = 7x - \frac{5}{12}$

g) $10t - \frac{2}{5} = \frac{3}{5}$ h) $7x + \frac{5}{4} = -3$ i) $11x - \frac{1}{2} = \frac{3}{2}$

j) $9t - \frac{3}{5} = -\frac{6}{5}$ k) $1 = \frac{1}{3} - 8s$ l) $1\frac{1}{6} + 8t = 2\frac{1}{6}$

3. Solve:

a) $\frac{1}{3}r - 3 = -6$ b) $\frac{1}{4}x + 6 = 10$ c) $\frac{1}{7}x - 1 = \frac{9}{7}$

d) $0 = \frac{6}{7}x - \frac{3}{7}$ e) $\frac{11}{12} = -\frac{1}{12}x + \frac{3}{12}$ f) $\frac{3}{4}y - 4 = -5$

g) $1.5x - 3 = -12$ h) $2.5y + 3 = -8$ i) $1 = 3.8x - 0.9$

j) $4.4y + 3 = 5.64$ k) $1.3w + 65 = 26$ l) $12.5z - 36 = 64$

4. Solve:

a) $4(x - 2) = 9$ b) $-3 = 5(z + 7)$

c) $-\frac{2}{3}(x + 1) = 6$ d) $-3(y - \frac{1}{2}) = \frac{1}{2}$

e) $0.02 = 0.8(y + 0.1)$ f) $1.2(t + 2.3) = 3.96$

g) $0.6(5s - 6) = 2.4$ h) $1.21 = 11(0.51 + 0.4x)$

B 5. Solve:

a) $12 = 9 - 2t$ b) $23 = 11 - 3r$ c) $8 - 3z = -19$

d) $5y + \dfrac{3}{5} = 3$ e) $7x - \dfrac{5}{9} = 3$ f) $\dfrac{5}{16} = \dfrac{11}{16} - \dfrac{1}{4}x$

g) $0.2x + 4 = 7$ h) $-1 = 5 - 0.3t$ i) $0.25p + 0.25 = 0.5$

6. Solve:

a) $\dfrac{2}{5}t + 3 = 11$ b) $\dfrac{n}{3} + 4 = -6$ c) $-\dfrac{3}{4}z + 5 = -1$

d) $1.2(2x - 3) = 7.2$ e) $-3.5(1 + 3r) = 7$

f) $3(5.6 + \dfrac{x}{3}) = 0$ g) $23.98 = 11(1.1w + 2.07)$

7. If $3w = 6 - 9z$, find

a) the values of w for the following values of z:

   i) 1   ii) 2   iii) -2 iv) -3 v) 0

b) the values of z for the following values of w:

   i) 0   ii) -1 iii) 2   iv) -19 v) 110

8. The cost, C, in dollars, of a daytime, operator-assisted, station-to-station telephone call from Toronto to Calgary lasting n minutes is given by the formula:
$C = 3.70 + 0.99(n - 3)$, where $n \geqslant 3$.

a) Find the cost of a call that lasts

   i) 4 min;   ii) 13 min;   iii) 15 min;   iv) 1 h.

b) How long was a call that cost

   i) $5.68?   ii) $7.66?   iii) $19.54?   iv) $23.50?

9. The cost, C, in dollars, of renting a car for a day from Ace Rentals is given by the formula: $C = 14.75 + 0.15x$, where x is the distance driven, in kilometres.

a) What is the cost of renting a car for a day and driving

   i) 100 km?   ii) 250 km?   iii) 800 km?

b) How far had the car been driven if the cost for the day was

   i) $31.00?   ii) $50.00?   iii) $78.05?

10. A loan company charges a flat rate of $50 to process a loan and $27\dfrac{1}{2}\%$ per year on the principal. The amount, A, in dollars, required to discharge a loan of P after 1 year is given by the formula: $A = 50 + 1.275P$.

a) Calculate the amount owed after 1 year on a loan of

   i) $100;   ii) $500;   iii) $3700.

b) Find the principal if the amount owed after 1 year is

   i) $305;   ii) $815;   iii) $2915.75.

4 - 3   Combining Terms Containing the Variable

We have seen that to solve an equation it is necessary to isolate the variable on one side of the equation and combine the numerical terms on the other. When several terms contain the variable, they too must be combined. Consider the following examples.

Example 1.   Solve: a) $6x - 5 = 2x + 7$;   b) $2 - 3y = 7 + y$.

Solution.   a)

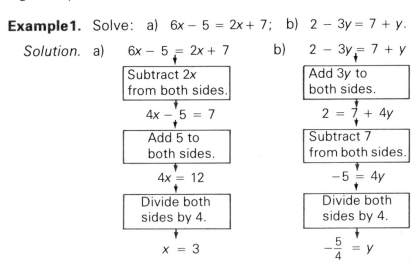

b)

a)
$$6x - 5 = 2x + 7$$

Subtract 2x from both sides.

$$4x - 5 = 7$$

Add 5 to both sides.

$$4x = 12$$

Divide both sides by 4.

$$x = 3$$

b)
$$2 - 3y = 7 + y$$

Add 3y to both sides.

$$2 = 7 + 4y$$

Subtract 7 from both sides.

$$-5 = 4y$$

Divide both sides by 4.

$$-\frac{5}{4} = y$$

The above examples show that it does not matter on which side of the equation you isolate the variable. The next two examples show that it is sometimes necessary to expand the expressions on both sides of an equation before you can combine terms and isolate the variable.

Example 2.   Solve: $5(y - 1) = 7(3 + y)$.

Solution.
$$5(y - 1) = 7(3 + y)$$
$$5y - 5 = 21 + 7y$$
$$-2y - 5 = 21$$
$$-2y = 26$$
$$y = -13$$

Example 3.   Solve: $3(y - 1) - 5y = 2y - (y - 2)$.

Solution.
$$3(y - 1) - 5y = 2y - (y - 2)$$
$$3y - 3 - 5y = 2y - y + 2$$
$$-3 - 2y = y + 2$$
$$-3y = 5$$
$$y = -\frac{5}{3}$$

Exercises 4 - 3

A 1. Solve:

a) $7x - 3 = 4x + 3$
b) $5y + 9 = 2y - 3$
c) $-4m + 2 = 6m + 12$
d) $-8t - 5 = -9t - 7$
e) $3r - 2 = -5r + 14$
f) $6p - 7 = -6p - 7$
g) $5 - y = 3 - 2y$
h) $9 - 2x = 6 - x$
i) $3 - 2t = 5 - 5t$
j) $4 - p = 5 - 3p$
k) $7 - 5p = 6 + p$
l) $8 - 3r = -6 + r$
m) $-11 + 6v = -6v + 11$
n) $-8w = -4 - 6w$

2. Solve:

a) $3(x - 1) = 12$
b) $5(x + 2) = 10$
c) $-14 = x - 3$
d) $x - 2 = 2(x - 1)$
e) $3y - 2 = y + 4$
f) $4y = -2(9 - y)$
g) $y + 7 = 3y - 9$
h) $9y - 3 = 3(y - 4)$
i) $2(t - 3) = -3(t - 1)$
j) $-3(r + 2) = -4(r - 1)$
k) $7(z + 3) = 5(z - 1)$
l) $4(2y - 1) = 5(3y + 1)$
m) $-3(4p + 2) = 4(2p - 2)$
n) $-2(1 - x) = -3(2 - x)$

B 3. Solve:

a) $6y - 2 = 5y + 4$
b) $3p + 2 = 5p - 7$
c) $4 - r = 3 - 2r$
d) $9 - 2p = -8 - p$
e) $5(x - 1) = 8(1 - x)$
f) $7(y - 2) = 13$
g) $-2(x - 1) = 3(x + 2)$
h) $-3(y + 1) = -2(y - 1)$
i) $r - 1 = 5r - 7$
j) $19t - 13 = 2t + 4$
k) $17y + 3 = 15y - 3$
l) $z - 2 = 2z - 2$
m) $x + 4 = 11x + 4$
n) $t - 8 = -12t + 18$

4. Solve:

a) $5x - 9 - 2x = 6 - 3x + 5$
b) $7y + 2 + 3y = 9 - 2y + 6$
c) $11 - 2(3 - s) = 13 - 3(1 - s)$
d) $2(y - 3) - 4y = 3y - (y - 5)$
e) $7(x - 2) - 3x = 4x - (x - 7)$
f) $-3(t - 2) + 5t = 7t - (t + 6)$
g) $3(1 - p) + p = -2(p - 1) + 7 - (2 - 3p)$
h) $7(t - 2) + 2t = 3(t + 3) - 5 - (2 - 5t)$
i) $4(x + 3) - 3(x - 2) + 5 = 2(x - 1) - 5 - 3(x + 2)$
j) $5(w - 1) - 4(w + 3) + 17 = 2(w + 3) - 4 - 6(w - 2)$

4 - 4 Equations Containing Fractions

When an equation contains fractions, multiply both sides of the equation by a common denominator of the fractions to obtain an equivalent equation without fractions. Then solve this equation by the method of Section 4-2.

Example 1. Solve: $\frac{a}{2} - 3 = \frac{1}{5}a + \frac{1}{2}$

Solution. $\frac{a}{2} - 3 = \frac{1}{5}a + \frac{1}{2}$

Multiply both sides by 10.

$$10(\frac{a}{2} - 3) = 10(\frac{1}{5}a + \frac{1}{2})$$

$$10 \times \frac{a}{2} - 10 \times 3 = 10 \times \frac{1}{5}a + 10 \times \frac{1}{2}$$

$$5a - 30 = 2a + 5$$

$$3a = 35$$

$$a = \frac{35}{3}, \text{ or } 11\frac{2}{3}$$

Example 2. Solve: $\frac{3x + 2}{2} - \frac{x + 1}{3} = x$

Solution. $\frac{3x + 2}{2} - \frac{x + 1}{3} = x$

Multiply both sides by 6.

$$6(\frac{3x + 2}{2}) - 6(\frac{x + 1}{3}) = 6x$$

$$3(3x + 2) - 2(x + 1) = 6x$$

$$9x + 6 - 2x - 2 = 6x$$

$$7x + 4 = 6x$$

$$x = -4$$

Now consider the problem on the first page of the chapter.

Example 3. A car uses gasoline at an average rate of 10.9 L/100 km. The rate of fuel consumption, y, in litres per 100 km, of this model is given by the formula: $y = \frac{-36}{5}x + \frac{29}{2}$, where x is the fraction of driving done on the highway. What fraction of the driving is on the highway?

Solution. $y = \frac{-36}{5}x + \frac{29}{2}$

$$10.9 = \frac{-36}{5}x + \frac{29}{2}$$

Multiply both sides by 10.

$$10(10.9) = 10(\frac{-36}{5}x + \frac{29}{2})$$

$$109 = 2(-36x) + 5(29)$$

$$109 = -72x + 145$$

$$72x = 145 - 109, \text{ or } 36$$

$$x = \frac{1}{2}$$

Half of the driving is on the highway.

Exercises 4 - 4

A 1. Solve:

a) $\frac{1}{2}x + \frac{1}{3}x = 10$ b) $\frac{1}{4}y - \frac{1}{2}y = 4$ c) $\frac{y}{3} - \frac{2}{3} = 4$

d) $\frac{a}{4} = \frac{1}{2}$ e) $\frac{x}{5} = -\frac{2}{3}$ f) $\frac{1}{3} = \frac{-2x}{5}$

g) $\frac{2}{5}a + \frac{a}{2} = a - 2$ h) $\frac{1}{2}n - \frac{2}{3}n + \frac{3}{4}n = -7$

i) $\frac{x}{3} + \frac{1}{2} = 2x$ j) $\frac{2x}{3} - \frac{1}{2} = \frac{1}{2} + \frac{1}{4}x$

2. Solve:

a) $\frac{a}{5} - a = \frac{1}{2}$ b) $\frac{2x}{3} = \frac{x}{2} - \frac{1}{4}$ c) $\frac{m}{6} - 5 = \frac{1}{2}m$

d) $\frac{3k}{4} + \frac{1}{2} = \frac{k}{3}$ e) $\frac{1}{3}(x + 1) = \frac{1}{2}(x - 2)$

f) $\frac{1}{5}(2n + 1) = \frac{2}{3}(n - 1)$ g) $\frac{a + 5}{3} = \frac{3 - a}{7}$

h) $\frac{1}{5}(\frac{1}{2}x + 4) = \frac{1}{3}(\frac{1}{4}x + 3)$ i) $\frac{1}{4}(2 - x) + \frac{1}{2} = \frac{1}{2}(\frac{1}{3}x + 7)$

B 3. If $y = \frac{-3}{5}x - \frac{1}{4}$, find

a) the values of y for these values of x:

   i) 0   ii) $\frac{1}{3}$ iii) $-\frac{1}{3}$ iv) 2   v) -2

b) the values of x for these values of y:

   i) 0   ii) $\frac{1}{2}$ iii) $-\frac{1}{2}$ iv) 1.5   v) -1.5

4. The rate of fuel consumption, y, in litres per 100 km, of Nicole's car is given by the formula $y = -\frac{25}{4}x + \frac{69}{5}$, where x is the fraction of driving done on the highway. If she is averaging 11.3 L/100 km, what fraction of Nicole's driving is on the highway?

C 5. Tap *A* can fill a tank in *a* min. Tap *B* can fill the tank in *b* min. Both taps together can fill the tank in *t* min, where $\frac{1}{t} = \frac{1}{a} + \frac{1}{b}$.

If tap *A* fills the tank in 8 min and both taps together fill the tank in 6 min, how long does it take tap *B* alone to fill the tank?

4 - 5 Checking Solutions

Solving an equation may require several steps. As the number of steps increases, the greater is the chance of an error. For this reason you should always check that your solution is correct. To do this:

- Substitute your solution for the variable in each side of the original equation.
- Simplify each side of the equation independently.

Your solution is correct if, and only if, each side of the equation simplifies to the same number.

Example 1.   Check that $y = -3$ is the solution of the equation $6y + 5 = 4y - 1$.

Check.   Substitute -3 for *y* in each side of the equation and simplify each side independently.

Left Side	Right Side
$6y + 5 = 6(-3) + 5$	$4y - 1 = 4(-3) - 1$
$= -18 + 5$	$= -12 - 1$
$= -13$	$= -13$

Each side simplifies to the same number. This means that $y = -3$ is the correct solution.

Example 2.   Solve and check: $3(x - 2) = 5(x + 6)$.

Solution.
$$3(x - 2) = 5(x + 6)$$
$$3x - 6 = 5x + 30$$
$$-2x - 6 = 30$$
$$-2x = 36$$
$$x = -18$$

Check.   Substitute -18 for *x* in each side of the equation.

$3(x - 2) = 3(-18 - 2)$	$5(x + 6) = 5(-18 + 6)$
$= 3(-20)$	$= 5(-12)$
$= -60$	$= -60$

Each side simplifies to the same number.
$x = -18$ is the correct solution.

Exercises 4 - 5

A 1. Solve and check:

a) $2(x + 1) = 3x$

b) $3(2 - m) = m + 2$

c) $3 - y = y - 7$

d) $4a + 6 = 2a - 2$

e) $2 - x = 5x + 8$

f) $3m - 5 = 2m + 1$

g) $3k - 5 = k - 4$

h) $\frac{1}{2}a + 3 = \frac{2}{3}a$

i) $\frac{1}{4}(c + 2) = \frac{1}{3}(c - 1)$

j) $\frac{1}{2}(b + 9) = \frac{2}{3}(b + 3)$

2. Solve and check:

a) $8(y - 5) = 7y$

b) $6(3 - x) = -9x - 6$

c) $9 - v = v - 9$

d) $8w - 4 = 4 - 8w$

e) $5t - 2 = 11t + 16$

f) $11 - p = 3p - 21$

g) $6(q + 1) = 3(q - 1)$

h) $3 - (6s - 15) = 4s + 8$

i) $3(2x - 1) - 7 = -41 - 2(x + 0.5)$

j) $\frac{1}{5}(y + 3) - \frac{5}{4} = \frac{1}{4}(y - 1)$

B 3. Consider the equation $3(x + 2) = x + 2(x + 3)$.

a) Check that $x = 5$, $x = 8$, and $x = -1$ are all solutions of the equation.

b) Choose any number you wish and check that it also is a solution of the equation.

c) Attempt to solve the equation. Can you suggest why every number is a solution of this equation?

4. Consider the equation $3(x + 2) = x + 2(x + 4)$.

a) Choose any number and show that it is *not* a solution of the equation.

b) Attempt to solve the equation. Can you explain why it does not have a solution?

C 5. Solve, if possible, and check:

a) $5(n + 2) = n + 6(n - 3)$

b) $2(p + 1) - 3 = 4(2p + 1) - 11$

c) $2(y + 1) = y + (1 + y)$

d) $5(x - 2) = x + 4(x - 3) + 2$

e) $4(q - 1) = q + 3(q + 3)$

f) $6(r + 2) = 3r + 3(r - 4)$

g) $4v + 3(20 - v) = 80$

h) $7(x - 3) = 3x + 4(x - 5) - 1$

i) $-3(1 - y) = -6(1 + y) + 3y$

j) $2(n - 2) = 6 + n - 3(1 + n)$

4 - 6   Working With Formulas

Formulas are equations that relate two or more variables using the basic operations of arithmetic. They are used to solve particular kinds of problems.

Science, engineering, and industry use a variety of formulas. Often, the values of all but one of the variables are known and it is necessary to solve an equation to find the value of the one that is not known.

Example 1. The mass, m, in grams, of a ball bearing is given by the formula: $m = 7.5v$, where v is its volume in cubic centimetres. What is the volume of a ball bearing that has a mass of 30 g?

Solution.

$$m = 7.5v$$

Substitute 30 for m. $$30 = 7.5v$$

Divide both sides by 7.5 $$\frac{30}{7.5} = v$$

$$4 = v$$

The volume of the ball bearing is 4 cm³.

Example 2. The annual interest, I, in dollars, on a sum of money, P, is given by the formula: $I = 0.095P$. What sum of money will earn \$807.50 interest in 1 year?

Solution.

$$I = 0.095P$$

Substitute 807.50 for I. $$807.50 = 0.095P$$

Divide both sides by 0.095. $$\frac{807.50}{0.095} = P$$

$$8500 = P$$

\$8500 will earn \$807.50 interest in 1 year.

Example 3. A science experiment shows that a rubber band stretches according to the formula: $I = 9.2 + 0.17m$, where I is its length, in centimetres, and m is the mass, in grams, suspended on one end.

a) Calculate I when the mass is 25 g.

b) What mass will stretch the band to 85.7 cm?

Solution. a) $I = 9.2 + 0.17\ m$

$$= 9.2 + 0.17 \times 25$$

$$= 9.2 + 4.25, \text{ or } 13.45$$

The length of the band is about 13.4 cm.

b) $l = 9.2 + 0.17\,m$
$85.7 = 9.2 + 0.17\,m$
$76.5 = 0.17\,m$
$m = \dfrac{76.5}{0.17}$, or 450.

A mass of 450 g will stretch the band to 85.7 cm.

Exercises 4 - 6

A 1. Use the formula $m = 7.5v$, where m is the mass in grams and v is the volume in cubic centimetres, to find
   a) the mass of a ball bearing that has a volume of 10 cm³;
   b) the volume of a ball bearing that has a mass of 165 g.

2. The length, l, in centimetres, of a rubber band suspending a mass of m g is given by the formula $l = 14.3 + 0.27m$.
   a) Calculate l when the suspended mass is 150 g.
   b) What mass stretches the band to a length of 98 cm?

3. The monthly interest, l, in dollars, on a loan is given by the formula $l = 0.0175P$, where P is the amount of the loan. What is the first month's interest on a loan of $1250?

4. The yearly interest, l, in dollars, on a loan is given by the formula $l = 0.11P$, where P is the principal.
   a) What is a year's interest on a loan of
       i) $100?   ii) $1000?   iii) $2500?   iv) $9000?
   b) What is the principal of a loan when a year's interest is
       i) $22?   ii) $55?   iii) $132?   iv) $159.50?

B 5. A person's maximum desirable pulse rate, m (in beats per minute) can be found from the formula $m = 220 - a$ if a, the person's age, is known.
   a) What is the maximum desirable pulse rate for a person who is
       i) 20 years old?   ii) 37 years old?   iii) 63 years old?
   b) How old is a person whose maximum desirable pulse rate is
       i) 170?   ii) 192?   iii) 141?

6. When the temperature at sea level is $t°C$, the temperature at altitude h, in metres, is $T°C$, where $T \doteq t - 0.0065h$.
   a) When it is 15°C at sea level, what is the temperature at
       i) Denver, Colorado, altitude 1600 m?
       ii) the top of Mt. Waddington, B.C., height 4000 m?

b) When t is 15°C, at what altitude will T be -8.4°C?

7. A company determines the age at which an employee can retire with full pension by the formula $a + b = 90$, where a is the employee's age and b is the number of years' service

a) How many years of service are required for retirement with full pension for an employee who is 60?   65?   70?

b) What is the minimum age for retirement with full pension after 30 years' service?   after 20 years' service?

8. The speed, s, of an object is given by the formula $s = \dfrac{d}{t}$, where t is the time taken to travel a distance d. What is the value of d for these values of s and t?

	a)	b)	c)	d)	e)
s	15	2	160	850	19
t	3	12	15	6	13

9. When an object falls freely from rest, its speed, v, in metres per second, after t s, is given by the formula $v = 9.8t$.

a) What is its speed after

i) 1 s?   ii) 5 s?   iii) 8 s?   iv) 10 s?

b) How long will it take the object to reach these speeds?

i) 54 m/s   ii) 81 m/s   iii) 343 m/s   iv) 1 km/s

10. When an object on the moon falls freely from rest, its speed, v, in metres per second, after t s, is given by the formula $v = 1.63t$.

a) What is its speed after

i) 1 s?   ii) 5 s?   iii) 8 s?   iv) 10 s?

b) How long will it take the object to reach these speeds?

i) 4.9 m/s   ii) 32.6 m/s   iii) 57 m/s   iv) 0.6 km/s

11. The time delay, t, in seconds, between seeing lightning and hearing the accompanying thunder is given by the formula $t = 3d$, where d is the distance, in kilometres, between the lightning and the observer. What is this distance when the time delay is

a) 7.2 s?   b) 4.5 s?   c) 3.9 s?   d) 2.7 s?

12. The mass, m, in grams, of a volume of air is given by the formula $m = 1.29v$, where v is the volume in litres. What is the volume of air that has a mass of

a) 19.35 g?   b) 303.15 g?   c) 22.575 kg?

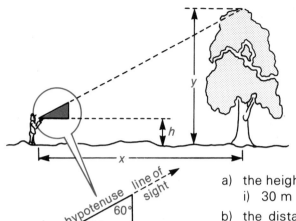

13. To find the height of a tall object:
    - Sight the top of the object along the hypotenuse of a $30° - 60° - 90°$ triangle.
    - Measure your distance, x, from the object.
    - Substitute your measured value for x in the formula $y = 0.577x + h$, and the height of your eyes above the ground for h. (x, y, and h must be in the same units.)

    Assume h to be 1.4 m and calculate

    a) the height for these values of x:
       i) 30 m   ii) 40 m   iii) 15 m   iv) 17.5 m

    b) the distances of a viewer from objects, when correctly sighted, whose heights are:
       i) 40 m,   ii) 55 m,   iii) 100 m,   iv) 81.5 m.

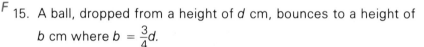

14. The sum of the angles of a triangle is $180°$. For $\triangle ABC$ and $\triangle DEF$

    a) find the values of p for these values of q:
       i) $20°$ ii) $36°$ iii) $70°$ iv) $80°$

    b) find the values of q for these values of p:
       i) $30°$ ii) $45°$ iii) $55°$ iv) $80°$

    c) What is the meaning of your answer to (b) (iv) for $\triangle DEF$?

15. A ball, dropped from a height of d cm, bounces to a height of b cm where $b = \frac{3}{4}d$.

    a) From what height was it dropped if the height of bounce is
       i) 90 cm?   ii) 75 cm?   iii) 60 cm?   iv) 39 cm?

    b) If it is dropped from a height of 160 cm, what is the height of its second bounce?   its third bounce?

C 16. The formula, $P = 35 \times D \times \frac{1}{S}$, relates a recommended per-cent mark up, P, of an article to the supply, S, and demand, D. Calculate

    a) P when the demand is for 40 articles and the supply is
       i) 20 articles;   ii) 40 articles;   ii) 80 articles;

    b) D when P is 70 and S is 30;

    c) S when P is 80 and D is 40.

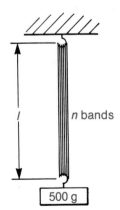

17. When a 500 g mass is suspended from a number of similar elastic bands, the length, l, in centimetres, to which they stretch is given by the formula $l = 9.2 + \frac{85}{n}$, where n is the number of bands. How many bands are used when the length is 23.4 cm?

4 - 7   Translating into the Language of Algebra

Equations are used to solve problems. Before we can write an equation, we must first translate the facts of the problem into the language of algebra.

Study these verbal expressions and their algebraic equivalents.

Verbal	Algebraic
4 more than a number	$x + 4$
a number increased by 8	$x + 8$
9 less than a number	$x - 9$
twice a number	$2x$
one-sixth of a number	$\frac{1}{6}x$ or $\frac{x}{6}$
5 more than 4 times a number	$4x + 5$
the product of 1 more than a number and 7	$7(x + 1)$
1 more than the product of a number and 7	$1 + 7x$
3 less than 5 times a number	$5x - 3$
the product of 6 less than a number and 8	$8(x - 6)$

Sometimes in a problem, two numbers are related in a certain way. We often represent one number by a variable, and express the other number in terms of the first. Study these examples.

Verbal	Algebraic
two consecutive integers	x and $x + 1$
two consecutive odd (or even) integers	x and $x + 2$
one number is six times another	x and $6x$
two numbers differ by five	x and $x + 5$
Mary's age now, and in six years	x and $x + 6$
a 4 m log is cut into two pieces	x and $4 - x$
the sum of two numbers is 45	x and $45 - x$

Exercises 4 - 7

A   1.   Write an algebraic equivalent for each verbal expression:
  a) five more than a number
  b) six less than a number
  c) eight times a number
  d) one-fifth of a number
  e) the product of a number and eight
  f) four more than five times a number
  g) two less than eight times a number
  h) the product of eight and two less than a number
  i) twelve less than three times a number
  j) three plus one-fourth of a number
  k) one-fourth of the sum of a number and three
  l) five plus four times a number
  m) the product of four and five more than a number

2.   For each of the following relationships, choose a variable to represent one quantity and express the other in terms of the first:
  a) A boy is twelve years older than his brother.
  b) A number is one-fifth of another number.
  c) The ages of John and Mary total 21.
  d) Two consecutive even numbers
  e) Joan's jump was 15 cm longer than Enid's.
  f) The Jaguar travelled 1.1 times as fast as the Mercedes.
  g) Roberta is three years younger than Rebecca.
  h) Sophia's age is four years less than twice Bertha's age.
  i) Bob has $3 more than four times the amount that Ian has.
  j) A 15 m log is cut into two pieces.
  k) The product of two numbers is 76.
  l) Three consecutive integers
  m) Three consecutive odd integers
  n) $750 is divided between two people.
  o) A number is 2 less then three times another number.
  p) Two numbers differ by 6.
  q) One-third the larger of two consective integers
  r) Three times the larger of two consecutive even integers
  s) The sum of two numbers is 81.

4 - 8 Writing Equations

When problems are solved by algebra, the first step is to translate the facts given into the language of algebra. The kinds of problems we are concerned with give two facts. One fact enables us to express each quantity in terms of the variable. The other fact enables us to write an equation.

Example 1. A number is 4 times another number. The sum of the numbers is 59. What are the numbers?

Solution. The facts of the problem are:

(1) A number is 4 times another number.

(2) The sum of the numbers is 59.

Because of the simplicity of the facts, we can use either one to express the numbers and the other to write the equation.

Let the smaller number be: x.

The larger number is:

$$4x \qquad (1) \quad \text{or} \qquad 59 - x. \qquad (2)$$

The equation is:

$$x + 4x = 59 \quad (2) \quad \text{or} \quad 59 - x = 4x. \quad (1)$$
$$5x = 59 \qquad\qquad\qquad 59 = 5x$$
$$x = 11.8 \qquad\qquad\qquad 11.8 = x$$

One number is 11.8 and the other is 47.2.

When one fact is more complicated than the other, use the complicated fact to write the equation.

Example 2. The sum of two numbers is 117. Five times the smaller number is nine more than three times the larger number. Find the numbers.

Solution. The simple fact is: The sum of two numbers is 117.

Let the smaller number be: x.

The larger number is: $117 - x$.

The more complicated fact is used to write the equation.

$$5x = 9 + 3(117 - x)$$
$$5x = 9 + 351 - 3x$$
$$8x = 360$$
$$x = 45$$

The two numbers are 45 and 72.

Exercises 4 - 8

Write an equation for each of the following exercises. Do not solve the equations at this time.

A 1. One number is one-fifth of another number. The sum of the numbers is 12. Find the numbers.

2. One number is seven times one-half of another number. The difference of the numbers is 35. What are the numbers?

3. Clyde is 12 years older than Bonnie. The sum of their ages is 42. How old is each?

4. Ian's mass is 1.5 kg less than that of Hamish. The sum of their masses is 118.5 kg. What is the mass of each?

5. The ages of John and Mary total 21. Mary's age plus twice John's age is 30. How old is each?

6. Joan's jump was longer than Enid's by 15 cm. Joan's jump was 1.04 times as long as Enid's. How far did each jump?

7. For two consecutive odd integers, the sum of the smaller and twice the larger is 19. What are the integers?

B 8. For two consecutive integers, the sum of twice the larger and three times the smaller is 242. Find the integers.

9. The Jaguar travelled 1.1 times as fast as the Mercedes. The difference in their speeds was 12 km/h. Find the speed of each car.

10. Roberta is three years younger than Rebecca. Eight years ago, Roberta was half Rebecca's age. How old is each girl now?

C 11. Sophia's age is four years less than twice Beryl's age. In two years, Beryl's age will be three-quarters Sophia's age. How old is each girl now?

12. A piece of string, 60 cm long, is cut into three pieces of different lengths. The middle-sized piece is 2 cm longer than the short piece and 2 cm shorter than the long piece. What are the lengths of the three pieces?

13. Girish has $2 more than three times the amount that Joseph had. Girish gave Joseph $5 who then had half as much as Girish. How much did each have at first?

4 - 9   Solving Problems Using Equations

Solving problems using equations involves three steps:

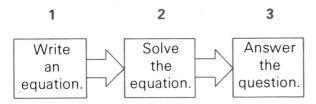

Sections 4-7 and 4-8 prepared you for Step 1. The earlier sections of the chapter prepared you for Step 2. Step 3 is merely taking the solution of the equation and answering the question the problem asks.

When you check your solution, do *not* substitute in the equation, substitute in the problem. You could have made a mistake in writing the equation.

Example 1.   A pair of skis and boots together cost $225. The skis cost $60 more than the boots. What is the cost of the skis?

Solution.   Step 1: Let x be the cost of the skis.
The cost of the boots is: $x - 60$.
The equation is: $x + (x - 60) = 225$.

Step 2:
$$x + (x - 60) = 225$$
$$2x - 60 = 225$$
$$2x = 285$$
$$x = 142.5$$

Step 3: The skis cost $142.50.

Check.   The skis cost $142.50.
The boots cost $142.50 − $60, or $82.50.
The total cost is $142.50 + $82.50, or $225.00.
The solution is correct.

Example 2.   A tree trunk 15 m long is cut into two pieces. One piece is four times as long as the other. How long is each piece?

Solution. Step 1: Let x be the length of the shorter piece.
Then $4x$ is the length of the longer piece.
The equation is: $x + 4x = 15$.

Step 2: $x + 4x = 15$
$5x = 15$
$x = 3$

Step 3: The shorter piece is 3 m long. The longer piece is 12 m long.

Check. The shorter piece is 3 m long.
The longer piece is 4×3 m long.
The total length is: 3 m + 12 m, or 15 m.
The solution is correct.

Example 3. A mother is three times as old as her daughter. Six years ago she was five times as old as her daughter. How old are mother and daughter now?

Solution. Step 1:

	Now	Then
daughter:	x	$x - 6$
mother:	$3x$	$3x - 6$
equation:	$3x - 6 = 5(x - 6)$	

Step 2: $3x - 6 = 5(x - 6)$
$3x - 6 = 5x - 30$
$24 = 2x$
$12 = x$

Step 3: The daughter is 12 years old now. The mother is 36 years old now.

Check. Six years ago, the daughter's age was $12 - 6$, or 6. The mother's age was then $36 - 6$, or 30.
$5 \times 6 = 30$
The solution is correct.

Remember: In reading the problem, look for the simple fact. Use it to express the relationship between the quantities involved. Use the more complicated fact to write the equation.

Exercises 4 - 9

A 1. Brian ran 2 km farther than Tom. They ran a total distance of 12 km. How far did each run?

2. Marie ran twice as far as Brenda. They ran a total distance of 12 km. How far did each run?

3. An airplane travels six times as fast as a train. The difference in their speeds is 500 km/h. How fast is each travelling?

4. The length of a rectangle is 5 cm longer than the width. The perimeter is 68 cm. Find the dimensions of the rectangle.

5. Find two consecutive numbers whose sum is 263.

6. Find three consecutive numbers whose sum is 141.

7. Find four consecutive numbers whose sum is 234.

8. Find two consecutive even numbers whose sum is 170.

9. One number is five times another number. If the two numbers total 18, find the numbers.

B 10. In a cross-country marathon, Jack and Jill ran a total of 73 km. Jill ran 5 km farther than Jack. How far did each run?

11. The combined mass of a dog and a cat is 24 kg. The dog is three times as heavy as the cat. What are their masses?

12. In a class of 33 students, there are 9 more girls than boys. How many girls are there?

13. Millie is four times as old as Marty. The sum of their ages is 65 years. How old are they?

14. Find two numbers that
   a) are consecutive and whose sum is 83;
   b) differ by 17 and whose sum is 39;
   c) differ by 36 and one is seven times the other.

15. The sum of two numbers is 36. Four times the smaller is 1 less than the larger. What are the numbers?

16. The difference of two numbers is 48. One number is nine times the other. What are the numbers?

17. One number is 0.25 less than another number. The sum of the numbers is 7.25. Find the numbers.

18. The sum of three numbers is 33. The second number is 7 less than the first, and the third number is 2 more than the second. What are the numbers?

19. The sum of three numbers is 75. The second number is 5 more than the first, and the third is three times the second. What are the numbers?

20. The least of three consecutive integers is divided by 10, the second is divided by 17, the greatest is divided by 26. What are the numbers if the sum of the quotients is 10?

21. James has two-fifths the amount that Lorna has, and Muriel has seven-ninths the amount that James has. Together they have $770. What do they each have?

22. Jeanne is twice as old as Michel. The sum of their ages 3 years ago was 45 years. What are their ages now?

23. Yvonne has equal numbers of nickels, dimes, and quarters in her purse. Their total value is $2.00. How many of each kind of coin does she have?

24. Roger has dimes and quarters with a total value of $2.50. If he has 3 more quarters than dimes, how many of each kind of coin does he have?

25. Find three consecutive odd numbers whose sum is 27.

C 26. At Happy Snack, a milk shake costs twice as much as an order of french fries. If two milkshakes and three orders of french fries cost $2.80, what is the cost of a milkshake?

27. The mass of a candy-bar wrapper is $\frac{1}{11}$ the mass of the wrapped bar. If the candy bar alone has a mass of 75 g, what is the mass of the wrapper?

28. A wine cellar contains port, sherry, claret, and champagne. One-fifth of all the wine is port and one-third is claret. If there are 180 bottles of sherry and 30 bottles of champagne, how much port and claret does the cellar contain?

29. Raymond has a box of candy bars. He gave Monique half of what he had plus half a bar. Then he gave Claude half of what he had left plus half a bar. After which he gave Laura half of what he had left plus half a bar. And, finally, he gave Alfred half of what he had left plus half a bar. He then had no bars left. How many candy bars did Raymond have to start with?

4 - 10   Solving Inequalities

An **inequality** is a mathematical sentence that uses the sign $>$ (is greater than) or the sign $<$ (is less than) to relate two expressions. We now check to see whether the rules for solving equations apply to inequalities as well.

Consider the inequality: $4 < 8$.

Operations on: $4 < 8$	Result	Correct or Incorrect
Add 3 to both sides.	$7 < 11$	Correct
Subtract 7 from both sides	$-3 < 1$	Correct
Multiply both sides by $+3$	$12 < 24$	Correct
Multiply both sides by -3	$-12 < -24$	Incorrect
Divide both sides by $+2$	$2 < 4$	Correct
Divide both sides by -2	$-2 < -4$	Incorrect

These results suggest that the rules for solving in-equalities are the same as those for solving equations with two exceptions:

> If both sides of an inequality are multiplied or divided by a *negative number,* the inequality sign must be reversed.

The method of solving inequalities is the same as for solving equations: (i) isolate the term containing the variable, (ii) isolate the variable.

Example 1.   Solve: $3x < 5x + 12$.

Solution. $3x < 5x + 12$
$-2x < 12$
$x > -6$

Note that $<$ becomes $>$.

In the above example, any number greater than -6 satisfies the inequality. On the number line, the solution is shown thus:

The open dot at -6 indicates that $x = -6$ is not part of the solution.

It is impossible to check the limitless number of solutions to an inequality. However, if one solution is selected and substituted in the inequality and the resulting statement is correct, we may reasonably conclude that our solution is correct.

Example 2. Solve, graph, and check: $3 - 2a > a + 9$.

Solution.
$$3 - 2a > a + 9$$
$$3 > 3a + 9$$
$$-6 > 3a$$
$$-2 > a$$

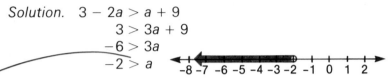

Note that $>$ remains $>$.

Check. The solution includes $a = -5$. (Since $-2 > -5$.)

When $a = -5$,   When $a = -5$,
$3 - 2a = 3 - 2(-5)$ $a + 9 = -5 + 9$
$\quad = 3 + 10$, or 13 $\quad = 4$

Since $13 > 4$, $a = -5$ satisfies the inequality and suggests that $a < -2$ is the correct solution.

Exercises 4 - 10

A   1. Solve and graph:

a) $x + 1 < 4$ b) $x - 1 < 3$ c) $x + 3 > 2$

d) $4 > 9 - x$ e) $-13 > x - 11$ f) $9 < 15 - x$

2. Solve and graph:

a) $7x < 14$ b) $5x > 10$ c) $9 > -2x$

d) $3y + 8 > 17$ e) $21 - 5z > 11$ f) $13.5 + 2y < 18.5$

g) $13 < 1 - \frac{3}{4}x$ h) $61 < 13w - 4$ i) $18 > 4.5 - 1.5a$

B   3. Solve, graph, and check:

a) $4x - 7 > 2x + 5$ b) $13 - 2y < 4y - 14$

c) $-25 + 11z < 30 - 11z$ d) $39 + 4w > 13 - 6w$

e) $3(x + 2) < 11$ f) $2(x + 8) < 4(3 + x)$

g) $5(2 - x) < 2(x + 7)$ h) $\frac{2}{3}(15 - 3x) > \frac{1}{2}(2 + 5x)$

4. Write the inequality represented by each graph:

a)   b)

c)   d)

Review Exercises

1. Solve:
   a) $5 + x = -11$ b) $y - 14 = 83$ c) $14 - z = -14$
   d) $8 - t = -2$ e) $w + 21 = -13$ f) $17 = 19 - t$

2. Solve:
   a) $5x = 45$ b) $-15 = -3n$ c) $\frac{1}{5}t = -3$

   d) $\frac{n}{14} = -7$ e) $\frac{s}{5} = \frac{1}{3}$ f) $8r = 56$

   g) $1.3x = 9.1$ h) $\frac{x}{17} = \frac{39}{51}$ i) $11 = 2.25w$

3. Solve:
   a) $6p - 3 = 15$ b) $13 = 4 + 3x$ c) $-6 - 2r = 8$

   d) $5 - 5y = 1$ e) $8p - 3 = 7$ f) $3x - \frac{1}{5} = 4$

   g) $3y - 7 = 14$ h) $\frac{1}{4}x - \frac{2}{3} = 2$ i) $2.7y - 3.1 = 5$

   j) $\frac{4}{5} - \frac{1}{3}x = \frac{1}{2}x$ k) $1.69 - 1.3x = 0$ l) $-64.5 + 2.5x = -2$

4. The cost, C, in cents, of making copies on a copying machine is given by the formula $C = 90 + 3n$, where n is the number of copies.
   a) What is the cost of making 200 copies?
   b) How many copies can be made for $6.00?

5. The cost, C, in dollars, of a telephone call from Vancouver, B.C., to St. John's, Newfoundland, is given by the formula $C = 1.20 + 0.95(n - 1)$, where n is the number of minutes the call lasts, and $n \geqslant 1$.
   a) Find the cost of a call that lasts
      i) 1 min;   ii) 3 min;   iii) 5 min.
   b) The charge for one call was $24. How long was the call?

6. Solve:
   a) $5y - 2 = 3y + 4$ b) $-7x + 6 = 2x - 3$
   c) $r - 3 = 2r + 4$ d) $11 - 1.3x = 4.7x - 7$
   e) $4(x - 3) = -2$ f) $-5(y + 3) = 14$
   g) $\frac{3}{8}(2 - 4x) = -1\frac{1}{4} + \frac{x}{2}$ h) $0.5(5x - 3) = 1.2$
   i) $3(1 - x) = -2(2 - x)$ j) $3(r - 1) = -2(r + 8)$

7. Solve:

   a) $-2z - 3 + 5z = 6 - z - 9$ b) $-8t + 7 - 5t = 9 - t - 11$

   c) $-13 - q - 9 = 5q - 9 - q$ d) $-19 - 3r + 6 = 7r - 11 - r$

   e) $4(w - 5) - w = -9 - w + 7$

   f) $-7(p - 3) + 11 = 3p - 12 - p$

   g) $t + 17 - 2t = -3(t - 1) - 3$

   h) $9 - 3(1 - q) = 7 - 4(2 - q)$

   i) $-9(r + 3) - 9r = -3r - (3 - r) + 8$

   j) $3(7v + 8) - 9 = 14 - 2v + 6(3v - 4)$

   k) $5(6w - 3) - 7 = 17 + 3w - 5(2w + 1)$

   l) $-2(5x - 1) - 4 = 22 - 7x + 8(3x - 1)$

8. Solve and check:

   a) $3(x + 1) = 2x$ b) $4n - 2 = n + 1$

   c) $8 - x = x - 8$ d) $12 - y = 3y - 14$

   e) $4(t + 1) = 2t + (1 - t)$ f) $6(w - 2) = 3w + 2(w + 1)$

   g) $7.2x - 7.5 - 1.7x = 4.6 + 4.4x$

   h) $15(0.3 - z) + 14.5z = 2(0.25z - 10)$

9. Express each of the following as an equation and solve:

   a) A number multiplied by 7 equals 56.

   b) When 13 is subtracted from a number the result is 18.

   c) When 16 is added to a number the result is 31.

   d) A number divided by 15 equals 75.

   e) When 29 is subtracted from a number the result is -2.

10. Find two numbers that differ by 6 and have a sum of 62.

11. Find two numbers that differ by 7 and have a sum of 35.

12. When a number is tripled and increased by 7 the result is the same as doubling the number. Find the number.

13. One number is 3 more than another number. The sum of the numbers is 1.5. Find the numbers.

14. A tree trunk, 16 m long, is cut into two pieces so that one piece is 8 m longer than the other. How long is each piece?

15. Finn is 12 years older than Haddy. Their ages total 32 years. How old is each?

16. A father is four times as old as his son. In 24 years time he will only be twice as old. What are their present ages?

17. One number is 2 less than three times another number. The sum of the numbers is -10. What are the numbers?

18. Two numbers differ by 7. The sum of the larger and one-half the smaller is 19. Find the numbers.

19. Nita swam twice as many lengths of the pool as Honor. If they swam a total of 42 lengths, how many lengths did each swim?

20. A jacket and slacks together cost $115. The jacket costs $35 more than the slacks. Find the cost of each.

21. The sum of the angles of a triangle is 180°. For the triangle shown, find
    a) the value of a when b is: i) 30°;   ii) 60°.
    b) the value of b when a is: i) 20°;   ii) 50°.

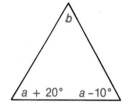

22. A company pays its employees a car allowance of $20 per month plus 13¢ for each kilometre driven on company business.
    a) Write an equation which shows the amount, A, that an employee receives for driving n km in a month.
    b) If an employee submits a claim for having driven 800 km, how much does she expect to receive?

23. Sheila calculated that the weekly expense of operating her car was $40 for the fixed costs of depreciation, insurance, license, and repairs, and 5¢/km for gasoline and oil.
    a) Write an equation for the cost of operating Sheila's car.
    b) What is the week's cost if the distance driven is
       i) 0 km?   ii) 100 km?   iii) 350 km?   iv) 1200 km?
    c) How far can Sheila drive if the weekly cost must be limited to
       i) $100?   ii) $65?   iii) $80   iv) $40?

24. In 1979, Calgary's population was ten times as great as that of Lethbridge. The populations of the two cities totalled 550 000. What was the population of each city?

25. Radio waves travel through space at approximately 300 000 km/s.
    a) If it takes 2.5 s for a radio signal to reach the moon and be reflected back, how far is the moon from Earth?
    b) When the distance between Earth and Venus is a minimum, it takes 280 s for a radio wave to go from one to the other and be reflected back. What is the minimum distance between Earth and Venus?
    c) How far do radio waves travel in 1 year?

Alberta

THE MATHEMATICAL MIND

SUMMING AN INFINITE SET OF FRACTIONS

Most ancient mathematicians were baffled by the idea of adding an infinite set of numbers. In fact, it was not until the nineteenth century that mathematicians learned to deal with infinite sums. These examples show how we can use our techniques for solving equations to sum certain infinite sets of fractions.

Farmer Brown planted

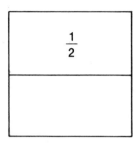

wheat in half his field,

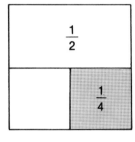

corn in half the remaining part,

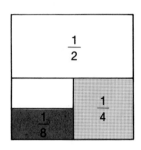

flax in half the remaining part,

and continued in this manner, each time planting a different crop in half the unplanted part of the field. If he could continue this process an unlimited number of times, what fraction of the field would be planted?

Let x represent the fraction of the field planted.

Then, $x = \frac{1}{2} + \frac{1}{4} + \frac{1}{8} + \frac{1}{16} + \cdots$

Multiply both sides of this equation by 2:

$$2x = (2 \times \tfrac{1}{2}) + (2 \times \tfrac{1}{4}) + (2 \times \tfrac{1}{8}) + (2 \times \tfrac{1}{16}) + (2 \times \tfrac{1}{32}) + \cdots$$

$$= 1 + \frac{1}{2} + \frac{1}{4} + \frac{1}{8} + \frac{1}{16} + \cdots$$

$$= 1 + x$$

That is, $2x = 1 + x$, or $x = 1$

This tells us what we intuitively knew to start with—eventually, the entire field would be planted. What is important is that we have discovered a new technique for finding the sum of certain infinite sets of fractions.

Example 1. Find the sum of: $\frac{1}{10} + \frac{1}{100} + \frac{1}{1000} + \frac{1}{10\,000} + \cdots$.

Solution. Let $x = \frac{1}{10} + \frac{1}{100} + \frac{1}{1000} + \frac{1}{10\,000} + \ldots$

Multiply both sides of the equation by 10:

$$10x = 10(\tfrac{1}{10}) + 10(\tfrac{1}{100}) + 10(\tfrac{1}{1000}) + 10(\tfrac{1}{10\,000}) + \ldots$$

$$= 1 + \underbrace{\frac{1}{10} + \frac{1}{100} + \frac{1}{1000} + \ldots}$$

$$= 1 + x$$

$$9x = 1, \quad \text{or} \quad x = \frac{1}{9}$$

$$\frac{1}{10} + \frac{1}{100} + \frac{1}{1000} + \frac{1}{10\,000} + \ldots = \frac{1}{9}$$

In the above result, if we write each common fraction as a decimal, we get:

$$0.1 + 0.01 + 0.001 + 0.0001 + \ldots = \frac{1}{9}$$

$$\text{or,} \quad 0.\overline{1} = \frac{1}{9}$$

The next example shows how we can use the above technique to express a repeating decimal as a common fraction without first expressing it as the sum of a set of fractions.

Example 2. Express as a common fraction: a) $0.\overline{4}$; b) $0.\overline{42}$.

Solution. a) $0.\overline{4}$ may be written 0.44. ...

$$\text{If } x = 0.44. \ldots \qquad \text{(i)}$$
$$10x = 4.44. \ldots \qquad \text{(ii)}$$

Subtract (i) from (ii).

$$9x = 4$$

Therefore $x = \frac{4}{9}$

$0.\overline{4}$ expressed as a common fraction is $\frac{4}{9}$.

b) $0.\overline{42}$ may be written 0.424 2.. ...

$$\text{If } x = 0.424\,2.. \ldots \qquad \text{(i)}$$
$$100x = 42.424.. \ldots \qquad \text{(ii)}$$

Subtract (i) from (ii).

$$99x = 42$$

Therefore $x = \frac{42}{99}$, or $\frac{14}{33}$

$0.\overline{42}$ expressed as a common fraction is $\frac{14}{33}$.

Use the technique above to express these repeating decimals as common fractions:

1. $0.\overline{5}$ 2. $0.\overline{71}$ 3. $3.\overline{27}$

PROBLEM
SOLVING
STRATEGY

Construct a Table

Once the essential information has been extracted from a problem, it is often useful to display it in a table. The remaining spaces in the table can then be filled in by calculations or reasoning. When the table is completed, the problem is usually easy to solve.

Example 1. An aircraft flies from city A to city B against the wind at an average speed of 600 km/h. On the return trip the average speed is 1000 km/h. What is the average speed for the round trip?

Table.

	Distance	Time	Speed
$A \to B$			600 km/h
$B \to A$			1000 km/h

Solution. To complete the table, let the distance from A to B be d km. Then: time (in hours) $= d \div$ speed

	Distance	Time	Speed
$A \to B$	d km	$\frac{d}{600}$ h	600 km/h
$B \to A$	d km	$\frac{d}{1000}$ h	1000 km/h

The solution can be found if it is remembered:

average speed $=$ total distance $\div$ total time

Exercises

1. A boat has an average speed of 50 km/h on the first lap of a two-lap race. On the second lap it averages only 30 km/h. What is its average speed for the whole race?

2. If Mr. Swan drives at an average speed of 100 km/h he arrives at work at 09:00. If he leaves home at the same time but averages 80 km/h, he arrives at 09:06. How far does he live from work?

3. A car's cooling system contains a 25% solution of antifreeze. Half the system is drained and then topped up with pure antifreeze. What is the strength of the antifreeze in the system now?

Tables are also useful when solving non-numerical problems.

Example 2. Three sports professionals, a tennis player, a golfer, and a skier, live in three adjacent houses on Oak Street. One of these athletes is a woman named Kim. The other two are Frank and Bill. From the information below determine the professional skier.

    A. Kim does not play tennis.

    B. Frank skis and plays tennis, but does not like golf.

    C. The golfer and the skier live beside each other.

    D. Three years ago, Bill broke his leg skiing and has not tried it since.

    E. Kim lives in the last house of the three.

    F. The golfer and the tennis player share a common backyard swimming pool.

Table.

	Tennis player	Golfer	Skier
Kim			
Frank			
Bill			

The above table displays all the possibilities.

Solution. Using the above information in A, B, and D, we can eliminate three possibilities.

	Tennis player	Golfer	Skier
Kim	X		
Frank		X	
Bill			X

The name of the professional skier can now be found by using C and F together with E to eliminate one more possibility.

4. Mr. Allen, Mr. Becker, and Mr. Carson each have a grown son. One of the sons is a teacher, one a banker, and the other is a lawyer. Determine which son is the lawyer from the following information.

    A. The lawyer often plays tennis with his father.

    B. Mr. Becker's son is friends with the teacher.

    C. The teacher's father plays golf every Sunday.

    D. Mr. Allen does not like sports of any kind.

CALCULATOR POWER

Solving Equations

Your calculator can be used to solve many equations quickly. You perform on your calculator the *inverses* of the operations indicated in the equation. That is, where the equation has multiplication, divide, and where it has add, subtract, and vice versa. Study these examples.

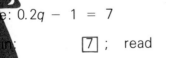

1. Solve: $0.2q - 1 = 7$

Key in:
[7] ; read 7
[+] [1] ; read 1
[÷] [.] [2] ; read 0.2
[=] ; read 40

$q = 40$

2. Solve: $3(x + 2) - 7 = 20$

Key in: [2] [0] ; read 20
[+] [7] ; read 7
[÷] [3] ; read 3
[-] [2] ; read 2
[=] ; read 7

$x = 7$

Exercises

1. Use a calculator to solve these equations:

   a) $3x + 6 = 27$ b) $4x + 7 = 39$ c) $5x - 3 = 102$
   d) $2x + 16 = 106$ e) $7x + 18 = 102$ f) $6x - 27 = 153$
   g) $97 = 6x - 23$ h) $0.5x - 12 = 36$ i) $63 = 0.3x - 27$
   j) $70 = 0.35x + 21$ k) $0.4x + 364 = 56$
   l) $0.75x + 4.5 = 10.5$ m) $1.69x + 52 = 221$
   n) $2.02 + 2.9x = 0.28$ o) $3.1 + 0.65x = 1.02$
   p) $2.9 = 19 + 0.46x$ q) $-3 = 0.41 + 0.37x$
   r) $\frac{13}{15}x + 218 = 23$ s) $\frac{13x}{16} - 9 = \frac{3}{4}$

2. Solve, using a calculator:

   a) $5(x + 2) = 45$ b) $7(x + 4) + 63 = 0$
   c) $8(x - 6) = -96$ d) $3(x - 2) = 48$
   e) $5(2x + 7) = 75$ f) $6(3x - 4) + 108 = 0$
   g) $2(x + 6) - 3 = 75$ h) $5(x - 4) + 7 = 57$
   i) $3(x + 8) - 9 = 51$ j) $8(x - 12) + 17 = 73$
   k) $4(x + 13) + 115 = 35$ l) $6(x - 6) + 38 = 140$
   m) $5(2x + 6) - 18 = 132$ n) $3(5x - 11) + 124 = 7$
   o) $3(4x - 5) + 15 = 81$ p) $2(7x + 11) + 144 = -18$

5 Powers and Roots

Keith's copy of this poster is 90 cm by 90 cm. He wants an enlargement made that has twice the area. What will be the length of each side of the enlargement? (See *Example 6* in Section 5-8.)

5 - 1   The Meaning of Exponents

To show a repeated addition, we use a **product**.

$$4 + 4 + 4 + 4 + 4 \text{ is written } 5 \times 4.$$

Each term in a product is called a **factor**.

To show repeated multiplication, we use a **power**.

$$4 \times 4 \times 4 \times 4 \times 4 \text{ is written } 4^5.$$

Read: 4 raised to the fifth power, or 4 to the fifth.

$$\text{Power} \longrightarrow 4^5 \begin{cases} \text{Exponent} \\ \text{Base} \end{cases}$$

The **base** is the number that is repeatedly multiplied. The **exponent** tells the number of times the base is a factor.

Example 1.   Write these products as powers:

a) $(-3)(-3)(-3)(-3)$

b) $\dfrac{3}{5} \times \dfrac{3}{5} \times \dfrac{3}{5} \times \dfrac{3}{5} \times \dfrac{3}{5}$

c) $\dfrac{1}{m} \times \dfrac{1}{m} \times \dfrac{1}{m}$

d) $a \cdot a \cdot a \cdot \ldots a$ (n factors)

Solution.   a) $(-3)(-3)(-3)(-3) = (-3)^4$

b) $\dfrac{3}{5} \times \dfrac{3}{5} \times \dfrac{3}{5} \times \dfrac{3}{5} \times \dfrac{3}{5} = (\dfrac{3}{5})^5$

c) $\dfrac{1}{m} \times \dfrac{1}{m} \times \dfrac{1}{m} = (\dfrac{1}{m})^3$

d) $a \cdot a \cdot a \cdot \ldots a = a^n$

Example 2. If $n = 3$, find the value of: a) n^5;   b) 5^n.

Solution.   a) $n^5 = 3^5$ b) $5^n = 5^3$

$\qquad\qquad\quad = 3 \times 3 \times 3 \times 3 \times 3 \qquad\quad = 5 \times 5 \times 5$

$\qquad\qquad\quad = 243 \qquad\qquad\qquad\qquad\quad = 125$

A power with a negative base simplifies to a positive number when the exponent is even and to a negative number when the exponent is odd.

Example. 3. Simplify: a) $(-5)^3$;   b) $(-5)^4$.

Solution.   a) $(-5)^3$ b) $(-5)^4$

$\qquad\qquad = (-5)(-5)(-5) \qquad\quad = (-5)(-5)(-5)(-5)$

$\qquad\qquad = -125 \qquad\qquad\qquad\quad = 625$

Evaluate powers before doing any operations unless brackets indicate otherwise.

Example 4. If $x = 5$ and $y = -3$, evaluate:

   a) $x^2 + y^2$;   b) $(x + y)^2$;   c) $3y^3$;   d) $(3y)^3$.

Solution.   a) $x^2 + y^2$

$\qquad = 5^2 + (-3)^2$

$\qquad = 25 + 9$

$\qquad = 34$

b) $(x + y)^2$

$\qquad = [5 + (-3)]^2$

$\qquad = [2]^2$

$\qquad = 4$

c) $3y^3 = 3(-3)^3$

$\qquad = 3(-27)$

$\qquad = -81$

d) $(3y)^3 = [3(-3)]^3$

$\qquad = [-9]^3$

$\qquad = -729$

When the base is a product, each factor is raised to the power.

Example 5. Simplify: a) $(3x)^3$;   b) $\left(\frac{y}{4}\right)^2$.

Solution.   a) $(3x)^3 = (3x)(3x)(3x)$

$\qquad = (3)(3)(3)(x)(x)(x)$

$\qquad = 27x^3$

b) $\left(\frac{y}{4}\right)^2 = \left(\frac{y}{4}\right)\left(\frac{y}{4}\right)$

$\qquad = \frac{y^2}{16}$

Some equations involving powers can be solved by inspection.

Example 6. Solve: a) $x^2 = 49$;   b) $y^3 = 0.125$

Solution.   a)   Since $7 \times 7 = 49$, and $(-7) \times (-7) = 49$, the equation, $x^2 = 49$, has two roots: 7 or -7. The solution is: $x = +7$ or $x = -7$.

b)   Since $0.5 \times 0.5 \times 0.5 = 0.125$, then $y = 0.5$

Exercises 5 - 1

A 1. Write as a power:

   a) $y \times y \times y \times y$

   b) $(-3)(-3)(-3)(-3)(-3)(-3)$

   c) $\frac{2}{5} \times \frac{2}{5} \times \frac{2}{5} \times \frac{2}{5} \times \frac{2}{5}$

   d) $\frac{-3}{8} \times \frac{-3}{8} \times \frac{-3}{8}$

   e) $(4a)(4a)(4a)(4a)(4a)$

   f) $2.9 \times 2.9 \times 2.9 \times 2.9$

   g) $m \times m \times m \times m$

   h) $(-6x)(-6x)(-6x)(-6x)(-6x)$

   i) $\pi \times \pi \times \pi \times \pi \times \pi \times \pi$

   j) $a \cdot a \cdot a \cdot a \cdot b \cdot b \cdot b$

2. Write as a power:

   a) six cubed

   b) 7 raised to the eighth power

   c) nine to the fourth

   d) twenty squared

   e) the fifth power of eleven

   f) five to the eleventh

   g) four to the nth power

   h) $2x$ to the tenth

   i) $x \times x \times x \times x \times x \times x \times x \times x \times x \times x \times x \times x \times x \times x \times x$

   j) $b \times b \times b \times \ldots$ (to n factors)

3. Simplify:

   a) 4^3

   b) 3^4

   c) $(-2)^5$

   d) $(-5)^2$

   e) 10^4

   f) $\left(\frac{1}{4}\right)^2$

   g) $(0.2)^3$

   h) $(2.1)^2$

   i) 3×2^4

   j) $2^3 \times \left(\frac{3}{4}\right)^2$

   k) $(3 \times 2)^4$

   l) $2 \times (-3)^4$

4. Simplify:

   a) $2^3 + 3^2$

   b) $3^2 + 4^2$

   c) $(3 + 4)^2$

   d) $5 \times (-4)^3$

   e) $3^3 - 5^2$

   f) $(-4)^2 - 7^2$

   g) $(-3)^3 - (-1)^3$

   h) $(-5)^3 - (-2)^5$

   i) $4(-2)^3 + 5(-3)^2$

5. Evaluate $2x^2 - 3x + 5$ for these values of x:

   a) 4

   b) -1

   c) -2

   d) 10

   e) -5

   f) $\frac{1}{2}$

   g) $\frac{-1}{3}$

   h) 1.5

   i) -0.4

   j) 100

6. If $x = -3$ and $y = 2$, evaluate:

   a) x^2

   b) y^3

   c) $-5x^3$

   d) $-x^4$

   e) $(-x)^4$

   f) $x^2 + y^2$

   g) $x^2 - y^2$

   h) $(x + y)^3$

   i) $x^3 + y^3$

   j) $x^3 - y^3$

   k) $(x + y)^9$

   l) $(3x - y)^2$

   m) $5(x^2 - 2y)$

   n) $(3x + y)^3$

   o) $4x^2 - 7y^2$

   p) $4x^2 + y^2$

7. Express as a power of 10:

   a) 1000

   b) 10 000

   c) 100

   d) 1 000 000

   e) 100 000

   f) 10 000 000

   g) 10

   h) 100 000 000

   i) 1 000 000 000 000

8. Evaluate for $n = 3$:

   a) n^3

   b) $n^2 - n^3$

   c) $4n^2$

   d) $(4n)^2$

   e) $(n + 2)^3$

   f) $(n - 2)^8$

   g) $(2 - n)^{15}$

   h) $(3n - 7)^4$

   i) $(4n - 2n^2)^4$

   j) $\left(\frac{n}{4}\right)^3$

   k) $\left(\frac{2}{n} - 1\right)^5$

   l) $(n - n^2)^3$

B 9. Arrange from greatest to least:
   a) 2^4, 3^2, 5^2, 2^3, 3^3
   b) $(-3)^4$, 4^3, 7^2, 2^5, 10^2
   c) $(1.2)^2$, $(1.1)^3$, $(1.05)^5$, $(1.15)^3$, $(1.3)^1$
   d) $(2.1)^4$, $(2.9)^3$, $(2.3)^2$, $(1.8)^7$, $(2.4)^5$
   e) $(0.3)^2$, $(0.2)^3$, $(0.2)^2$, $(0.3)^3$, $(0.4)^2$

10. Solve for x:
   a) $x^2 = 25$ | b) $x^2 = 81$
   c) $x^2 = 121$ | d) $x^2 = 169$
   e) $x^2 = 0.01$ | f) $x^2 = 0.36$
   g) $x^3 = 8$ | h) $x^3 = 64$
   i) $x^3 = -216$ | j) $4x^2 = 144$
   k) $10x^2 = 1.6$ | l) $16x^3 = -128$

C 11. Which is greater,
   a) 3^{22} or 3^{25}? | b) $3x^2$ or $(3x)^2$?
   c) $(-5n)^3$ or $5n^3$? | d) $(-2)^{16}$ or $(-2)^{19}$?
   e) $(0.9)^{14}$ or $(0.9)^{11}$? | f) $(\frac{-3}{4})^{10}$ or $(\frac{-3}{4})^{7}$?

12. a) For what values of y is $y^2 < y$?
   b) For what values of x is $x^3 < x^4$?

13. Solve for n:
   a) $2^n = 8$ | b) $3^n = 81$
   c) $10^n = 1\,000\,000$ | d) $3 \times 2^n = 48$
   e) $2 \times 5^n = 50$ | f) $10 \times 3^n = 810$

14. Identify the greater number in each pair:
   a) 2^5, 5^3 | b) 3^4, 4^3 | c) 10^4, 2^{10}
   d) 6^4, 11^{11} | e) 6^3, 3^6 | f) 9^4, 3^8

15. Express the first number as a power of the second:
   a) 16, 4 | b) 27, 3 | c) 64, 2
   d) 625, 5 | e) 16, -2 | f) -243, -3
   g) 343, 7 | h) 256, -4 | i) 81, -3
   j) 6561, 9 | k) 7776, 6 | l) 1.4641, 1.1

5 - 2 Exponents in Formulas

The formulas for area, volume, distance fallen by an object, and compound interest are just a few of the many that involve exponents. Study the useful formulas for area and volume given below.

Circular and spherical measurement, whether of area or volume, involve π, the ratio of circumference to diameter. An adequate approximation of the value of π is 3.14.

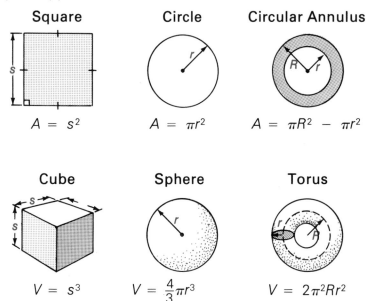

Square	Circle	Circular Annulus
$A = s^2$	$A = \pi r^2$	$A = \pi R^2 - \pi r^2$

Cube	Sphere	Torus
$V = s^3$	$V = \dfrac{4}{3}\pi r^3$	$V = 2\pi^2 R r^2$

Example 1. Find the area of the shaded part of each figure to one decimal place.

a)   b)   c)

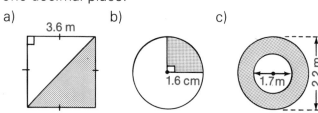

Solution. a) Area of square, $A = s^2$

$$= (3.6)^2$$

Area of shaded part, $\dfrac{1}{2}A = \dfrac{1}{2}(3.6)^2$

$$= 6.48$$

The area of the shaded part of the square is approximately 6.5 m².

b) Area of circle, $A = \pi r^2$

　　Area of shaded part, $\frac{1}{4} A = \frac{1}{4} \pi r^2$

$$\doteq \frac{1}{4}(3.14)(1.6)^2$$

$$\doteq 2.0096$$

The area of the shaded part of the circle is approximately 2.0 cm².

c) Area of annulus, $A = \pi R^2 - \pi r^2$

$$\doteq 3.14(1.1)^2 - 3.14(0.85)^2$$

$$\doteq 3.7994 - 2.268\,65$$

$$\doteq 1.530\,75$$

The area of the annulus is approximately 1.5 m².

Example 2. Find the volume of each 3-dimensional figure to one decimal place.

a) Cube　　　b) Sphere　　　c) Torus

r = 2 mm

3.0 m

1.2 cm

R = 8 mm

Solution. a) Volume of cube, $V = s^3$

$$= (1.2)^3$$

$$= 1.728$$

The volume of the cube is approximately 1.7 cm³.

b) Volume of sphere, $V = \frac{4}{3} \pi r^3$

$$\doteq \frac{4}{3}(3.14)(3.0)^3$$

$$\doteq 113.04$$

The volume of the sphere is approximately 113.0 m³.

c) Volume of torus, $V = 2\pi^2 R r^2$

$$\doteq 2(3.14)^2(8)(2)^2$$

$$\doteq 631.0144$$

The volume of the torus is approximately 631.0 mm³.

Example 3. The distance, d, in metres, an object falls from rest in t s is given by the formula: $d = 4.9t^2$.

a) If a pebble dropped from the top of a building takes 3.5 s to reach the ground, how tall is the building?

b) If a pebble is dropped from a height of 313.6 m, how long will it take to reach the ground?

Solution. a) Substitute 3.5 for t in the formula.
$$d = 4.9 \times (3.5)^2$$
$$= 60.025$$
The building is approximately 60 m tall.

b) Substitute 313.6 for d in the formula.
$$313.6 = 4.9t^2$$
$$64 = t^2$$
$$8 = t$$
The pebble takes 8 s to reach the ground.

Example 4. A $500 Canada Saving Bond (CSB) paying 9% interest annually allows the interest to compound (earn more interest). Its value, V, after n years is given by
$$V = 500 \times (1.09)^n.$$

a) What is the CSB worth after 2 years, and after 5 years?

b) How long does the bond take to double in value?

Solution. a) After 2 years,   After 5 years,
$$V = 500 \times (1.09)^2 \qquad V = 500 \times (1.09)^5$$
$$= 500 \times 1.1881 \qquad \doteq 500 \times 1.538\,62$$
$$= 594.05 \qquad\qquad \doteq 769.31$$
After 2 years, the CSB is worth $594.05, and after 5 years it is worth $769.31.

b) When the $500 CSB doubles in value, the formula becomes: $1000 = 500 \times (1.09)^n$.
$$2 = (1.09)^n$$
Using a calculator, we find that $(1.09)^8$ is 1.992 56 and $(1.09)^9$ is 2.171 89. $(1.09)^8$ is closer to 2 than $(1.09)^9$. That is, $n \doteq 8$.
The CSB doubles in value in approximately 8 years.

Exercises 5 - 2

A 1. Give the area of each figure using exponents:

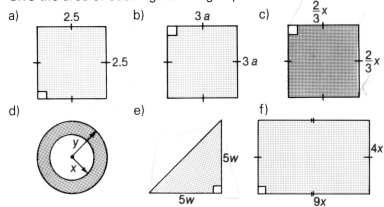

a) 2.5   2.5

b) 3 a   3 a

c) $\frac{2}{3}x$ $\frac{2}{3}x$

d) y x

e) 5w   5w

f) 4x   9x

2. Give the volume of each of these figures using exponents:

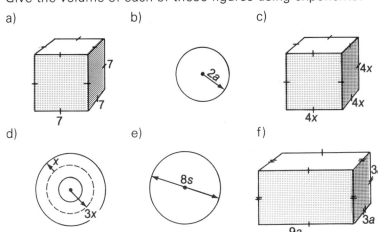

a) 7   7   7

b) 2a

c) 4x   4x   4x

d) x 3x

e) 8s

f) 3a   3a   9a

3. Find the areas of squares with these side lengths:
   a) 5 cm   b) 9 m   c) 1.5 cm   d) 13.7 m
   e) 0.6 cm   f) 2.6 m   g) $4a$ cm   h) $5x$ m

4. Find the lengths of the sides of squares with these areas:
   a) 36 cm^2 b) 400 cm^2 c) 6.25 m^2 d) 2.56 cm^2
   e) 0.16 cm^2 f) 0.09 m^2 g) $81x^2$ m^2 h) $1.44y^2$ cm^2

5. Find the volumes of cubes having edge lengths the same as those given in Exercise 3.

6. Find the lengths of the edges of cubes with these volumes:
   a) 125 cm^3 b) 216 m^3 c) 1000 cm^3 d) 0.008 m^3

7. $200 in a savings account that pays 8% interest annually grows to $200(1.08)^n$ in n years.
   a) What does it grow to in
      i) 2 years?   ii) 5 years?   iii) 7 years?
   b) How long does it take to double in value?

B 8. Find the areas of the shaded regions:

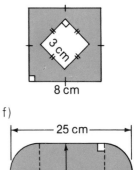

   a)   b)   c)

   2 cm
   7 cm   5 mm   3 cm
                                               8 cm

   d)   e)   f)

                         4 mm   ⟵— 25 cm —⟶
                         4 mm
   12 cm   8 mm   10 cm
                         8 mm

─7 cm─

10 cm   SUPER
        SoUP

9. A path, 2 m wide, is to enclose a circular lawn that has a 25 m radius. What will be the total cost if the cost per square metre is $3.00?

10. The label just covers the curved surface of a soup tin. What is its area?

11. A punch bowl is hemispherical (a half sphere) and 50 cm in diameter. How many litres of punch can it hold? ($1 L = 1000 cm^3$)

12. How many bouillon cubes with an edge length of 2 cm can be packed into a cubic box with an edge length of 0.5 m?

13. Find the volume of air contained in a spherical balloon with a radius of 12 cm.

⟵————— x —————⟶

C 14. A class found, by measuring, that the relationship of the area, A, of the maple leaf on the Canadian flag to the flag's length, x, is $A = 0.072x^2$.
   a) Find the area of the maple leaf on a flag of length
      i) 20 cm;   ii) 40 cm;   iii) 80 cm;   iv) 1.6 m.
   b) Find the length of the flag that has a maple leaf with an area of
      i) 583.2 cm²;   ii) 1036.8 cm²;   iii) 0.45 m².

Mathematics Around Us

How Bacteria Grow and Multiply

part 1

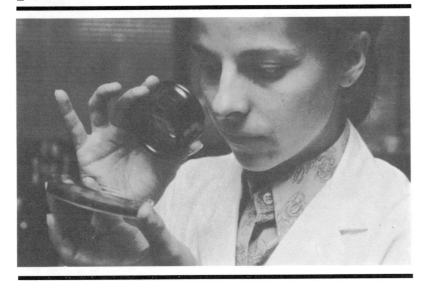

Questions

1. Express as powers of 2 the thousands of bacteria in the culture
   a) at 01:00;   d) at 04:00;
   b) at 02:00;   e) at 06:00;
   c) at 03:00;   f) at 10:00;

2. How many bacteria are there n h after midnight?

3. At 01:30 there are about 2800 bacteria in the culture. About how many would there be 1 h later?   2 h later?   1 h earlier?

4. Explain why powers of 2 are involved in the growth of bacteria.

5. If a petri dish is half-covered by bacteria at midnight, when will it be completely covered?

Scientists use bacteria in medical research. They grow bacteria in dishes called petri dishes, after Julius Petri, a noted bacteriologist. The bacteria "garden" is called a "culture". By counting the bacteria in the culture at regular intervals of time, scientists can learn how bacteria grow under controlled conditions.

The table shows a typical bacteria count every hour starting with a bacteria count of 1000 at midnight. The number of bacteria doubles every hour.

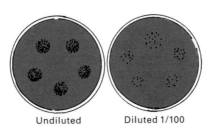

Cultures in Petri Dishes

Undiluted   Diluted 1/100

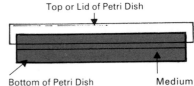

Cross Section of a Petri Dish

Top or Lid of Petri Dish

Bottom of Petri Dish   Medium

Time	Number of bacteria	Number of bacteria (thousands)
midnight	1000	1
01:00	2000	2
02:00	4000	4
03:00	8000	8

5 - 3 Operations With Powers

Three kinds of operations with powers are examined in this section. All three are based on the meaning of an exponent:

$$x^n = x \cdot x \cdot x \ldots \text{ to } n \text{ factors.}$$

Example 1. Simplify: a) $5^3 \times 5^4$;   b) $x^2 \times x^3$.

Solution. a) $5^3 \times 5^4 = 5 \times 5 \times 5 \quad \times \quad 5 \times 5 \times 5 \times 5$
$$= 5^7$$

b) $x^2 \times x^3 = x \times x \quad \times \quad x \times x \times x$
$$= x^5$$

Conclusion 1. For the product of powers with the same base, x, having exponents m and n that are positive integers:
$$x^m \cdot x^n = x^{m+n}$$

Example 2. Simplify: a) $7^5 \div 7^3$;   b) $z^7 \div z^3$.

Solution. a) $7^5 \div 7^3 = \dfrac{7^5}{7^3}$

$$= \frac{7 \times 7 \times \overset{1}{\cancel{7}} \times \overset{1}{\cancel{7}} \times \overset{1}{\cancel{7}}}{\underset{1}{\cancel{7}} \times \underset{1}{\cancel{7}} \times \underset{1}{\cancel{7}}}$$

$$= 7^2$$

b) $z^7 \div z^3 = \dfrac{z^7}{z^3}$

$$= \frac{z \times z \times z \times z \times \overset{1}{\cancel{z}} \times \overset{1}{\cancel{z}} \times \overset{1}{\cancel{z}}}{\underset{1}{\cancel{z}} \times \underset{1}{\cancel{z}} \times \underset{1}{\cancel{z}}}$$

$$= z^4$$

Conclusion 2. For the quotient of powers with the same base, x ($x \neq 0$), having exponents m and n that are positive integers, $m > n$:
$$x^m \div x^n = x^{m-n}$$

Example 3. Simplify: a) $(3^2)^4$;   b) $(y^3)^5$.

Solution. a) $(3^2)^4 = 3^2 \times 3^2 \times 3^2 \times 3^2$
$$\doteq 3^{2+2+2+2} \qquad \text{by Conclusion 1}$$
$$= 3^8$$

b) $(y^3)^5 = y^3 \times y^3 \times y^3 \times y^3 \times y^3$
$$= y^{3+3+3+3+3} \qquad \text{by Conclusion 1}$$
$$= y^{15}$$

Conclusion 3. When a power, x^m, is raised to a power n, for all positive integral values of m and n:

$$(x^m)^n = x^{mn}$$

The next example illustrates the application of these conclusions.

Example 4. Simplify: a) $3a^4 \times 2a^3$; b) $\dfrac{12n^5}{6n^2}$;

c) $\dfrac{5(b^2)^2 \times 3b^4}{2b^3}$.

Solution. a) $3a^4 \times 2a^3$ b) $\dfrac{12n^5}{6n^2}$

$= 3 \times 2 \times a^{4+3}$ $= 2n^{5-2}$

$= 6a^7$ $= 2n^3$

c) $\dfrac{5(b^2)^2 \times 3b^4}{2b^3} = \dfrac{5 \times 3}{2} \times \dfrac{b^4 \times b^4}{b^3}$

$= \dfrac{15}{2} \times \dfrac{b^{4+4}}{b^3}$

$= 7.5 \times b^{4+4-3}$

$= 7.5b^5$

Exercises 5 - 3

A 1. Simplify:

a) $3^4 \times 3^6$ b) $7^4 \times 7^7$ c) $(-5)^{16}(-5)^9$

d) $(2.1)^5 \times (2.1)^{11}$ e) $(-8)^2(-8)^3(-8)$

f) $(-1.7)^4(-1.7)^2(-1.7)$ g) $(\frac{2}{5})^8(\frac{2}{5})^{14}$

h) $(\frac{3}{11})^{21}(\frac{3}{11})^{15}$ i) $(-2\frac{1}{4})(-2\frac{1}{4})^6(-2\frac{1}{4})^{17}$

2. Simplify:

a) $x^7 \cdot x^4$ b) $k^3 \cdot k^9$ c) $n^6 \cdot n^{17}$

d) $s^4 \cdot s^5 \cdot s^2$ e) $v^{12} \cdot v^5 \cdot v$ f) $y^7 \cdot y \cdot y^2$

g) $(-a)^4(-a)^6$ h) $(-c)^7(-c)$ i) $t^{10} \cdot t^7 \cdot t^4$

3. Simplify:

a) $3^8 \div 3^3$ b) $2^{16} \div 2^7$ c) $m^{20} \div m^5$

d) $\dfrac{s^{18}}{s^6}$ e) $\dfrac{14z^{12}}{-2z^4}$ f) $\dfrac{24r^{24}}{8r^8}$

g) $\dfrac{6^8}{6^2}$ h) $\dfrac{(-2)^7}{(-2)^3}$ i) $\dfrac{(-10)^{11}}{-10}$

4. Simplify:

a) $\dfrac{2^3 \times 2^5}{2^6}$

b) $\dfrac{3 \times 3^7}{3^2 \times 3^2}$

c) $\dfrac{m^4 \times m^3}{m^2}$

d) $\dfrac{b^4 \times b}{b^2}$

e) $\dfrac{(-a)^5(-a)}{(-a)^2}$

f) $\dfrac{x^{12} \times x^6}{x^5 \times x^4}$

g) $\dfrac{c^8 \times c^6}{c^2 \times c^9}$

h) $\dfrac{(-5)^{41} \times (-5)^{19}}{(-5)^{50}}$

i) $\dfrac{7^{14}}{7^3 \times 7^4}$

5. Simplify:

a) $3a^2 \times 5a^3$

b) $2m^3 \times 9m^5$

c) $4x^4 \times 9x^9$

d) $6y^5(-3y^7)$

e) $5(3)^8 \times 6(3)^4$

f) $8(-7)^4 \times 4(-7)^{11}$

g) $3x \times 19x^{10}$

h) $2p^5 \times 5p^2 \times 3p^3$

i) $4s^5 \times 7s^{10} \times 3s$

6. Simplify:

a) $\dfrac{20d^5}{4d^2}$

b) $\dfrac{36a^{12}}{-4a^3}$

c) $\dfrac{-42z^3}{-7z^2}$

d) $15m^9 \div 5m^3$

e) $-32x^{12} \div 8x^4$

f) $50a^{20} \div 20a^5$

g) $12(2)^7 \div 4(2)^3$

h) $-24(3)^{18} \div 4(3)^6$

i) $\dfrac{4n^{12} \times 5n^3}{10n^6}$

j) $\dfrac{3c^6 \times 2c}{4c^2}$

k) $\dfrac{18m^{14} \times 5m^7}{10m^3}$

l) $\dfrac{-9a^7 \times (-8)a^9}{-18a^8}$

7. Simplify:

a) $(m^4)^5$

b) $[(-t)^3]^5$

c) $(a^7)^7$

d) $(2^3)^4$

e) $(12^5)^7$

f) $(10^2)^6$

g) $[(-5)^4]^3$

h) $(z^9)^3$

i) $[(-11)^4]^4$

8. Simplify:

a) $(6m^2)^3$

b) $(4x^5)^2$

c) $(13a^7)^4$

d) $(-3p^2)^8$

e) $(-6c^4)^6$

f) $(-3x^5)^2$

g) $\dfrac{(2k^2)^2}{k^3}$

h) $\dfrac{(3n^4)^3}{(2n)^2}$

i) $\dfrac{(4a^5)^3(4a^6)^2}{(4a^4)}$

B 9. Evaluate the following when $a = 5$:

a) $a \times a^3$

b) $a^2 \times a^4$

c) $a^9 \div a^7$

d) $\dfrac{a^3 \times a^5}{a^4}$

e) $\dfrac{15a^3}{5a}$

f) $\dfrac{(-6a^2)(8a^3)}{(4a^2)^2}$

g) $(a^2)^3$

h) $(3a^2)^2$

i) $(2a^3)(4a^2)$

10. Evaluate Exercise 9 for these values of a:

a) -5

b) 2

c) -2

11. Explain why the following are incorrect:

a) $3^2 \times 3^3 = 3^6$

b) $5^7 \times 5^4 = 25^{11}$

c) $10^8 \div 10^2 = 10^4$

d) $12^{20} \div 12^5 = 1^{15}$

e) $2^7 \times 3^5 = 6^{35}$

f) $6^6 \times 3^3 = 2^3$

g) $(5x^2)^7 = 5x^{14}$

h) $(3m^3)^5 = 243m^8$

i) $(-2y^4)^2 = -4y^8$

j) $(-4z)^2 \div (-2z) = 2z$

C 12. a) Make a table of powers of 2 up to 2^{24}.

b) Use your table of powers of 2 to evaluate the following without actually doing the arithmetic:

i) 16×64

ii) 32×512

iii) $65\,536 \div 2048$

iv) $\dfrac{128 \times 4096}{32}$

v) 64^3

vi) 4^5

13. Fold a piece of paper in half (giving 2 layers of paper). Fold it in half again (giving 4 layers). Fold it in half again (giving 8 layers), and continue folding it in half in this manner. Suppose you can do this 20 times.

a) How many layers of paper would there be?

b) About how thick would the resulting wad of paper be?

c) Find out how many times you can actually fold a piece of paper in this way.

14. Astronomers estimate that there are about 10^{11} galaxies in the universe, and that each galaxy contains about 10^{11} stars. About how many stars are there in the universe?

5 - 4 The Exponent Laws

Conclusion 2 of the previous section concerns the quotients of powers:

$$x^m \div x^n = x^{m-n}$$

where m and n are positive integers, and $m > n$. We want this to be true for $m = n$ and $m < n$.

Let us see what happens when $m = n$, as in the example: $x^3 \div x^3$. Using the meaning of an exponent:

$$x^3 \div x^3 = \frac{x \cdot x \cdot x}{x \cdot x \cdot x} \qquad (x \neq 0)$$
$$= 1$$

Assuming *Conclusion 2* to be true for $m = n$:

$$x^3 \div x^3 = x^{3-3}$$
$$= x^0$$

For *Conclusion 2* to apply when $m = n$, x^0 must equal 1.

Conclusion 4. For any number x, $x \neq 0$: $x^0 = 1$

Now let us see what happens when $m < n$, as in the example: $x^3 \div x^7$. Using the meaning of an exponent:

$$x^3 \div x^7 = \frac{x \cdot x \cdot x}{x \cdot x \cdot x \cdot x \cdot x \cdot x \cdot x} \qquad (x \neq 0)$$
$$= \frac{1}{x^4}$$

Assuming *Conclusion 2* to be true for $m < n$:

$$x^3 \div x^7 = x^{3-7}$$
$$= x^{-4}$$

For *Conclusion 2* to apply when $m < n$, x^{-4} must equal $\frac{1}{x^4}$.

Conclusion 5. For any integer, m, and $x \neq 0$: $x^{-m} = \frac{1}{x^m}$

We can now remove some of the restrictions on our earlier conclusions and refer to them as **exponent laws**. Thus:

For all integers m and n:
1. $x^m \times x^n = x^{m+n}$
2. $x^m \div x^n = x^{m-n}$ $(x \neq 0)$
3. $(x^m)^n = x^{mn}$
4. $x^{-m} = \dfrac{1}{x^m}$ $(x \neq 0)$
5. $x^0 = 1$ $(x \neq 0)$

The pattern of this table suggests that the results we have obtained for powers having exponents that are negative integers and zero are reasonable.

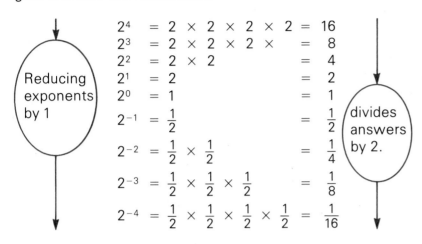

$$
\begin{aligned}
2^4 &= 2 \times 2 \times 2 \times 2 = 16 \\
2^3 &= 2 \times 2 \times 2 \times = 8 \\
2^2 &= 2 \times 2 = 4 \\
2^1 &= 2 = 2 \\
2^0 &= 1 = 1 \\
2^{-1} &= \frac{1}{2} = \frac{1}{2} \\
2^{-2} &= \frac{1}{2} \times \frac{1}{2} = \frac{1}{4} \\
2^{-3} &= \frac{1}{2} \times \frac{1}{2} \times \frac{1}{2} = \frac{1}{8} \\
2^{-4} &= \frac{1}{2} \times \frac{1}{2} \times \frac{1}{2} \times \frac{1}{2} = \frac{1}{16}
\end{aligned}
$$

Reducing exponents by 1

divides answers by 2.

Example 1. Simplify: a) 5^{-2}; b) $\left(\frac{1}{4}\right)^{-3}$; c) $(3^0 - 3^{-1})^{-2}$.

Solution. a) $5^{-2} = \frac{1}{5^2}$ b) $\left(\frac{1}{4}\right)^{-3} = \frac{1}{\left(\frac{1}{4}\right)^3}$

$\phantom{a) 5^{-2}} = \frac{1}{25}$

$\phantom{b) \left(\frac{1}{4}\right)^{-3}} = \frac{1}{\frac{1}{64}}$, or 64

c) $(3^0 - 3^{-1})^{-2} = \left(1 - \frac{1}{3}\right)^{-2}$

$\phantom{c) (3^0 - 3^{-1})^{-2}} = \left(\frac{2}{3}\right)^{-2}$

$\phantom{c) (3^0 - 3^{-1})^{-2}} = \frac{1}{\left(\frac{2}{3}\right)^2}$

$\phantom{c) (3^0 - 3^{-1})^{-2}} = \frac{1}{\frac{4}{9}}$, or $\frac{9}{4}$

Example 2. Simplify:
a) $(x^{-2})(x^5)(x^0)$;   b) $x^3 \div x^{-5}$;   c) $(x^2)^{-3}$

Solution. a) $(x^{-2})(x^5)(x^0)$ b) $x^3 \div x^{-5}$
$ = x^{-2+5+0}$ $ = x^{3-(-5)}$
$ = x^3$ $ = x^8$

c) $(x^2)^{-3} = x^{2(-3)}$
$\phantom{c) (x^2)^{-3}} = x^{-6}$

Example 2 shows how the exponent laws are used when some of the exponents are negative integers or zero. We often use these laws in numerical computations.

Example 3. Simplify: a) $4^{-3} \times 4^2 \times 4^{-1}$;
b) $(-2)^{-4} \div (-2)^{-1}$; c) $(3^{-1})^{-2}$.

Solution. a)
$$4^{-3} \times 4^2 \times 4^{-1}$$
$$= 4^{-3+2-1}$$
$$= 4^{-2}$$
$$= \frac{1}{4^2}$$
$$= \frac{1}{16}$$

b)
$$(-2)^{-4} \div (-2)^{-1}$$
$$= (-2)^{-4-(-1)}$$
$$= (-2)^{-3}$$
$$= \frac{1}{(-2)^3}$$
$$= -\frac{1}{8}$$

c)
$$(3^{-1})^{-2}$$
$$= 3^{(-1)(-2)}$$
$$= 3^2$$
$$= 9$$

Exercises 5 - 4

A 1. Express each power with a positive exponent, and then simplify:

a) 2^{-1} b) 5^{-1} c) 3^{-2} d) 2^{-3}

e) 5^{-3} f) 10^{-2} g) 12^{-3} h) 10^{-4}

i) $(\frac{1}{2})^{-1}$ j) $(\frac{1}{4})^{-2}$ k) 10^{-5} l) $\frac{1}{5^{-1}}$

m) $\frac{1}{2^{-5}}$ n) $\frac{3}{4^{-2}}$ o) $(\frac{3}{4})^{-2}$ p) $(\frac{1}{10})^{-1}$

q) $(0.1)^{-3}$ r) $(0.5)^{-2}$ s) $(2.5)^{-3}$ t) $\frac{2}{3^{-4}}$

2. Simplify:

a) $3^2 + 3^{-2}$ b) $3^2 - 3^{-2}$ c) $3^{-2} - 3^2$

d) $3^2 \times 3^{-2}$ e) $3^2 \div 3^{-2}$ f) $3^{-2} \div 3^2$

3. Simplify:

a) $2^3 - 2^{-1}$ b) $5^2 + 5^{-1}$ c) $7^{-2} - 7$

d) $(2 \times 3)^{-2}$ e) $4^2 + 4^0$ f) $3^{-1} + 3^{-2}$

g) $6^2 + 6^0 + 6^{-2}$ h) $(2^2 - 1)^{-2}$ i) $3^{-2} - 2^{-4}$

4. Simplify:

a) $(-2)^3$

b) $(-2)^{-3}$

c) $-(-2)^{-3}$

d) $(-5)^0$

e) $-(5^0)$

f) $(6 - 4)^{-3}$

g) $(5 - 8)^{-1}$

h) $(\frac{1}{4} - \frac{1}{4^2})^{-2}$

i) $[(-3)^{-2} + (-3)^{-1}]^{-1}$

5. Express the following as powers of 2, and then arrange in order from greatest to least:

$$\frac{1}{32} \qquad 16 \qquad 128 \qquad \frac{1}{64} \qquad 1 \qquad \frac{1}{2}$$

6. Express the following as powers of 3, and then arrange in order from least to greatest:

$$\frac{1}{9} \qquad \frac{1}{243} \qquad \frac{1}{81} \qquad 27 \qquad \frac{1}{729} \qquad 1$$

7. Express the following as powers with positive exponents:

a) 49

b) $\frac{1}{100}$

c) $\frac{1}{343}$

d) $\frac{1}{-32}$

e) $\frac{1}{1\,000\,000}$

f) 0.25

g) 0.001

h) 0.125

8. Express the following as powers with negative exponents other than -1:

a) $\frac{1}{121}$

b) $\frac{1}{169}$

c) 0.01

d) 0.16

e) 0.000 01

f) 0.008

g) 0.0081

h) $\frac{1}{1728}$

9. Simplify:

a) $5^{-1} \div 3^{-2}$

b) $(3^{-1} - 3^{-2})^{-1}$

c) $(\frac{1}{4})^{-1} - (\frac{1}{3})^{-2}$

d) $(\frac{1}{-2})^{-3} + (\frac{1}{2})^{-2}$

e) $(\frac{2}{3^{-1}})^{-3}$

f) $(0.5)^{-3} + (0.5)^0$

g) $\frac{4}{4^{-1} + 4^0}$

h) $\frac{2^{-1}}{2^{-2} - 2^{-3}}$

i) $[47(5)^{-2}]^0$

B 10. Simplify:

a) $x^{-9} \times x^{-4}$

b) $2^{-5} \times 2^{11}$

c) $3^4 \div 3^{-7}$

d) $m^{-14} \div m^5$

e) $s^5 \div s^{-8}$

f) $(-5)^{-11} \div (-5)^{19}$

g) $(-3)^{-6} \times (-3)^4 \div (-3)^3$

h) $y^{-5} \div y^9 \times y^4$

i) $(x^{-2})^3 \div (x^3)^{-2}$

j) $a^{-3} \div a^6 \div a^{-9}$

11. Simplify:
    a) $5n^{-4} \times 2n^{-17}$
    b) $12t^4 \div 3t^{-3}$
    c) $60x^5 \div 12x^{-5}$
    d) $7a^{-4} \times (-4)a^{-2}$
    e) $16w^{-8} \div 4w^{-2}$
    f) $15s^{-15} \div 3s^5$
    g) $45b^{-3} \div 5b^5 \times 3b^{-7}$
    h) $4x^{-7} \times 12x^{-5} \div (-8)x^{-4}$

12. Evaluate for $a = -3$, $b = 2$, and $c = -1$:
    a) a^{-1}
    b) $-a^{-1}$
    c) $a^{-1} + b^{-1}$
    d) $(a + b + c)^{-1}$
    e) $a^{-1} + b^{-1} + c^{-1}$
    f) a^b
    g) $\left(\dfrac{a}{b-c}\right)^{-2}$
    h) $\left(\dfrac{2a}{b+4c}\right)^{-3}$

13. Evaluate $3a^{-2} + b^c$ for the following values of a, b, and c:
    a) $a = 4$, $\quad b = 3$, $\quad c = 0$
    b) $a = 3$, $\quad b = 2$, $\quad c = -1$
    c) $a = -1$, $b = -2$, $c = 3$
    d) $a = 2$, $\quad b = 5$, $\quad c = -1$

14. Evaluate the following for (i) $x = 2$, (ii) $x = -2$:
    a) x^3
    b) $(-x)^3$
    c) $-x^3$
    d) $-(-x)^3$
    e) x^{-3}
    f) $(-x)^{-3}$
    g) $-x^{-3}$
    h) $-(-x)^{-3}$

C 15. Solve by inspection:
    a) $5^x = 1$
    b) $2^x = \dfrac{1}{2}$
    c) $(-3)^x = \dfrac{1}{9}$
    d) $x^{-3} = \dfrac{1}{125}$
    e) $2^x = \dfrac{1}{32}$
    f) $x^2 = \dfrac{1}{25}$
    g) $4^{x-1} = \dfrac{1}{64}$
    h) $10^{2-x} = 0.001$

Mathematics Around Us

How Bacteria Grow and Multiply
part 2

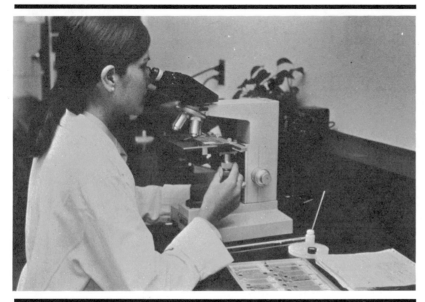

If, at midnight, there were 1000 bacteria in a culture, the number of bacteria n h later would be 1000×2^n. Does this statement make sense if n is negative?

If n were -3, it would be like asking: "How many bacteria were there -3 h after midnight?" -3 h after midnight is 3 h *before* midnight. To find the number of bacteria 3 h before midnight multiply 1000 by 2^{-3}.

$$1000 \times 2^{-3} = 1000 \times \frac{1}{8}$$
$$= 125$$

There were about 125 bacteria in the culture 3 h before midnight.

Questions

1. About how many bacteria are in the culture
    a) 1 h before midnight?
    b) 2 h before midnight?

2. At 00:30, there are about 1400 bacteria in the culture. About how many were there
    a) 1 h earlier?
    b) 2 h earlier?
    c) 3 h earlier?

3. At 03:45, there are about 13 500 bacteria in the culture. About how many were there
    a) 3 h earlier?
    b) 4 h earlier?
    c) 5 h earlier?

4. State the number of bacteria after
    a) 24 h
    b) 72 h
    c) 29 h

    Solution to (a).
    After 24 h, the number of bacteria is 1000×2^{24}.

    An approximate value for 2^{24} is found as follows:
    $$\begin{aligned} 2^{24} &= 2^{10} \times 2^{10} \times 2^4 \\ &= 1024 \times 1024 \times 16 \\ &\doteq 1000 \times 1000 \times 16, \\ & \text{or } 10^6 \times 16 \end{aligned}$$

    The number of bacteria after 24 h is approximately
    $$1000 \times 16 \times 10^6$$
    $$= 16 \times 10^9.$$

    Now complete the question using this method.

Night Sky:
double cluster in Perseus

5 - 5 Scientific Notation

Scientists tell us that there are about 120 000 000 000 stars in our galaxy, the Milky Way. If you saw a falling star and then said there are now only 119 999 999 999 stars left, you would be wrong on two counts. First, a falling, or shooting, star is only a meteor burning up as it passes through Earth's atmosphere. Second, there are not exactly 120 000 000 000 stars. The first two digits are the only ones with any degree of accuracy. The zeros are merely place holders to show the position of the decimal point.

Large numbers like this are awkward to write, difficult to read, and liable to be interpreted as exact numbers. The same is true of very small numbers. A much better way to express very large and very small numbers is to express them as a product of

- a number equal to or greater than 1 but less than 10, and
- a power of 10 (to locate the decimal point).

For the number: 120 000 000 000 1.2×10^{11}

place the decimal point after the first non-zero digit. As its true position is 11 places to the right (the positive direction), the exponent is positive and the number becomes

For the number: 0.000 000 000 167 1.67×10^{-10}

place the decimal point after the first non-zero digit. As its true position is 10 places to the left (the negative direction), the exponent is negative and the number becomes

Numbers expressed in this form are said to be written in **scientific notation**. Study these further examples:

$$9\,000\,000\,000\,000\,000 = 9 \times 10^{15}$$
$$24\,900\,000\,000\,000 = 2.49 \times 10^{13}$$
$$0.000\,000\,000\,000\,075 = 7.5 \times 10^{-14}$$
$$0.000\,000\,000\,000\,000\,000\,81 = 8.1 \times 10^{-19}$$

Example 1. Simplify: $\dfrac{24\,000\,000\,000 \times 0.000\,02}{3200}$

Solution. Rewrite the expression using scientific notation.

$$\frac{24\,000\,000\,000 \times 0.000\,02}{3200} = \frac{2.4 \times 10^{10} \times 2 \times 10^{-5}}{3.2 \times 10^{3}}$$
$$= \frac{2.4 \times 2}{3.2} \times \frac{10^{10} \times 10^{-5}}{10^{3}}$$
$$= 1.5 \times 10^{2}, \quad \text{or} \quad 150$$

Example 2. It has been estimated that if the average mass of an automobile was reduced from 1500 kg to 1000 kg, the amount of oil Canada would save would be 30 000 m³/day.

a) About how much oil would Canada save in a year?

b) If there are 9 000 000 cars on Canadian roads, and 3 m³ of oil yield 2 m³ of gasoline, what would be the approximate annual saving per car with gasoline at 35¢/L?

Solution. a) Cubic metres of oil saved per year:
$$365 \times 30\,000$$
$$= 3.65 \times 10^{2} \times 3 \times 10^{4}$$
$$= 1.095 \times 10^{7}$$
The annual saving of oil would be about 1.1×10^{7} m³.

b) Cubic metres of gasoline saved per year:
$$\frac{2}{3} \times 1.095 \times 10^{7}$$

Annual saving per car:
$$\frac{2}{3} \times 1.095 \times 10^{7} \div 9\,000\,000$$

$$35¢/L = \$350/m^{3}$$

Dollar saving per car: $\dfrac{2}{3} \times \dfrac{1.095 \times 10^{7} \times 350}{9 \times 10^{6}}$

$$\doteq 28.4 \times 10, \text{ or } 284$$

The annual saving per car would be about $280.

Exercises 5 - 5

A   1. Write the following in scientific notation:

     a) 1000   b) 100 000 000   c) 100
     d) 750   e) 1100   f) 3 700 000
     g) 0.0001   h) 0.000 000 1   i) 0.000 001
     j) 0.000 85   k) 0.000 092   l) 0.000 000 008 2
     m) 85   n) 0.038   o) 9900

2. Write the following in scientific notation:

     a) Speed of light   300 000 km/s
     b) Distance to the sun   150 000 000 km
     c) Distance to the nearest star   40 000 000 000 000 km
     d) Distance to Andromeda galaxy
                      14 000 000 000 000 000 000 km
     e) World population in 1980   4 470 000 000
     f) Mass of the Earth   5 980 000 000 000 000 000 000 000 kg
     g) Time of fastest camera exposure   0.000 000 1 s
     h) Mass of the ball in a ball-point pen   0.004 g
     i) Mass of a hydrogen atom
                  0.000 000 000 000 000 000 000 001 67 g
     j) Radius of an atom's nucleus   0.000 000 000 000 1 cm
     k) Density of hydrogen   0.0838 g/L

3. Copy and complete this table:

	Physical Quantity	Decimal Notation	Scientific Notation
a)	A light-year		9.5×10^{12} km
b)	Temperature of sun's interior	1 300 000°C	
c)	Thickness of plastic film	0.000 01 m	
d)	Mass of an electron		9.2×10^{-28} g
e)	Number of stars in our galaxy		1.2×10^{11}
f)	Estimated age of Earth	4 500 000 000 yrs.	
g)	Diameter of a hydrogen atom	0.000 000 0113 cm	
h)	Land area of Earth		1.5×10^{8} km²
i)	Ocean area of Earth		3.6×10^{8} km²
j)	Mass of the Earth		5.9×10^{24} kg
k)	Cost of a Concorde aircraft	8 500 000 000 F	

4. Write the following in scientific notation:

a) 32×10^4 b) 247×10^8 c) 49.2×10^7

d) 685×10^{10} e) 0.387×10^4 f) 0.087×10^3

g) 672×10^{-5} h) 43.7×10^{-6} i) 0.841×10^{-2}

j) 0.49×10^{-7} k) 125×10^0 l) $1.85 \div 10^{-2}$

Give your answers to the following exercises in scientific notation rounded to two decimal places.

5. Simplify:

a) $349\,000 \times 2650 \times 120\,000$

b) $8600 \times 1\,500\,000 \times 0.0003$

c) $27\,000\,000\,000 \times 6\,800\,000 \times 0.000\,000\,025$

d) $\dfrac{480\,000 \times 62\,000\,000}{300\,000}$ e) $\dfrac{850\,000 \times 400\,000}{6\,200\,000}$

f) $\dfrac{0.000\,006 \times 54\,000}{0.000\,009}$ g) $\dfrac{2\,400\,000 \times 0.000\,000\,000\,8}{0.000\,000\,4 \times 12\,000}$

B   6. Find n:

a) $1265 = 1.265 \times 10^n$ b) $76.3 = 7.63 \times 10^n$

c) $0.0041 = 4.1 \times 10^n$ d) $0.860 = 8.60 \times 10^n$

e) $0.005 = 5 \times 10^n$ f) $0.000\,056\,3 = 5.63 \times 10^n$

g) $1150 = 1.150 \times 10^n$

h) $4\,961\,000\,000 = 4.961 \times 10^n$

i) $7\,430\,000 = 7.43 \times 10^n$

j) $0.000\,000\,583\,1 = 5.831 \times 10^n$

7. Californium 252, one of the world's rarest metals, is used in treating cancer. If one-tenth of a microgram costs $100, what is its cost per gram?

8. It is estimated that the total volume of water in the world is 1.4×10^9 km³. It is distributed as follows:

    97.3% is in the oceans and is unfit for human use;

    2.1% is locked in glaciers and polar ice;

    0.3% is trapped far underground;

    0.2% is in the atmosphere.

The rest of the water is in the lakes and rivers and is easily available for human use.

a) What percent of the world's water is available for human use?

b) How many cubic metres of water are available for human use?

Rhône Glacier

9. The measured daily deposit of the pollutant sulphur dioxide on Metropolitan Toronto is approximately 4.8×10^{-6} g/cm^2. If Metropolitan Toronto has an area of about 620 km^2, and the pollutant is distributed evenly, calculate the amount of sulphur dioxide that falls on the city

   a) in 1 day;   b) in 1 year;   c) in your lifetime.

10. If it takes 1200 silkworm eggs to balance the mass of 1 g, what is the mass of one silkworm egg?

11. The volume of water in the oceans is an estimated 1.35×10^{18} m^3. If the density of sea water is 1025 kg/m^3, what is the mass of the oceans?

12. A faucet is leaking at the rate of one drop of water per second. If the volume of one drop is 0.1 cm^3, calculate

   a) the volume of water lost in a year;

   b) how long it would take to fill a rectangular basin 30 cm by 20 cm by 20 cm.

C 13. In 1974, a Canadian astronomer was reported to have discovered the largest known object in the universe. It stretches 16.3 million light-years from end to end. If a light-year is the distance that light, with a speed of 300 000 km/s, travels in a year, calculate

   a) the approximate number of kilometres in a light-year;

   b) the length, in kilometres, of the astronomer's discovery.

14. It has been estimated that the insect population of the world is at least 1 000 000 000 000 000 000. The scientist who made this estimate also reckoned that if the average mass of an insect is 2.5 mg, then the total mass of insects on Earth is twelve times greater than the total mass of all human beings.

   a) Estimate the total mass of Earth's insect population.

   b) Estimate the total mass of Earth's human population.

   c) Use the population estimate in Exercise 2 and your answer to part (b) to calculate the average mass per person. Is this average reasonable?

Cluster of butterflies

15. A drop of oil, with a volume of 1 mm^3, spreads out on the surface of some water until it is a film one molecule thick. If the film has an area of 1 m^2, what is the thickness of an oil molecule?

16. Googol is the name given to the number 10^{100}. To give you an idea how large a number this is, think of the visible universe as a cube with edge of length 2×10^{25} m. Suppose this cube is filled with cubical grains of sand with edges of length 2×10^{-4} m. How many grains of sand would be needed?

17. Express the distances and measurements on this page in scientific notation.

From Galactic Distances to Atomic Measurements, All in Metres

100 000 000 000 000 000 000 000 000	Distance to quasars
14 000 000 000 000 000 000 000	Distance to nearest galaxy
760 000 000 000 000 000 000	Diameter of our galaxy
41 000 000 000 000 000	Distance to nearest star
12 000 000 000 000	Diameter of the solar system
150 000 000 000	Distance to the sun
380 000 000	Distance to the moon
13 000 000	Diameter of Earth
8 800	Height of Mount Everest
2	Height of man
Size of an insect	0.005
Diameter of a grain of sand	0.000 1
Size of bacteria	0.000 001
Diameter of an atom	0.000 000 000 1
Diameter of a nucleus	0.000 000 000 000 01
Diameter of a proton	0.000 000 000 000 000 1
Wavelength of cosmic rays	0. 000 000 000 000 000 000 01

Mathematics Around Us

GOLD IN THE OCEANS

The oceans contain many dissolved salts and minerals, including gold. 1 km³ of sea water contains about 4 kg of gold. The immensity of the oceans means they have within them a vast quantity of gold. However, it will most likely stay there owing to the difficulty of recovering it.

Questions

1. If the total volume of the oceans is approximately 1.4×10^9 km³, how many grams of gold will they contain?

2. The price of gold is still quoted in troy ounces (about 31.2 g). Look in your newspaper to get the current price of gold. Then calculate the total value of the gold in the oceans.

3. If all the gold in the oceans were recovered and divided equally among the world's population (estimated to be 4 470 000 000 in 1980), what would be the value of each person's share?

4. All the gold ever mined has a volume equal to that of a cube with an edge of 1650 cm. The density of gold is 19 300 kg/m³. What is the total mass of the gold that has been mined?

5. If the mined gold were collected and shared equally among the world's population, what would be the value of each person's share?

6. What would happen to the price of gold if all the gold dissolved in the oceans were to be recovered?

5 - 6   Square Roots

The square roots of 9 are 3 and -3 because
$$3 \times 3 = 9 \qquad \text{and} \qquad (-3) \times (-3) = 9.$$
Any number, n, can be written as the product of two factors:
$$(\quad) \times (\quad) = n$$
When the two factors are the same, the factor is a **square root** of n.

Because $7 \times 7 = 49$, and $(-7) \times (-7) = 49$,
  7 and -7 are the square roots of 49.

Because $0.2 \times 0.2 = 0.04$, and $(-0.2) \times (-0.2) = 0.04$,
  0.2 and -0.2 are the square roots of 0.04.

Because $\frac{3}{2} \times \frac{3}{2} = \frac{9}{4}$, and $(-\frac{3}{2}) \times (-\frac{3}{2}) = \frac{9}{4}$,
  $\frac{3}{2}$ and $-\frac{3}{2}$ are the square roots of $\frac{9}{4}$.

Positive numbers always have two square roots, one positive the other negative. The symbol, $\sqrt{}$, is known as the **radical sign**, and it always denotes the positive square root. Thus:

$$\sqrt{49} = 7 \qquad \sqrt{0.04} = 0.2 \qquad \sqrt{\frac{9}{4}} = \frac{3}{2}$$

$\sqrt{n}$ means the positive square root of n.

Example 1. Simplify: $-3\sqrt{6.25}$

  Solution. $-3\sqrt{6.25} = -3 \times 2.5$
  $\phantom{-3\sqrt{6.25}} = -7.5$

Radical signs are usually treated like brackets. Operations within them are performed first.

Example 2. Simplify: a) $\sqrt{9 + 16}$;
  b) $\sqrt{64} + \sqrt{36}$.

  Solution. a) $\sqrt{9 + 16} = \sqrt{25}$
  $\phantom{\sqrt{9 + 16}} = 5$

  b) $\sqrt{64} + \sqrt{36} = 8 + 6$
  $\phantom{\sqrt{64} + \sqrt{36}} = 14$

Example 3. Simplify: a) $-3\sqrt{400}$;

b) $5\sqrt{4} - 3(\sqrt{121} - \sqrt{81})$.

Solution. a) $-3\sqrt{400} = -3 \times 20$

$= -60$

b) $5\sqrt{4} - 3(\sqrt{121} - \sqrt{81})$

$= 5 \times 2 - 3(11 - 9)$

$= 10 - 3 \times 2$

$= 4$

Equations in the forms $ax^2 + b = c$ and $ax^2 - b = c$ can be solved by isolating the variable in the usual way and then taking the square root.

Example 4. Solve and check: $2x^2 - 3 = 15$

Solution. $2x^2 - 3 = 15$

$2x^2 = 18$

$x^2 = 9$

$x = 3$, or $x = -3$

Check. When $x = 3$, $\qquad$ When $x = -3$,

$2x^2 - 3 = 2(3)^2 - 3 \qquad 2x^2 - 3 = 2(-3)^2 - 3$

$= 18 - 3 \qquad\qquad\qquad = 18 - 3$

$= 15 \qquad\qquad\qquad\quad = 15$

For both values of x, the left side of the equation is equal to the right side. The solutions are correct.

Exercises 5 - 6

A 1. Find the square roots of these numbers:

a) 10 000 $\quad$ b) 81 $\qquad$ c) 1.69 $\qquad$ d) 0.25 $\qquad$ e) $\frac{1}{100}$

f) 0.04 $\qquad$ g) $\frac{1}{64}$ $\qquad$ h) $1\frac{9}{16}$ $\qquad$ i) 900 $\qquad$ j) $2\frac{7}{9}$

2. Simplify:

a) $\sqrt{49}$ $\qquad$ b) $\sqrt{400}$ $\qquad$ c) $\sqrt{\frac{36}{121}}$ $\qquad$ d) $\sqrt{2\frac{1}{4}}$

e) $-\sqrt{0.16}$ $\qquad$ f) $\sqrt{0.0004}$ $\qquad$ g) $-\sqrt{225}$ $\qquad$ h) $-\sqrt{30\frac{1}{4}}$

i) $\sqrt{6400}$ $\qquad$ j) $-\sqrt{\sqrt{16}}$ $\qquad$ k) $\sqrt{10^{12}}$ $\qquad$ l) $\sqrt{10^{-16}}$

3. Simplify:

a) $\sqrt{64+36}$
b) $\sqrt{16} + \sqrt{9}$
c) $2\sqrt{16} - 3\sqrt{4}$

d) $3\sqrt{36} + 2\sqrt{25}$
e) $5\sqrt{100-36}$
f) $\sqrt{1+3+5+7+9}$

g) $2\sqrt{81} - 7\sqrt{49}$
h) $4\sqrt{289-225}$
i) $2\sqrt{\sqrt{81}}$

4. Evaluate for $a = 5$ and $b = -3$:

a) $\sqrt{20a}$
b) $\sqrt{9a^2}$
c) $\sqrt{\dfrac{125}{a}}$

d) $\sqrt{2a - 13b}$
e) $\sqrt{-12b}$
f) $-\sqrt{3a - 3b + 1}$

g) $\sqrt{a^2 + 3b}$
h) $\sqrt{7a - 8b + 5}$
i) $-4\sqrt{11a - 3b}$

B 5. Evaluate for $x = 3$, $y = -4$, and $z = -7$:

a) $-\sqrt{12x}$
b) $\sqrt{z^2 + 6y}$
c) $4\sqrt{x^2 + y^2}$

d) $\sqrt{6x - z}$
e) $5\sqrt{15x - y}$
f) $-\sqrt{7x - 4y - 1}$

g) $2\sqrt{x - 2y - 2z}$
h) $-\sqrt{2z^2 + 5y + x}$
i) $6\sqrt{3y^2 + x^0}$

6. Solve and check:

a) $x^2 = 64$
b) $a^2 = 225$
c) $y^2 = 1.96$

d) $m^2 + 3 = 12$
e) $s^2 - 4 = 12$
f) $2n^2 = 50$

g) $3c^2 + 1 = 28$
h) $9k^2 - 49 = 0$
i) $4x^2 - 70 = 155$

7. From the area of each of these squares, calculate
   i) the length of a side;   ii)   the perimeter.

a)
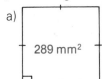
289 mm²

b)

4.41 cm²

c)

0.0064 m²

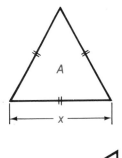

A

C 8. For an equilateral triangle of side length x, the area, A, is given
   by $A \doteq 0.433x^2$. For equilateral triangles with these areas,
   calculate: i)   the length of a side; ii)   the perimeter.

a) 10.825 m²
b) 27.712 m²
c) 0.350 73 m²

d) 389.7 cm²
e) 0.004 33 km²
f) 97.425 km²

9. In a right-angled triangle with side lengths as shown,
   $z = \sqrt{x^2 + y^2}$. Calculate z for these values of x and y.

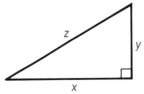

z

y

x

	x	y		x	y
a)	6 mm	8 mm	b)	8 mm	15 mm
c)	0.5 m	1.2 m	d)	24 mm	7 mm
e)	0.03 km	0.04 km	f)	2.8 km	9.6 km

5 - 7 Solving an Equation Involving the Square Root of a Variable

In Chapter 4, we learned that if the same operation was applied to both sides of an equation the equation remained true. The operations we used were addition, subtraction, multiplication, and division. To these we can now add squaring.

Example 1. Solve and check: $3\sqrt{5x} + 2 = 47$

Solution.

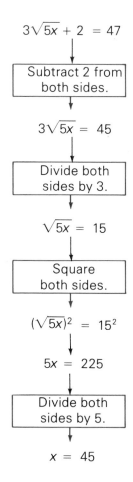

$$3\sqrt{5x} + 2 = 47$$

Subtract 2 from both sides.

$$3\sqrt{5x} = 45$$

Divide both sides by 3.

$$\sqrt{5x} = 15$$

Square both sides.

$$(\sqrt{5x})^2 = 15^2$$

$$5x = 225$$

Divide both sides by 5.

$$x = 45$$

Check. If $x = 45$, $3\sqrt{5x} + 2 = 3\sqrt{5 \times 45} + 2$
$$= 3\sqrt{225} + 2$$
$$= 3 \times 15 + 2$$
$$= 47$$

The left side of the equation equals the right side. $x = 45$ is the correct solution.

Example 2. Solve and check: $2\sqrt{x-1} = 10$.

Solution.
$$2\sqrt{x-1} = 10$$
$$\sqrt{x-1} = 5$$
$$(\sqrt{x-1})^2 = 5^2$$
$$x - 1 = 25$$
$$x = 26$$

Check. If $x = 26$, $2\sqrt{x-1} = 2\sqrt{26-1}$
$$= 2\sqrt{25}$$
$$= 2 \times 5, \text{ or } 10$$

The left side of the equation equals the right side. $x = 26$ is the correct solution.

Exercises 5 - 7

1. Solve:
   a) $\sqrt{x} = 3$ b) $\sqrt{y} = 13$ c) $\sqrt{z} = 2.5$
   d) $3\sqrt{x} = 15$ e) $5\sqrt{y} = 35$ f) $9\sqrt{z} = 11.7$

2. Solve:
   a) $\sqrt{3a} = 6$ b) $\sqrt{5b} = 1$ c) $\sqrt{7c} = 7$
   d) $\sqrt{25a} = 5$ e) $\sqrt{65b} = 13$ f) $\sqrt{0.1c} = 11$
   g) $7\sqrt{8a} = 56$ h) $12\sqrt{5b} = 60$ i) $13\sqrt{14c} = 91$

3. Solve and check:
   a) $\sqrt{r} - 3 = 4$ b) $\sqrt{s} + 10 = 11$ c) $\sqrt{t} + 25 = 30$
   d) $-17 + \sqrt{r} = -8$ e) $26 - \sqrt{s} = 9$ f) $\sqrt{0.1t} - 10 = 0$
   g) $2\sqrt{m} + 1 = 7$ h) $6\sqrt{n} - 2 = 40$ i) $8\sqrt{p} + 11 = 99$
   j) $3\sqrt{5m} - 5 = 7$ k) $5 + 9\sqrt{9n} = 59$ l) $25 + 7\sqrt{10p} = 39$

4. Solve and check:
   a) $\sqrt{c - 2} = 8$ b) $\sqrt{k + 7} = 11$
   c) $\sqrt{m + 1} = 1$ d) $\sqrt{5a - 4} = 6$
   e) $\sqrt{9b - 9} = 9$ f) $\sqrt{6c + 25} = 25$
   g) $3\sqrt{x - 1} + 5 = 14$ h) $5\sqrt{2z + 3} - 2 = 43$
   i) $13\sqrt{3m + 19} + 9 = 100$ j) $11\sqrt{4n + 21} - 15 = 106$

5 - 8   Estimating and Calculating Square Roots

Some numbers have exact square roots:

$$\sqrt{25} = 5; \qquad \sqrt{289} = 17; \qquad \sqrt{4.41} = 2.1.$$

Most numbers do not have exact square roots:

$$\sqrt{23} \doteq 4.796; \qquad \sqrt{290} \doteq 17.029; \qquad \sqrt{4.5} \doteq 2.121.$$

If you have a calculator with a key marked $\boxed{\sqrt{}}$, the approximate square root of a positive number can be found easily and quickly. Even if your calculator lacks this key, the approximate square root of a positive number can be found quite readily by other methods.

Systematic Trial Method

Let the number whose square root is to be found be *n*.

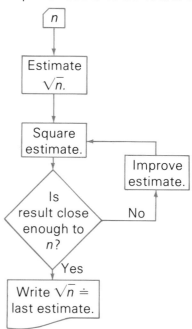

Example 1.   Calculate $\sqrt{27}$ to three decimal places.

Solution.   Estimate: $\sqrt{27} \doteq 5.1$

$$(5.1)^2 = 26.01 \text{ too small}$$

Try 5.2. $\qquad (5.2)^2 = 27.04 \text{ too big}$

Try 5.19. $\qquad (5.19)^2 = 26.9361$

Try 5.195. $\quad (5.195)^2 \doteq 26.9880$

Try 5.197. $\quad (5.197)^2 \doteq 27.0088$

Try 5.196. $\quad (5.196)^2 \doteq 26.9984$

Of the last two estimates, $(5.196)^2$ is closer to 27.
Therefore, $\sqrt{27} \doteq 5.196$.

Newton's Method of Finding Square Roots

Let the number whose square root is to be found be n.

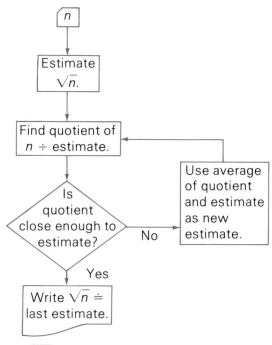

Example 2. Calculate $\sqrt{27}$ to three decimal places.

Solution.   Estimate: $\sqrt{27} \doteq 5.1$ Quotient: $\frac{27}{5.1} \doteq 5.294$

Average (new estimate): $\frac{5.294 + 5.1}{2} = 5.197$

Quotient: $\frac{27}{5.197} \doteq 5.195$

Average (new estimate): $\frac{5.195 + 5.197}{2} = 5.196$

Quotient: $\frac{27}{5.196} \doteq 5.196$

Quotient and last estimate are almost the same.
Therefore, $\sqrt{27} \doteq 5.196$.

In *Example 2*, try using 5, 5.3, and even 6 as your initial
estimate of $\sqrt{27}$. You will find that in only a few steps you will
obtain the same answer.

Example 3. Find $\sqrt{0.763}$ by Newton's method to two decimal places.

Solution. $\sqrt{0.763}$ lies between 0.8 and 0.9. Try 0.85 as the first estimate.

Quotient: $\dfrac{0.763}{0.85} \doteq 0.898$.

Average (new estimate): $\dfrac{0.898 + 0.85}{2} = 0.874$

Quotient: $\dfrac{0.763}{0.874} \doteq 0.873$

Quotient and last estimate are almost the same.

Therefore, $\sqrt{0.763} = 0.87$, to two decimal places.

Square Roots From Tables

The approximate square root of a positive number can be found from a table of square roots, such as the one on page 420. The main disadvantage of a table is that numbers must sometimes be rounded to the nearest entry in the table, or an interpolation made.

Example 4. Use the table of square roots on page 420 to find
  a) $\sqrt{3.4}$;   b) $\sqrt{47}$;   c) $\sqrt{67.3}$.

Solution. a) $\sqrt{3.4} \doteq 1.844$ b) $\sqrt{47} \doteq 6.856$
  c) Since 67.3 is not in the table, we can use the closest value, 67.
$$\sqrt{67} \doteq 8.185$$
Or, we can get a better approximation by interpolating between the reading for $\sqrt{67}$ and $\sqrt{68}$.
$$\sqrt{68} - \sqrt{67} \doteq 8.246 - 8.185, \text{ or } 0.61.$$
$$\sqrt{67.3} \doteq 8.185 + 0.3 \times 0.061$$
$$\doteq 8.185 + 0.018$$
$$\doteq 8.203$$

Numbers greater than 100 or less than 1 must first be written as the product of a number between 1 and 100 and an *even* power of 10.

Example 5.   Use the table of square roots to find

a) $\sqrt{240}$;

b) $\sqrt{2400}$;

c) $\sqrt{0.35}$;

d) $\sqrt{0.035}$.

Solution.

a) $240 = 2.4 \times 10^2$

$\sqrt{240}$

$= \sqrt{2.4 \times 10^2}$

$= \sqrt{2.4} \times \sqrt{10^2}$

$\doteq 1.549 \times 10$

$\doteq 15.49$

b) $2400 = 24 \times 10^2$

$\sqrt{2400}$

$= \sqrt{24 \times 10^2}$

$= \sqrt{24} \times \sqrt{10^2}$

$\doteq 4.899 \times 10$

$\doteq 48.99$

c) $0.35 = 35 \times 10^{-2}$

$\sqrt{0.35}$

$= \sqrt{35 \times 10^{-2}}$

$= \sqrt{35} \times \sqrt{10^{-2}}$

$\doteq 5.916 \times 10^{-1}$

$\doteq 0.5916$

d) $0.035 = 3.5 \times 10^{-2}$

$\sqrt{0.035}$.

$= \sqrt{3.5 \times 10^{-2}}$

$= \sqrt{3.5} \times \sqrt{10^{-2}}$

$\doteq 1.871 \times 10^{-1}$

$\doteq 0.1871$

Example 6.   What will be the length of each side of a square poster that is twice the area of the one shown?

Solution.   Area of first poster: $90 \text{ cm} \times 90 \text{ cm} = 8100 \text{ cm}^2$
Area of enlarged poster:

$$2 \times 8100 \text{ cm}^2 = 16\,200 \text{ cm}^2$$

Let x be the length, in centimetres, of each side of the large poster.

Then, $x^2 = 16\,200$

and $x = \sqrt{16\,200}$

$\sqrt{16\,200}$ can be found by any one of these methods:

Using a calculator with a $\boxed{\sqrt{}}$ key:

$\sqrt{16\,200} \doteq 127.27922$

By Systematic Trial and Newton's method:

$\sqrt{16\,200} \doteq 127.3$

By the table, with interpolation:

$\sqrt{16\,200} \doteq 127.3$

—90 cm—

—?—

The length of each side of the poster is approximately 127.3 cm.

This solves the problem posed at the beginning of the chapter. Note that though the area has doubled, the side length has not.

Exercises 5 - 8

A   1. Estimate the following to one decimal place. Square your estimate, using a calculator, to check how good your estimate is.

a) $\sqrt{29}$ b) $\sqrt{290}$ c) $\sqrt{14}$ d) $\sqrt{6.5}$ e) $\sqrt{175}$

f) $\sqrt{640}$ g) $\sqrt{887}$ h) $\sqrt{0.83}$ i) $\sqrt{43.5}$ j) $\sqrt{0.435}$

2. Use the Systematic Trial method to calculate the following
square roots to two decimal places.

a) $\sqrt{34}$ b) $\sqrt{18}$ c) $\sqrt{28.5}$ d) $\sqrt{150}$ e) $\sqrt{3.84}$

f) $\sqrt{0.29}$ g) $\sqrt{0.62}$ h) $\sqrt{521}$ i) $\sqrt{0.05}$ j) $\sqrt{7.4}$

3. Use Newton's method to calculate the following square roots to
three decimal places.

a) $\sqrt{51}$ b) $\sqrt{22.6}$ c) $\sqrt{321}$ d) $\sqrt{117}$ e) $\sqrt{4.92}$

f) $\sqrt{15.3}$ g) $\sqrt{0.06}$ h) $\sqrt{0.45}$ i) $\sqrt{0.121}$ j) $\sqrt{205}$

4. Find the approximate values of the following by using the table
of square roots:

a) $\sqrt{8.6}$ b) $\sqrt{86}$ c) $\sqrt{860}$ d) $\sqrt{8600}$ e) $\sqrt{86\,000}$

f) $\sqrt{58}$ g) $\sqrt{5.8}$ h) $\sqrt{0.58}$ i) $\sqrt{0.058}$ j) $\sqrt{0.0058}$

5. Calculate, giving your answer to three decimal places:

a) $\sqrt{2} + \sqrt{3}$ b) $(\sqrt{2} + \sqrt{3})^2$ c) $(\sqrt{2})^2 + (\sqrt{3})^2$

d) $3\sqrt{5} - \sqrt{2}$ e) $2\sqrt{5} + \sqrt{6}$ f) $5\sqrt{6} - 2\sqrt{10}$

g) $(\sqrt{2.1} + \sqrt{7.9})^2$ h) $\sqrt{23.5} - \sqrt{11.1}$ i) $8\sqrt{0.7} + 3\sqrt{7}$

B   6. Evaluate to two decimal places for $x = 2$, $y = -3$, $z = -7$:

a) $\sqrt{x + y - z}$ b) $\sqrt{3x - 2y}$ c) $\sqrt{x^2 + y^2}$

d) $\sqrt{x^2 + y^2 + z^2}$ e) $\sqrt{2x^2 - y^3}$ f) $\sqrt{5x - 2y - z}$

g) $3\sqrt{2x - z}$ h) $5\sqrt{3x - y + z}$ i) $2\sqrt{3x} + 5\sqrt{zy}$

7. Solve, rounding your answer to two decimal places:

a) $x^2 = 65$ b) $a^2 = 2.4$ c) $m^2 = 1.88$

d) $y^2 = \dfrac{5}{12}$ e) $x^2 + 5 = 12$ f) $a^2 - 5 = 12$

g) $3m^2 = 39$ h) $2x^2 + 1 = 13$ i) $a^2 - 37 = 0$

8. The perimeter, P, of a square of area A is given by the formula
$$P = 4\sqrt{A}.$$

   a) Find the perimeters of the squares with these areas:
     i) 25 cm²   ii) 64 cm²   iii) 78 m²   iv) 3.8 m²

   b) Find the areas of the squares with these perimeters:
     i) 12 cm   ii) 30 cm   iii) 5.7 m   iv) 7.4 m

   c) Explain why the formula is correct.

9. The diameter, d, of a circle of area A is given by the formula

$$d \doteq 1.13\sqrt{A}.$$

   a) Find the diameters of the circles with these areas:
     i) 22 cm²   ii) 33 cm²   iii) 40 cm²   iv) 5 m²

   b) Find the areas of the circles with these diameters:
     i) 10 cm   ii) 20 cm   iii) 31.8 cm   iv) 4 m

10. The distance, d, in kilometres, of the horizon from an observer at a height h m is given by the formula $d \doteq 3.6\sqrt{h}$.

   a) To the nearest kilometre, how far is the horizon from
     i) the 266 m observation level (above Lake Ontario) of the Skylon?
     ii) the 457 m observation level of the CN Tower?
     iii) eye level, 1.5 m, when you are standing on a beach?

   b) From what height would the distance to the horizon be
     i) 36 km?   ii) 100 km?

11. An object falls through a distance d m when falling from rest for t s. The relationship between d and t is given by the formula $d = 4.9t^2$. If it takes 3 s to fall a certain distance, how long will it take to fall

Skylon

   a) twice as far?   b) five times as far?

   c) half as far?   d) ten times as far?

12. For any planet, its year is the time it takes to circle the sun once. The number of Earth days in its year, N, is given by the formula $N \doteq 0.2(\sqrt{R})^3$, where R is its mean distance from the sun, in millions of kilometres. Find the number of Earth days in the year of each of these planets:

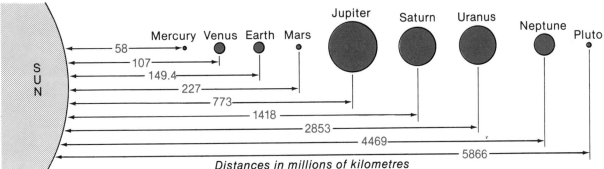

Distances in millions of kilometres

13. The length, d, of the diagonal of a rectangle of length l and width w is given by the formula $d = \sqrt{l^2 + w^2}$.

a) Calculate the lengths of the diagonals of rectangles with these dimensions:

   i) 8 cm by 3 cm   ii) 2.3 cm by 1.2 cm

   iii) 10 m by 7 m   iv) 2 m by 9 m

b) A rectangle is 9.2 cm long and its diagonal measures 11.5 cm. What is its width?

14. The period of a pendulum, T, in seconds, is given by the formula $T = 2\pi\sqrt{\dfrac{l}{9.8}}$, where $\pi \doteq 3.14$ and l is the length of the pendulum in metres.

a) Find the period of a pendulum whose length is

   i) 2.45 m;   ii) 0.5 m.

b) How long is a pendulum whose period is 1 s?

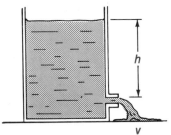

A period is the time taken for one complete swing to and fro.

15. The velocity, v, in metres per second, with which liquid discharges from a small hole in a container is given by the formula $v = \sqrt{19.6h}$, where h is the height of the liquid above the hole, in metres.

a) What is the velocity of discharge for these values of h?

   i) 1.0 m   ii) 0.5 m   iii) 10 cm

b) What height of liquid gives a discharge velocity of 2 m/s?

16. In winter, you will have noticed that the stronger the wind the colder it feels. Winter weather forecasts often give "wind-chill" temperatures. For wind speeds of 10 km/h and higher, the following formula gives an approximate value for the wind-chill temperature:

$$w = 33 - (0.23\sqrt{v} + 0.45 - 0.01v)(33 - T)$$

where w is the wind-chill temperature, in degrees Celsius,
      v is the wind speed in kilometres per hour, and
      T is the still-air temperature in degrees Celsius.

For a still-air temperature of $-18°C$, calculate the wind-chill temperatures at these wind speeds.

a) 100 km/h   b) 80 km/h   c) 30 km/h   d) 15 km/h

anemometer(for measuring wind speeds)

17. Find which of the following expressions is the closest approximation to π by calculating each to four decimal places.

a) $\sqrt{10}$ b) $\sqrt{\dfrac{2500}{253}}$ c) $\sqrt{51} - 4$

d) $\dfrac{13\sqrt{146}}{50}$ e) $\sqrt{13.5 - \sqrt{12}}$ f) $\sqrt{\sqrt{\dfrac{2143}{22}}}$

Mathematics Around Us

Saskatchewan produces more than half of Canada's wheat crop

If Canada's wheat production is represented by a 5 cm square, Saskatchewan's production may be shown as a smaller square inside it. The side length of the smaller square may be found as follows:

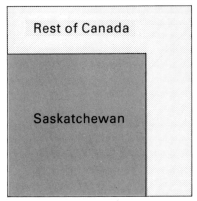

Wheat Production

Area of large square: 25 cm²
Therefore, area of small square is:
0.57 × 25 cm² = 14.25 cm²

Wheat Production of the Provinces as a Percent of the Canadian Total	
Saskatchewan	57%
Alberta	25%
Manitoba	12%
Other Provinces	6%
Total	**100%**

Let the side length of the smaller square be x cm ($x > 0$).

Then $x^2 = 14.25$
$x = \sqrt{14.25}$
$\doteq 3.77$

The side length of the square representing Saskatchewan's wheat production is approximately 3.8 cm.
A graph drawn in this way is called a **box graph**.

Questions

1. a) What would be the side length of the square representing the wheat production of i) Alberta? ii) Manitoba?

    b) Draw the graphs for (a).

2. Use the data in the table to draw box graphs to represent the population of:
    a) Saskatchewan;
    b) Alberta;
    c) Manitoba.

Population of Canada (1979)	
Saskatchewan	960 000
Alberta	2 030 800
Manitoba	1 126 400
Other Provinces	19 724 800
Total	**23 842 000**

*5 - 9 Cube Roots

The cube root of 27 is 3 because $\quad 3 \times 3 \times 3 = 27$

The cube root of -64 is -4 because
$$(-4) \times (-4) \times (-4) = -64$$

A number, x, is the cube root of a number, n, if $x^3 = n$

We write: $\sqrt[3]{27} = 3$, $\quad \sqrt[3]{-64} = -4$, and $\quad \sqrt[3]{n} = x$.

The cube root of a positive number is positive and the cube root of a negative number is negative.

Example 1. Simplify: a) $\sqrt[3]{8}$; b) $\sqrt[3]{-125}$; c) $\sqrt[3]{\dfrac{216}{343}}$

Solution. a) $\sqrt[3]{8} = 2$, because $2 \times 2 \times 2 = 8$

b) $\sqrt[3]{-125} = -5$, because
$$(-5)(-5)(-5) = -125$$

c) $\sqrt[3]{\dfrac{216}{343}} = \dfrac{6}{7}$, because $\dfrac{6}{7} \times \dfrac{6}{7} \times \dfrac{6}{7} = \dfrac{216}{343}$

Cube root signs are usually treated like brackets. Operations within them are performed first.

Example 2. Simplify: a) $2\sqrt[3]{-1000} + \sqrt[3]{729}$;

b) $\sqrt[3]{216 + 343 - 47}$

c) $5\sqrt[3]{27} - 2(\sqrt[3]{64} - \sqrt[3]{-8})$

Solution. a) $\quad 2\sqrt[3]{-1000} + \sqrt[3]{729}$
$$= 2(-10) + 9$$
$$= -20 + 9$$
$$= -11$$

b) $\quad \sqrt[3]{216 + 343 - 47}$
$$= \sqrt[3]{512}$$
$$= 8$$

c) $\quad 5\sqrt[3]{27} - 2(\sqrt[3]{64} - \sqrt[3]{-8})$
$$= 5(3) - 2[4 - (-2)]$$
$$= 15 - 12$$
$$= 3$$

When the cube root of a number cannot be found by inspection, the systematic trial method can be employed.

Let the number whose cube root is to be found be n.

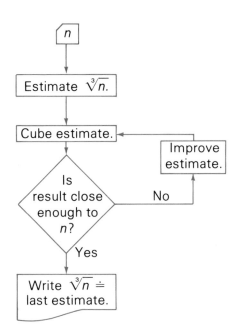

Example 3. Calculate $\sqrt[3]{15}$ to three decimal places.

Solution. Estimate: $\sqrt[3]{15} \doteq 2.5$.

$$(2.5)^3 = 15.625 \text{ too big}$$

Try 2.4. $(2.4)^3 = 13.824 \text{ too small}$

Try 2.46. $(2.46)^3 \doteq 14.8869$

Try 2.466. $(2.466)^3 \doteq 14.9961$

$\sqrt[3]{15} \doteq 2.466$

Example 4. The volume of a spherical balloon is 14 130 cm³.
What is its radius?

Solution. Volume of a sphere, $V = \frac{4}{3}\pi r^3$

$$14\,130 \doteq \frac{4}{3}(3.14)r^3$$

$$r^3 \doteq \frac{3 \times 14\,130}{4 \times 3.14}$$

$$\doteq 3375$$

$$r \doteq \sqrt[3]{3375}, \text{ or } 15$$

The radius of the balloon is approximately 15 cm.

Exercises 5 - 9

A 1. Find the cube root of each of the following:

a) 1000   b) −125   c) 0.008

d) $\dfrac{1}{27}$ e) $\dfrac{1}{1\,000\,000}$ f) −0.216

2. Simplify:

a) $\sqrt[3]{0.125}$ b) $\sqrt[3]{-27}$ c) $\sqrt[3]{-0.512}$

d) $\sqrt[3]{1.728}$ e) $\sqrt[3]{-\dfrac{27}{64}}$ f) $\sqrt[3]{-\dfrac{729}{343}}$

3. Calculate to two decimal places:

a) $\sqrt[3]{10}$ b) $\sqrt[3]{100}$ c) $\sqrt[3]{1.23}$

4. Simplify:

a) $\sqrt[3]{216} + \sqrt[3]{343}$ b) $2\sqrt[3]{-64} + \sqrt[3]{512}$

c) $\sqrt[3]{125 + 216} + 2$ d) $\sqrt[3]{729 - 8 + 343 - 64}$

5. Find the radius of a sphere which has a volume of

a) 113.04 cm³;   b) 904.32 m³;   c) 381 510 mm³.

B 6. What are the dimensions of the rectangular bar shown if its volume is 1000 cm³?

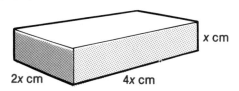

7. Solve:

a) $x^3 = -3.375$ b) $27x^3 = 1$ c) $-7x^3 = 448$

Review Exercises

1. Simplify:

a) $3^2 + 2^3$ b) $3^2 \times 2^3$ c) $(1.6)^2$

d) 2×4^2 e) $3^2 + 4$ f) $(-2)^3 + (-3)^2$

g) $(2 + 3)^3$ h) $(-4)^2 + (-2)^4$ i) $\left(-\dfrac{3}{7}\right)^3$

2. If $x = -2$ and $y = 3$, evaluate:
   a) $-5x^2$
   b) $(-5x)^2$
   c) $y^2 - x^2$
   d) $(x + y)^3$
   e) $4x^2 + 3y^2$
   f) $(y - x)^2$
   g) $\dfrac{x^2 - y^2}{x - y}$
   h) $2(x + y)^2$
   i) $\dfrac{y^2 - x^3}{y - x}$

3. Solve for x:
   a) $x^2 = 49$
   b) $x^3 = -8$
   c) $4x^2 = 100$
   d) $3x^2 = 108$
   e) $7x^2 = 1008$
   f) $69.5x^2 = 10\,008$

4. Find the lengths of the sides of squares with these areas:
   a) 64 mm^2
   b) 0.81 m^2
   c) 49 cm^2
   d) 2.25 cm^2

5. Simplify:
   a) $x^4 \times x^5$
   b) $x^{36} \div x^{12}$
   c) $x^{12} \times x^6$
   d) $(x^4)^7$
   e) $3x^2 \times 5x^4$
   f) $2^9 \div 2^4$
   g) $9m^4 \times 3m^2$
   h) $(-3)^{12} \div (-3)^4$
   i) $(5x)^2$
   j) $5(x^2)^3$
   k) $(3x^2)^3$
   l) $-42y^{12} \div 6y^8$

6. Simplify:
   a) $\dfrac{18x^4 \times 5x^2}{15x^3}$
   b) $\dfrac{120x^5}{-15x} \times \dfrac{15x^4}{5x^2}$

7. Evaluate when $x = 3$:
   a) $x^2 \times x^4$
   b) $(2x^2)^2$
   c) $(2x^2)(3x^3)$

8. Simplify:
   a) 5^{-3}
   b) $\left(\dfrac{1}{2}\right)^{-1}$
   c) $2^{-3} - 4^{-1}$
   d) $7^0 + 2^{-2}$
   e) $3^2 - 3^{-2}$
   f) $\left(\dfrac{2}{3}\right)^{-1} + \left(\dfrac{2}{3}\right)^0$

9. Evaluate for $a = -2$, $b = 2$, $c = -1$:
   a) $a^{-1} + b^{-1}$
   b) $(a + b - c)^{-1}$
   c) a^b
   d) $a^{-1} + b^{-1} + c^2$

10. Simplify:
    a) $w^8 \div w^{-4}$
    b) $w^{-9} \div w^{-12}$
    c) $15x^4 \div (-3x^{-4})$
    d) $-24y^4 \div 3y^{-2}$
    e) $\dfrac{10}{y^4} \div y^2$
    f) $16z^{-2} \div (2z)^2$

11. Write in scientific notation:
    a) $10\,000$
    b) $740\,000$
    c) $0.000\,01$
    d) 0.057

12. Express in scientific notation, then simplify:
    a) $49\,000\,000 \times 730\,000$
    b) $26\,500\,000 \times 7900 \times 0.0046$
    c) $\dfrac{320\,000 \times 64\,000\,000}{9\,600\,000}$

13. Simplify:
    a) $\sqrt{36}$
    b) $\sqrt{2\tfrac{7}{9}}$
    c) $\sqrt{14\,400}$
    d) $\sqrt{0.0081}$
    e) $3\sqrt{25} - 5\sqrt{9}$
    f) $\sqrt{225} + \sqrt{49}$
    g) $4\sqrt{169} - \sqrt{25}$

14. Solve and check:
    a) $5m^2 = 180$
    b) $4x^2 - 2 = 79$
    c) $2\sqrt{a} = 14$

15. Calculate to three decimal places:
    a) $\sqrt{28}$
    b) $\sqrt{17.4}$
    c) $\sqrt{250}$
    d) $\sqrt{0.44}$

16. A square, with a side length of 10 cm, is to be reduced in area by one-half. Find the length of the side of the reduced square.

Apollo XI Lunar Module in lunar orbit

17. a) If $y = 1.2x^2$, find y for these values of x:
    i) 4   ii) 0.5   iii) -3
    b) If $y = 1.2x^2$, find x for these values of y:
    i) 10.8   ii) 7.5   iii) 1

18. The approximate velocity, v, in metres per second, of an orbiting satellite is given by the formula $v = \sqrt{9.8r}$, where r is the distance of the satellite, in metres, from the centre of the Earth. Find the velocity of the satellite if
    a) $r = 2 \times 10^7$ m;
    b) $r = 5 \times 10^7$ m.

19. When a ball is dropped from a height of 2 m, the height, h, in metres, to which it bounces is given by $h = 2(0.8)^n$, where n is the number of bounces.
    a) To what height does the ball bounce after
    i) the first bounce?   ii) the third bounce?
    b) After how many bounces will it rise to
    i) about 1.3 m?   ii) about 80 cm?

20. Solve by inspection:
    a) $2\sqrt[3]{x} + 1 = 7$
    b) $2^x = 16$
    c) $3x^3 = -24$
    d) $4^x = 1$
    e) $3^x = \dfrac{1}{9}$
    f) $x^{-2} = 0.01$

6

Ratio
Proportion
and Rate

In a move to conserve Earth's resources, there is now an international standard of paper sizes. The largest, A0, has an area of 1 m², and the ratio of the lengths of its sides are such that cutting it in half results in the next largest size, A1, with side lengths in exactly the same ratio. The next size, A2, is obtained by cutting an A1 sheet in half, and so on. For every size the length-to-width ratio is the same.

There is only one length-to-width ratio that keeps yielding similar rectangles in this way. Can you find what it is? (See *Example 4,* Section 6-3.)

6 - 1 Ratios

80 out of every 100 Canadians buy lottery tickets.

CP. A survey...

A **ratio** is a comparison of quantities measured in the same units.

The newspaper clipping says: "80 out of 100 Canadians buy lottery tickets." Thus, 20 out of 100 do not.

The ratio of those who buy to those who do not is written: **80 : 20**, and is read, "eighty to twenty".

The numbers 80 and 20 are the **terms** of the ratio. The ratio 80 : 20 can also be written $\frac{80}{20}$. Multiplying or dividing each term of a ratio by the same non-zero number produces **equivalent ratios**.

$$\frac{80}{20} = \frac{800}{200} = \frac{8}{2} = \frac{4}{1}$$

The ratio 4 : 1 is in lowest terms. A ratio is in **lowest terms** when the only integral factor the terms have in common is 1.

Example 1. Write each ratio in lowest terms:

a) 9 : 6   b) 16 : 12   c) 25 : 20   d) $\frac{60}{15}$

Solution. a) 9 : 6
  Divide by 3:
  9 : 6 = 3 : 2

b) 16 : 12
  Divide by 4:
  16 : 12 = 4 : 3

c) 25 : 20
  Divide by 5:
  25 : 20 = 5 : 4

d) $\frac{60}{15}$
  Divide by 15:
  $\frac{60}{15} = \frac{4}{1}$

Name	Fights	Knock-outs
Doug James	28	7
Mike Seiling	24	8
Gary Burgess	20	5
Dan Stoffer	24	4

Example 2. The table shows the knockout records of four boxers in the Winnipeg Boxing Club. Which two boxers have the same ratio of knockouts to fights?

Solution. The knockouts-to-fights ratios are:
James: 7 : 28 = 1 : 4;   Seiling 8 : 24 = 1 : 3
Burgess: 5 : 20 = 1 : 4;   Stoffer 4 : 24 = 1 : 6
Doug James and Gary Burgess have the same knockouts-to-fights ratio.

Example 3. A parking lot contains domestic and foreign cars in the ratio 7:4. If there are 77 cars in the lot, how many of them are foreign?

Solution. The ratio of domestic to foreign is 7:4. Then 7 out of 11 cars are domestic and 4 out of 11 cars are foreign.

The number of foreign cars is: $\frac{4}{11} \times 77 = 28$.

There are 28 foreign cars in the lot.

Example 4. Fuel X is composed of ingredients A and B in the ratio 3:5. Fuel Y is composed of ingredients A and B in the ratio 4:7. Which fuel is richer in ingredient A?

Solution. $\frac{3}{8}$ of fuel X is ingredient A.

$\frac{4}{11}$ of fuel Y is ingredient A.

$\frac{3}{8} = \frac{33}{88}$ and $\frac{4}{11} = \frac{32}{88}$ $\frac{33}{88} > \frac{32}{88}$

Therefore, fuel X is richer in ingredient A.

Exercises 6 - 1

A   1. Give the meaning of each statement without using the word "ratio".

a) Mr. Adams and Mr. Becker divided the profits in the ratio 3:2.

b) The ratio of girls to boys in the class is 7:5.

c) Mrs. Arbor's chain saw runs on a 25:1 mixture of gasoline and oil.

d) The scale of a map is 1:250 000.

e) Brass is an alloy of copper and zinc in the ratio 3:2.

2. Write each ratio in lowest terms:

a) 40:12   b) 5:65   c) 28:8   d) 32:52

e) 12:72   f) 50:250   g) $\frac{60}{12}$ h) $\frac{144}{9}$

3. Which is the greater ratio,

a) $\frac{5}{8}$ or $\frac{11}{15}$?   b) 3:7 or 9:22?

c) 6:5 or 22:19?   d) 8:3 or 41:17?

4. "Dad's" cookies have raisins and chocolate chips in the ratio 3:7. "Mum's" cookies have raisins and chocolate chips in the ratio 5:11. Which brand has the greater ratio of raisins to chocolate chips?

5. Air consists of oxygen and nitrogen in the approximate ratio 1:4. What fraction of air is oxygen?   nitrogen?

6. Sterling silver is an alloy of silver and copper in the ratio 37:3.
   a) What fraction of a sterling silver fork is silver?
   b) A sterling silver ingot with a mass of 500 g will contain how many grams of silver?

7. Write an equivalent ratio with a second term of 1:
   a) 5:2   b) 2:0.5   c) 3:10   d) 4:0.8

8. Write an equivalent ratio with a second term of 24:
   a) 5:6   b) 8:48   c) 27:36   d) 5:0.6

B   9. At a school dance, there are 15 teachers, 275 girls, and 225 boys. Express the following ratios in lowest terms:
   a) girls to boys   b) boys to girls
   c) teachers to girls   d) students to teachers

10. A newspaper costs 20¢ each day from Monday to Friday and 45¢ on Saturday. What is the ratio of
   a) the cost on Saturday to the cost for one week?
   b) the cost on Monday to the cost for one week?

11. If the ratio of domestic cars to foreign cars in Metropolitan Toronto is 9:5, how many domestic cars might you expect to find in a lot holding 247 cars? 280

12. The length and width of a rectangle are in the ratio 9:7. If the perimeter is 256 cm, what are the dimensions of the rectangle?

13. The front gears of a ten-speed bicycle have 40 and 52 teeth. The back gears have 14, 17, 20, 24, and 28 teeth.
   a) Write the ten different gear ratios (front:back).
   b) Arrange the ten gear ratios in order from lowest to highest.

C   14. In an isosceles triangle, the two different sizes of angle are in the ratio 2:5. What are the angles of the triangle?

15. The ratio of the mass of a hydrogen atom to the mass of an average man (70 kg) is about the same as the ratio of the mass of an average man to the mass of the sun. The mass of a hydrogen atom is about 1.7×10^{-29} kg. What is the approximate mass of the sun?

6 - 2   Applying Ratios

In many applications of ratios only one term is stated. Since the second term is always the same in each type of application it is omitted. The purity of gold, Consumer Price Index, Mach numbers, and Intelligence Quotients are all examples of ratios in which only the first term is stated.

Example 1. The amount of gold in jewellery and coins is measured in karats (K) with 24 K representing pure gold. The mark, 14 K, on a ring means the ratio of the mass of gold in the ring to the mass of the ring is 14 : 24. A gold bracelet is marked 18 K:

a) Express the gold content as a ratio in lowest terms.

b) If the mass of the bracelet is 52 g and the value of pure gold is $25.50/g, what is the value of the gold in the bracelet?

Solution. a) $\dfrac{18}{24} = \dfrac{3}{4}$

b) Mass of gold in bracelet: $\dfrac{3}{4} \times 52 \, g = 39 \, g$

Value of gold in bracelet: $39 \times \$25.50$
$= \$994.50$

Example 2. The Consumer Price Index (CPI) is a measure of the change in the cost of living. The CPI at any given time is the ratio of the price of 300 selected items at that time to the price of the same items in a base year, currently 1971.

$$\frac{\text{Present CPI}}{100} = \frac{\text{Present cost of the 300 items}}{\text{Cost of the 300 items in 1971}}$$

If the CPI in 1981 is 217.4, how much will a family spend to maintain the same standard of living as they did in 1971 when they spent $6500?

Solution. The cost of living for the family in 1981, in dollars:

$\dfrac{217.4}{100} \times 6500 = 2.174 \times 6500$
$= 14\,131$

To maintain the same standard of living in 1981 as in 1971, the family will have to spend approximately $14 100.

When we wish to compare more than two quantities, we use ratios with more than two terms. For example:

$$6:8:12 = 3:4:6$$

This means: $\quad \dfrac{6}{3} = \dfrac{8}{4}; \qquad \dfrac{8}{4} = \dfrac{12}{6}; \qquad \dfrac{6}{3} = \dfrac{12}{6}$

Example 3. The gravities of Mars, Earth, and Jupiter are in the ratio $2:5:13$. How many times heavier would you be on Jupiter than on Mars?

Solution. Let your weights on Mars, Earth, and Jupiter be m, e, and j respectively.

$$m:e:j = 2:5:13$$
$$\frac{m}{2} = \frac{j}{13}$$
$$2j = 13m$$
$$j = \frac{13}{2}m$$

Your weight on Jupiter would be $\dfrac{13}{2}$, or 6.5, times your weight on Mars.

Example 4. Profits in a business are to be shared by the three partners in the ratio $2:3:5$. If the profit for a year was $100 000, what was each partner's share?

Solution. Let the shares, in dollars, be $2x$, $3x$, and $5x$. Since the total profit was $100 000,

$$2x + 3x + 5x = 100\,000$$
$$10x = 100\,000$$
$$x = 10\,000$$

The shares were: $20 000, $30 000, and $50 000.

Exercises 6 - 2

A   1. Express the gold content of the following as a ratio in lowest terms:

    a) a 22 K gold coin   b) a 16 K gold pin

    c) a charm marked 9 K, the legal minimum for an article to be called gold.

2. A 14 *K* gold ring has a mass of 24.7 g.

   a) Find the mass of the gold in the ring.

   b) At $25.50/g, find the value of the gold in the ring.

3. In 1980, the Consumer Price Index rose to 204.6. How much did it cost a family in 1980 for goods that cost them $8700 in 1971?

4. Three people chipped in to buy a lottery ticket. They contributed to the cost of the ticket in the ratio $2:5:3$. If the ticket wins $25 000, how should the prize be divided?

5. Four partners in a business agreed to share the profits in the ratio $4:2:3:6$. The first year's profits were $84 000. Calculate each partner's share.

6. A Mach number is the ratio of the speed of an aircraft to the speed of sound at the same altitude and temperature. At an altitude where the speed of sound is 1085 km/h,

   a) what is the Mach number of an aircraft whose speed is
       i) 3255 km/h?   ii) 1302 km/h?   iii) 1000 km/h?

   b) what is the speed, in kilometres per hour, of
       i) a Concorde flying at Mach 2?
       ii) the North American Aviation X-15A-2 flying at Mach 6.72 — the highest achieved by a fixed-wing aircraft?

Concorde

7. A child's Intelligence Quotient (IQ) is the first term of a ratio with a second term of 100. The ratio of mental age to physical age is IQ : 100.

   a) Find the IQ of
       i) a twelve-year old with a mental age of 12.5;
       ii) a seven-year old with a mental age of 6.8.

   b) Find the mental age of
       i) a nine-year old with an IQ of 100;
       ii) a six-year old with an IQ of 150 (genius level).

8. a) A 1 cent coin minted before 1860 had a mass of 4.5 g and contained copper, tin, and zinc in the ratio $95:4:1$. What mass of tin did each coin contain?

   b) From 1876 to 1920, the mass of each 1 cent coin minted was 5.67 g, and the ratio of copper to tin to zinc was $95.5:3:1.5$. What mass of copper did each coin contain?

   c) After 1942, each 1 cent coin minted had a mass of 3.24 g and contained copper, tin, and zinc in the ratio $98:0.5:1.5$. What mass of zinc did each coin contain?

9. The frequencies of the notes in the musical scale of C major are related by an eight-term ratio as follows:

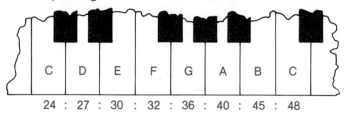

24 : 27 : 30 : 32 : 36 : 40 : 45 : 48

Musical instruments are usually tuned so that A in the scale has a frequency of 440 Hz (cycles per second). Find the frequencies of the other notes in the scale.

10. The gravities on Mars, Earth, and Jupiter are in the ratio 2 : 5 : 13. How many times heavier would you be on Jupiter than on Earth?

C 11. Chemical fertilizers usually contain nitrogen, phosphorus, and potassium. The amount of each nutrient present is expressed as a percent of the total mass of the fertilizer in a three-term ratio:

% nitrogen : % phosphorus : % potassium.

A, *B*, and *C* are 10 kg bags of three kinds of fertilizer.

 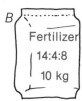 

a) Which fertilizer, *A*, *B*, or *C*, contains the most nitrogen?

b) Which fertilizer has the greatest ratio of nitrogen to phosphorus?

c) Which fertilizer contains the most nutrients?

d) Why are the ratios not expressed in lowest terms?

12. The ratio of the approximate distances of Earth and Uranus from the Sun is 4 : 77, and that of Mars and Uranus from the Sun is 2 : 25.

a) What is the ratio of the approximate distances of Earth and Mars from the Sun?

b) How far is Mars from the Sun if Earth's distance is 150 Gm?

13. The angles of a triangle are in the ratio 2 : 3 : 4. What are their measures?

14. The angles of a quadrilateral are in the ratio 2 : 3 : 3 : 4. What are their measures?

6 - 3 Proportions

A statement that two ratios are equivalent is called a **proportion**. A proportion may be written in two ways:

$$\text{i)}\quad a:b = c:d \qquad\text{and}\qquad \text{ii)}\quad \frac{a}{b} = \frac{c}{d}$$

Multiplying both sides of the proportion in (ii) by the product of the denominators, bd, produces a simple equation.

$$bd \times \frac{a}{b} = bd \times \frac{c}{d}$$

$$ad = bc \qquad\qquad bd \neq 0$$

This result can be obtained more simply as follows:

$$\frac{a}{b} \diagdown\kern-1.2em=\kern-1.2em\diagup \frac{c}{d}$$

$$\boldsymbol{ad = bc}$$

This technique enables us to find the unknown term of a proportion when the other three terms are known.

Example 1. Find the missing terms:

$$\text{a)}\quad \frac{8}{x} = \frac{6}{30} \qquad\qquad \text{b)}\quad \frac{3}{19} = \frac{7}{x}$$

Solution.

$$\text{a)}\quad \frac{8}{x} = \frac{6}{30} \qquad\qquad \text{b)}\quad \frac{3}{19} = \frac{7}{x}$$

$$240 = 6x \qquad\qquad\qquad 3x = 133$$

$$40 = x \qquad\qquad\qquad\quad x \doteq 44.3$$

Example 2. When a photograph is enlarged, the ratio of any length on the enlargement to the matching length on the original is always the same. Find the height of the mast of the sailboat in the enlargement shown.

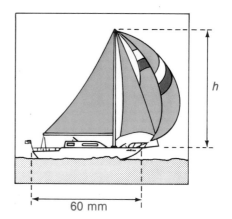

Solution. Let h represent the height of the mast in the enlargement. Then we can write the proportion:

$$\frac{\text{height of mast on enlargement}}{\text{height of mast on original}} = \frac{\text{length of boat on enlargement}}{\text{length of boat on original}}$$

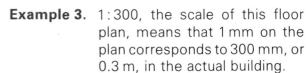

$$\frac{h}{50} = \frac{60}{36}$$
$$36h = 3000$$
$$h \doteq 83.3$$

On the enlargement, the height of the mast is approximately 83 mm.

Example 3. 1:300, the scale of this floor plan, means that 1 mm on the plan corresponds to 300 mm, or 0.3 m, in the actual building.

a) What are the dimensions of the living room in metres?

b) What is the total floor area in square metres?

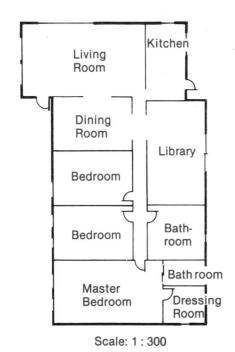

Scale: 1 : 300

Solution. a) By measurement, the dimensions of the living room on the plan are 33 mm by 20 mm. Let l be the length and w the width of the actual room. Then:

$$\frac{l}{33} = \frac{300}{1} \quad \text{and} \quad \frac{w}{20} = \frac{300}{1}$$
$$l = 9900 \qquad\qquad w = 6000$$

The living room measures 9.9 m by 6.0 m.

b) The floor plan can be divided into two rectangles. One rectangle comprises the living room and kitchen and measures 45 mm by 20 mm. The other rectangle comprises all the other rooms and measures 60 mm by 42 mm.

45 mm by 20 mm corresponds to 13 500 mm by 6000 mm, or 13.5 m by 6.0 m.
60 mm by 42 mm corresponds to 18 000 mm by 12 600 mm, or 18.0 m by 12.6 m.

Total floor area in the actual building, in square metres:

13.5 × 6.0 + 18.0 × 12.6 = 81.0 + 226.8
= 307.8

The floor area is approximately 308 m².

Example 4. A rectangular sheet of paper is 1 m wide. When folded in half, each half has the same length-to-width ratio as the unfolded sheet. How long is the sheet?

Solution. Let x be the length of the sheet, in metres.

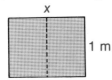

The folded length is 1 m and the folded width is 0.5x.

$$\frac{\text{original length}}{\text{original width}} = \frac{\text{folded length}}{\text{folded width}}$$

$$\frac{x}{1} = \frac{1}{0.5x}$$

$$0.5x^2 = 1$$

$$x^2 = 2$$

$$x = \sqrt{2}, \text{ or approximately } 1.41$$

The length of the sheet of paper is approximately 1.41 m.

This last example solves the problem at the beginning of the chapter. By starting with a sheet of paper where

$$\text{length} : \text{width} = \sqrt{2} : 1,$$

we can produce, with one cut, two sheets with measurements in the same ratio. Subsequent cuts yield smaller and smaller sheets having the same ratio of length to width.

Exercises 6 - 3

A 1. Solve for the missing terms:

a) $\frac{3}{8} = \frac{m}{24}$ b) $\frac{12}{16} = \frac{n}{8}$ c) $\frac{a}{12} = \frac{15}{36}$

d) $\frac{x}{18} = \frac{9}{54}$ e) $\frac{90}{b} = \frac{30}{11}$ f) $\frac{72}{x} = \frac{360}{15}$

g) $\frac{9}{8} = \frac{144}{d}$ h) $\frac{27}{5} = \frac{81}{x}$ i) $\frac{7}{15} = \frac{n}{105}$

2. Solve:

a) $\dfrac{11}{16} = \dfrac{n}{8}$ b) $\dfrac{x}{3} = \dfrac{2}{7}$ c) $\dfrac{5}{8} = \dfrac{9}{x}$

d) $\dfrac{7}{11} = \dfrac{9}{x}$ e) $\dfrac{t}{4} = \dfrac{5}{7}$ f) $\dfrac{9}{b} = \dfrac{5}{6}$

g) $\dfrac{3}{5} = \dfrac{w}{7}$ h) $\dfrac{5}{13} = \dfrac{10}{y}$ i) $\dfrac{19}{25} = \dfrac{57}{m}$

3. In a photograph of a mother and son standing together, the son measures 27 mm and the mother 63 mm. If the mother is actually 180 cm tall, how tall is her son?

4. At a given time of day, the ratio of the height of a tree to the length of its shadow is the same for all trees. A tree 12 m tall casts a shadow 5 m long. How tall is a tree that casts a shadow 3 m long?

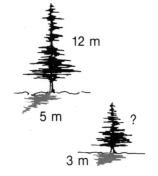

12 m

5 m   ?

3 m

5. The ratio of a man's height to his arm span is 24 : 23. How tall is he if his arm span is 184 cm?

6. A metre stick held vertically casts a shadow 0.7 m long. How tall is a tree that casts a shadow 12.7 m long?

5 out of 6 smokers want to quit

An extensive survey made in the factories of

B 7. If the headline statement is correct,

a) how many out of 30 smokers want to quit?

b) how many out of 600 smokers want to quit?

c) how many smokers were polled if 200 want to quit?

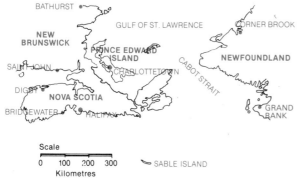

8. Find the actual dimensions and area of the master bedroom shown in the floor plan of *Example 3*.

9. Find these distances from the map:

a) Halifax, N.S., to Sable Island

b) Bathurst, N.B., to Corner Brook, Nfld.

c) The width of Cabot Strait

d) The length of Prince Edward Island

e) Bridgewater, N.S., to Grand Bank, Nfld.

f) Digby, N.S., to St. John, N.B.

10. The distance between two towns is 180 km. How far apart will they be on a map drawn to a scale of 1 : 1 500 000?

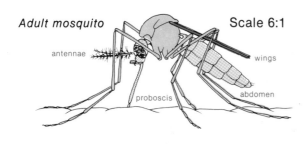

Adult mosquito   Scale 6:1

antennae   wings

proboscis   abdomen

11. Find the approximate lengths of these parts of the mosquito:

a) abdomen   b) antennae   c) wings

Mathematics Around Us

SCALES OF MODEL RAILROADING

The following diagram illustrates the scales that are available to model railroaders.

N	**TT**	**HO**	**S**	**O**
1:160	*1:120*	*1:87*	*1:64*	*1:48*

Questions

1. Find the lengths, to the nearest millimetre, of the models of these items of rolling stock:
   a) a boxcar 12.2 m long, in the HO scale
   b) a locomotive 15.5 m long, in the N scale
   c) a passenger car 25.0 m long, in the O scale

2. Find the lengths, to the nearest decimetre, of the actual pieces of rolling stock when the lengths of the models are:
   a) a baggage car in the S scale, 333 mm;
   b) a flat car in the HO scale, 175 mm;
   c) a caboose in the TT scale, 86 mm.

3. The width of the track in the N scale is 9 mm.
   a) What are the track widths in the other four scales?
   b) What is the track width of the actual railroad?

4. When the track plan in the diagram is used for N scale models, it measures 2.4 m by 1.8 m.
   a) Find the dimensions of the track plan if it is used for S scale models.
   b) For the S and N scale track plans, find
      i) the ratio of their lengths;
      ii) the ratio of their areas.

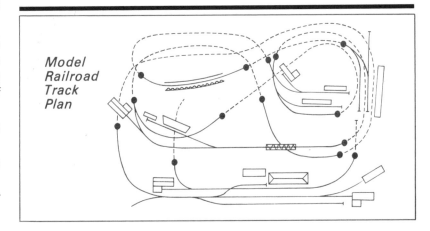

Model Railroad Track Plan

Fuel Economy
5.5 L/100 km

6 - 4 Rates

In the advertisement, two quantities having different units are compared. The amount of fuel consumed (in litres) is compared to the distance the car travels (in kilometres). Such a comparison is called a **rate**. In this case the rate is 5.5 L *per* 100 km.

50 km/h Begins

Rates are all around us in our daily lives. Kilometres per hour, heartbeats per minute, kilograms per cent, dollars per day, goals per game are just some examples. You can supply many more.

3 kg for 89¢

Problems involving rates can be solved using proportions. We need only be sure the quantities being compared appear in the same order on both sides of the equation.

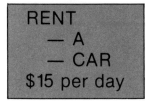

RENT
— A
— CAR
$15 per day

Example 1. If a car's rate of fuel consumption is 5.5 L/100 km,

    a) how much fuel is needed to go 275 km?

    b) how far will the car travel on 48 L?

Solution.   a) Let x be the number of litres needed.

$$\text{Then, } \frac{x}{275} = \frac{5.5}{100}$$
$$100x = 5.5 \times 275, \text{ or } 1512.5$$
$$x = 15.125$$

Approximately 15.1 L of fuel are needed to go 275 km.

    b) Let z be the number of kilometres on 48 L.

$$\text{Then, } \frac{48}{z} = \frac{5.5}{100}$$
$$5.5z = 48 \times 100, \text{ or } 4800$$
$$z \doteq 872.7$$

The car will travel nearly 873 km on 48 L of fuel.

Example 2. In one season, Casey batted at the rate of 2 hits for every 5 official times at bat. At this rate,

    a) how many hits should he get in 400 times at bat?

    b) how many times at bat should he have to get 180 hits?

Solution.   a) Let x be the number of hits Casey should get.

Then, $\dfrac{x}{400} = \dfrac{2}{5}$

$$5x = 2 \times 400, \text{ or } 800$$
$$x = 160$$

Casey should get approximately 160 hits in 400 times at bat.

    b) Let z be the number of times at bat to get 180 hits.

Then, $\dfrac{180}{z} = \dfrac{2}{5}$

$$2z = 5 \times 180, \text{ or } 900$$
$$z = 450$$

Casey should bat about 450 times to get 180 hits.

    To solve some problems involving rates, you may find it helpful to draw up a table. The equation can then be written from the table.

Example 3. It took 3.5 h to drive 225 km to a cottage. Part of the trip was a detour where the average speed was 30 km/h. The rest of the trip was on a highway at an average speed of 80 km/h.

    a) How much time was spent on the detour?

    b) How long was the detour?

Solution.   a) Let t be the time on the detour, in hours. Recall that *distance* equals *average speed* times *time.*

	Distance km	Time h	Average Speed km/h
Detour	$30t$	t	30
Highway	$80(3.5 - t)$	$3.5 - t$	80

The total distance was 225 km.

Therefore, $30t + 80(3.5 - t) = 225$

$$30t + 280 - 80t = 225$$
$$30t - 80t = 225 - 280$$
$$-50t = -55$$
$$t = 1.1$$

The time spent on the detour was 1.1 h.

b) The length of the detour, in kilometres:
$$30 \times 1.1, \text{ or } 33$$
The length of the detour was 33 km.

Check. Time on highway: $3.5 \text{ h} - 1.1 \text{ h} = 2.4 \text{ h}$

Distance on highway:
$$80 \text{ km/h} \times 2.4 \text{ h} = 192 \text{ km}$$

Total distance travelled:
$$192 \text{ km} + 33 \text{ km} = 225 \text{ km}$$

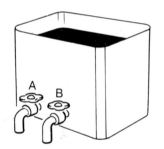

Example 4. The tank empties in 30 min when only tap *A* is open. It empties in 20 min when only tap *B* is open. How long does it take to empty with both taps open?

Solution. In 1 min, tap *A* empties $\frac{1}{30}$ of the tank.

In 1 min, tap *B* empties $\frac{1}{20}$ of the tank.

In 1 min, the fraction of the tank emptied with both taps open is:
$$\frac{1}{30} + \frac{1}{20} = \frac{5}{60}, \text{ or } \frac{1}{12}.$$

Since, with both taps open, $\frac{1}{12}$ of the tank empties in 1 min, the whole tank will empty in 12 min.

Check. Amount drained in 12 min by —

tap *A*: $\frac{1}{30} \times 12 = \frac{2}{5}$ of the whole tank;

tap *B*: $\frac{1}{20} \times 12 = \frac{3}{5}$ of the whole tank;

both taps together:
$$\frac{2}{5} + \frac{3}{5} = 1 \text{ (the whole tank).}$$

Exercises 6 - 4

1. Dale drives 45 km on 3 L of gasoline.
   a) How far would she drive on 2 L of gasoline?   on 5 L?
   b) How much gasoline does she need to drive 60 km?   to drive 270 km?

2. A car uses fuel at the rate of 7.2 L/100 km.
   a) How much fuel is needed to travel 360 km?
   b) How far will the car travel on a full tank of 85 L?

3. The new "miracle" car uses fuel at the rate of 2.8 L/100 km.
   a) How much fuel does it need to go 850 km?
   b) How far can it go on 80 L of fuel?

4. Bob can type at the rate of 30 words per minute.
   a) How long would it take him to type 20 words?   to type 100 words?
   b) How many words can he type in 4 min?   in 6.5 min?

5. An electronic typewriter can type 540 words per minute.
   a) How long will it take to type 1000 words?   to type 1 000 000 words?
   b) How many words can it type in 1 h?   in 1 week?

6. In the first 20 games of the baseball season, Reggie Jackson hit 12 home runs. If he continued at this rate, how many home runs would he hit in 160 games?

7. In the first 6 games of the football season, Dave Cutler scored 83 points. If he continued scoring at this rate, how many points would he score in 14 games?

8. If 18 houses are built in 45 days, how long at this rate would it take to build 63 houses?   144 houses?

9. A car travelled 520 km in 7.2 h. Part of the trip was at an average speed of 50 km/h and the remainder at 90 km/h.
   a) How much time was spent at each speed?
   b) How far did the car travel at each speed?

10. Marie drove her 18-wheeler the 1280 km from Calgary to Winnipeg in 15.2 h. Part of the trip she drove on icy roads at an average speed of 60 km/h. The rest of the time she drove at 100 km/h. How far did she drive on icy roads?

11. A tank with two taps, *A* and *B*, can be drained in 4 h by tap *A* alone, and in 6 h by tap *B* alone. How long will it take to drain the tank with both taps open?

12. A boat has two motors, one large and one small. If only the large motor is running, a tank of fuel lasts 3 h. If only the small motor is running, the tank of fuel lasts 5 h. How long will the fuel last if both motors are running?

13. Two girls, 60 km apart, start cycling toward each other at the same time. One girl cycles at 18 km/h. What must be the speed of the other girl if they meet in 1.5 h?

14. Car *A* and car *B* leave Halifax on the same road 1 h apart. Car *A* leaves first and travels at a steady 80 km/h. What must be car *B*'s speed in order to overtake car *A* in 4 h?

15. Machine *X* makes 200 boxes in 3 min and machine *Y* makes 200 boxes in 2 min. With both machines working, how long will it take to make 200 boxes?

16. A worker is paid $8.60/h for a 40 h week and time-and-a-half for overtime. How many hours are worked to earn $414.95 in one week?

6 - 5 Percent

**LOANS
at only
14.5%**

**First
Mortgages**
available at
$13\frac{3}{4}$ %

Percent has many applications — interest rates, discount, inflation, the chance of rain are just a few examples from the world around us.

A **percent** is another way of writing a ratio whose second term is 100. The word "percent" means "per hundred" For example:

$$75:100$$
is written
$$75\%.$$

GRAPHO
250 mL added to your oil will improve your car's performance by up to 28%.

**End-of-Line
Sale
Discounts**
up to **40%**

Example 1. Write the following ratios as percents:

a) $\frac{27}{100}$ b) $6:100$ c) $175:100$ d) $0.5:100$

Solution.   a) 27%   b) 6%   c) 175%   d) 0.5%

Ratios whose second terms are not 100 are changed to percents by using proportions.

Example 2. Change the following ratios to percents:

a) $2:5$ b) $3.5:4$ c) $19:6$

Solution. a) $2:5$

$$\frac{2}{5} = \frac{n}{100}$$
$$5n = 200$$
$$n = 40$$

$2:5$ is 40%.

b) $3.5:4$

$$\frac{3.5}{4} = \frac{n}{100}$$
$$4n = 350$$
$$n = 87.5$$

$3.5:4$ is 87.5%.

c) $19:6$

$$\frac{19}{6} = \frac{n}{100}$$
$$6n = 1900$$
$$n = 316\frac{2}{3}$$

$19:6$ is $316\frac{2}{3}\%$.

Decimals are quickly written as percents, and percents as decimals using the following procedures:

To change
decimals to percents,
$0.76 \longrightarrow 76\%$
multiply by 100.

To change
percents to decimals,
$37.5\% \longrightarrow 0.375$
divide by 100.

Example 3. Express the following as percents:

a) 0.64 b) 0.018 c) 0.0073 d) 3.15

Solution. a) $0.64 = 64\%$ b) $0.018 = 1.8\%$

c) $0.0073 = 0.73\%$ d) $3.15 = 315\%$

Example 4. Express the following as decimals:

a) 28%   b) 7.5%   c) 0.9%   d) 156%

Solution. a) 28% = 0.28   b) 7.5% = 0.075
c) 0.9% = 0.009   d) 156% = 1.56

It is sometimes convenient to express a percent as a ratio in lowest terms.

Example 5. Express these percents as ratios in lowest terms:

a) 18%   b) 12.5%   c) 175%   d) 0.4%

Solution. a) $18\% = \dfrac{18}{100}$ b) $12.5\% = \dfrac{12.5}{100}$

$= \dfrac{9}{50}$ $= \dfrac{1}{8}$

c) $175\% = \dfrac{175}{100}$ d) $0.4\% = \dfrac{4}{1000}$

$= \dfrac{7}{4}$ $= \dfrac{1}{250}$

Exercises 6 - 5

A   1. Write these ratios as percents:

a) 7 : 100   b) 18.5 : 100   c) 57 : 100
d) 0.8 : 100   e) 365 : 100   f) 36.5 : 100
g) 540 : 100   h) 1875 : 100   i) 101 : 100

2. Change each ratio to a percent:

a) 1 : 4   b) 5 : 8   c) 7 : 10
d) 11 : 20   e) 5 : 6   f) 3 : 5
g) 8 : 5   h) 2 : 3   i) 13 : 10
j) 31 : 40   k) 20 : 3   l) 19 : 50
m) 1 : 25   n) 7 : 40   o) 11 : 200

3. Express the following as percents:

a) 0.38   b) 0.57   c) 0.81
d) 0.06   e) 0.035   f) 0.072
g) 0.091   h) 0.007   i) 0.0086
j) 0.0051   k) 0.0007   l) 3.6
m) 3.06   n) 3.006   o) 30.6

4. Express these percents as decimals:

   a) 24%   b) 39%   c) 57.4%

   d) 3%   e) 5.8%   f) 11.5%

   g) 1.6%   h) 0.9%   i) 137%

   j) 264%   k) 375%   l) 375.8%

5. Express the following as ratios in lowest terms:

   a) 26%   b) 35%   c) 64%

   d) 75%   e) 62.5%   f) 125%

   g) $83\frac{1}{3}$%   h) $16\frac{2}{3}$%   i) 185%

   j) 360%   k) 0.8%   l) 0.125%

Mathematics Around Us

Election to the Baseball Hall of Fame

To be elected to the Baseball Hall of Fame a player must receive at least 75% of the votes cast by all eligible sportswriters. To be eligible, sportswriters must have 10 years experience. Each can vote for up to 10 players.

Questions

1. In 1979 Willie Mays was elected to the Hall of Fame with the highest percentage of votes since 1936, the year the Hall of Fame started. He received 409 votes out of a total of 432 ballots.

   a) On what percent of the ballots did Willie Mays' name appear?

   b) What was the minimum number of votes needed for election in 1979?

2. In 1980, there were 385 ballots. Al Kaline received 88% and Duke Snider received 86%. How many votes did each get?

3. Three players were elected to the Hall of Fame in 1936: Ty Cobb with 98.2%, and Babe Ruth and Honus Wagner each with 95.1%. If Ty Cobb received 222 votes,

   a) how many were cast that year?

Willy Mays, elected to Hall of Fame in 1979

   b) how many votes did each of the other two players receive?

6 - 6 Working With Percent

Businesses frequently run sales. Such sales are usually a reduction, by some percent, in the normal costs of their goods. As the sign states, the car dealer is offering buyers a 15% reduction on the usual price.

> **TAKE 15% OFF**
> THE STICKER PRICE OF ANY CAR ON THE LOT.

Example 1. What is the sale price of a car that has a sticker price of $3750?

Solution. Since the reduction is 15%, you will pay:
100% − 15%, or 85%, of the sticker price.
If the sale price is $x,

$$\text{then } \frac{85}{100} = \frac{x}{3750}$$
$$100x = 85 \times 3750$$
$$x = 0.85 \times 3750, \text{ or } 3187.50$$

The sale price of the car is $3187.50.

Example 2. a) Calculate 16% of 85.

b) 18% of a number is 50. What is the number?

c) What percent of 65 is 13?

d) 12 is what percent of 80?

Solution. a) If x is 16% of 85,

$$\text{then } \frac{16}{100} = \frac{x}{85}$$
$$100x = 16 \times 85$$
$$100x = 1360$$
$$x = 13.6$$

b) If 18% of y is 50,

$$\text{then } \frac{18}{100} = \frac{50}{y}$$
$$18y = 5000$$
$$y \doteq 277.7$$

c) If 13 is z% of 65,

$$\text{then } \frac{z}{100} = \frac{13}{65}$$
$$65z = 1300$$
$$z = 20$$

d) If 12 is w% of 80,

$$\text{then } \frac{w}{100} = \frac{12}{80}$$
$$80w = 1200$$
$$w = 15$$

> **GAS FIRM WANTS 6% INCREASE IN RATES**

Example 3. Your gas bill last year was $625. What will be your bill for this year if you use the same amount of gas and the gas company gets the 6% increase?

Solution. Your gas bill will be 106% of last year's bill.

$$\$625 \times \frac{106}{100} = \$662.50$$

This year's bill will be $662.50.

Example 4. Which game has the greater percent reduction?

Table Tennis Set	Dart Game
4⁴⁴	3⁹⁸
Reg. $5.49	Reg. $4.89

Solution. For each game, we express the ratio of the *change in price* to the *original price* as a percent. Let x and y be the percent reductions in the prices of the table tennis set and dart game.

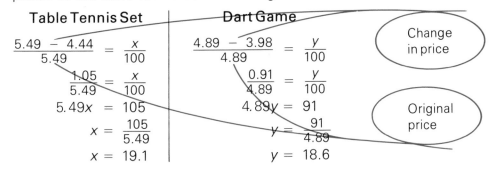

Table Tennis Set

$$\frac{5.49 - 4.44}{5.49} = \frac{x}{100}$$

$$\frac{1.05}{5.49} = \frac{x}{100}$$

$$5.49x = 105$$

$$x = \frac{105}{5.49}$$

$$x = 19.1$$

Dart Game

$$\frac{4.89 - 3.98}{4.89} = \frac{y}{100}$$

$$\frac{0.91}{4.89} = \frac{y}{100}$$

$$4.89y = 91$$

$$y = \frac{91}{4.89}$$

$$y = 18.6$$

Change in price

Original price

The table tennis set has the greater percent reduction.

Example 5. What was the regular price of this radio?

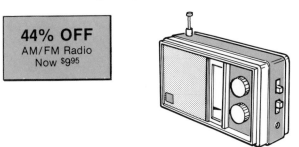

44% OFF
AM/FM Radio
Now $9⁹⁵

Solution. Let the regular price, in dollars, be x. Since the reduction is 44%, $9.95 must represent $100\% - 44\%$, or 56% of x.

$$\frac{56}{100} = \frac{9.95}{x}$$

$$56x = 995$$

$$x = \frac{995}{56}, \text{ or } 17.77$$

The regular price of the radio was $17.77.

Interest is the money paid for the use of money. If you have a savings account, the bank pays you interest for the use of your savings. If you borrow money from a bank, the bank charges you interest for the use of its money.

Simple interest is computed using the formula: $I = p \times r \times t$, where I is the interest, p the principal (the money saved or loaned), r the rate of interest, and t the time in years.

Example 6. Find the interest on
  a) a $324 credit-card bill for 1 month at 21% per annum;
  b) $6500 in a daily-interest account for 90 days at $11\frac{1}{2}$% per annum.

Solution. a) $I = prt$

$$= 324 \times 0.21 \times \frac{1}{12}$$

$$= 5.67$$

The interest is $5.67.

b) $I = prt$

$$= 6500 \times 0.115 \times \frac{90}{365}$$

$$\doteq 184.31$$

The interest is $184.31.

Exercises 6 - 6

A 1. Find:
   a) 25% of 40   b) 20% of 40   c) 60% of 150
   d) 5% of 35   e) 100% of 75   f) 4% of 150
   g) 7% of 95   h) 65% of 18   i) 135% of 50

2. Determine the number in each of the following:
   a) 50% of a number is 10.   b) 20% of a number is 3.
   c) 40% of a number is 10.   d) 75% of a number is 30.
   e) 60% of a number is 42.   f) 15% of a number is 15.
   g) $66\frac{2}{3}$% of a number is 18.   h) 10% of a number is 8.
   i) 104% of a number is 26.   j) 130% of a number is 91.

3. A ten-speed bicycle regularly sells for $227.50. What will it cost during a "15% off" sale?

4. a) What percent of 80 is 16?

   b) What percent of 135 is 15?

   c) What percent of 75 is 125?

   d) What percent of 50 is 45?

   e) 18 is what percent of 144?

   f) 27 is what percent of 81?

   g) 19 is what percent of 1900?

   h) 3 is what percent of 6000?

5. Skis are being sold at a discount of 45%. What will be the cost of a pair of skis that regularly sell for $180?

6. If food costs for the coming year are estimated to rise by 11.7%, what will be the coming year's food costs for a family that spent $8400 last year?

7. During a "20% off" sale, a clock radio is priced at $29.95. What is its regular price?

8. A hand-calculator is priced at $9.98 during a "25% off" sale. What is its regular price?

9. What is the rate of reduction on:

   a) a TV set regularly priced at $540 and selling for $499?

   b) an overcoat regularly priced at $195 and selling for $156?

10. Find the simple interest on:

    a) $1200 at 16% for 3 years;

    b) $15 000 at 12.5% for 2 years;

    c) $8500 at 18% for $1\frac{1}{2}$ years;

    d) $3000 at 15% for 3 months.

11. Long distance telephone rates are reduced 35% between 6 p.m. and midnight, and 60% between midnight and 8 a.m.

    a) If a 3 min call from Thunder Bay to Calgary at 10 a.m. costs $2.70, calculate the cost of a 3 min call at 8 p.m.;   at 7 a.m.

    b) If a 3 min call from Regina to Vancouver at 9 p.m. costs $1.25, what is the cost of a 3 min call at noon?   at 2 a.m.?

12. Sterling silver is an alloy of silver and copper in the ratio 37:3.

    a) What percent of a sterling silver bracelet is pure silver?

    b) If the bracelet has a mass of 30 g, what mass of silver does it contain?

Checks nab 187

Police spot-check crews stopped 860 vehicles yesterday and found 187 with defective equipment in the holiday traffic safety blitz across the city.

A total of 28 759 cars and trucks have been stopped for safety checks since the campaign began.

13. a) In the news item, what percent of the cars stopped yesterday had defective equipment?

   b) Estimate the number of cars found with defective equipment since the campaign began. What assumption are you making?

14. If a worker receives a cut of 20% in salary, what percent increase must he get to regain his original salary?

15. A pair of skis is priced at $185. What is their total cost
   a) when a 7% sales tax is added?
   b) with a 15% discount and a 7% sales tax?

16. A dress that sells for $80 is placed on sale at "15% off". After two weeks on the rack, the current selling price is reduced another 10%. What does it sell for now?

17. Specifications for the alloy, phosphor bronze, call for 85% copper by mass, 7% tin, 0.06% iron, 0.2% lead, 0.3% phosphorus, and the remainder zinc.
   a) The zinc content is what percent of phosphor bronze?
   b) How many kilograms of each element are needed to make 1 t of phosphor bronze?

C 18. In a recent year, Statistics Canada reported that 8.9 million persons were employed and 810 000 were unemployed.
   a) What was the rate of unemployment?
   b) If the size of the work force does not change, how many of the unemployed need to find jobs to bring the unemployment rate down to 4%?

19. Energy-conservation experts report that there is a 4.5% reduction in home-heating costs for every 1°C reduction in house temperature. If you have been keeping the house temperature at a constant 20°C day and night, what percent reduction in heating costs should you get if you keep the house at 18°C from 07:00 to 22:00 and 15°C from 22:00 to 07:00?

20. A company's profit is 5.4% of its sales. It must pay 48% corporate taxes on its profit. It always pays 60% of its after-taxes profit to its stockholders as dividends, and retains the balance as a reserve. How much did it retain as a reserve in a year when its sales were $10 000 000?

Mathematics Around Us

Reducing Reductions Gives Many Sizes of Photocopies

Some copying machines allow several different sizes of reduction to be made. By making a copy of the copy, an even wider range of reductions can be made. On one such machine, four different sizes of reduction can be selected, as follows:

1	85%	3	65%
2	77%	4	61.5%

Thus, if 1 is selected for a drawing 12 cm by 10 cm, the dimensions of the copy will be:

length: 0.85 × 12 cm, or 10.2 cm

width: 0.85 × 10 cm, or 8.5 cm

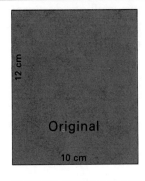
12 cm / Original / 10 cm

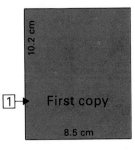
1 → First copy / 10.2 cm / 8.5 cm

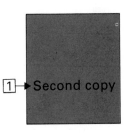
1 → Second copy

Questions

Give answers correct to one decimal place.

1. If 1 gives a copy 10.2 cm by 8.5 cm, what size copy does 1 again give? What is the size of reduction of the second copy?

2. What size of reduction is obtained by using
   a) 2 twice?
   b) 1 and then 2 ?
   c) 3 three times?

3. What combination of two reductions gives
   a) a 50% size of reduction?
   b) a 40% size of reduction?

4. For each of the following selections express the ratio as a percent:

   area of copy
   area of original

   a) 1   f) 2 twice
   b) 2   g) 1 then 2
   c) 3   h) 2 then 3
   d) 4   i) 3 then 4
   e) 1 twice   j) 3 three times

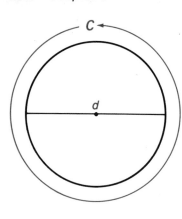

6 - 7 Ratios in Geometry

One ratio, above all others, has captured the interest of mathematicians through the ages—the ratio of the circumference, C, of a circle to its diameter, d. The value of the ratio is the same for all circles and is denoted by the Greek letter π.

For any circle,

$$C : d = \pi : 1 \quad \text{or} \quad \frac{C}{d} = \pi \quad \text{or} \quad C = \pi d.$$

By means of a computer, the value of π has been calculated to over one million decimal places. As it is not a rational number, the decimal never repeats. Its value to 18 decimal places is given on page 204. A value of 3.14 is usually sufficiently accurate.

Example 1. A tree trunk with a circular cross section has a circumference of 172.7 cm. What is the diameter of the trunk?

Solution.
$$C = \pi d$$
$$C \doteq 3.14d$$
$$172.7 \doteq 3.14d$$
$$d \doteq \frac{172.7}{3.14}, \text{ or } 55$$

The diameter of the trunk is about 55 cm.

For a circle of radius r and area A,

$$A : r^2 = \pi : 1 \quad \text{or} \quad \frac{A}{r^2} = \pi \quad \text{or} \quad A = \pi r^2.$$

Example 2. A pipe has a circular cross section with an area of 18 cm². What is the radius of the cross section?

Solution.
$$A = \pi r^2$$
$$18 \doteq 3.14r^2$$
$$r^2 \doteq \frac{18}{3.14}, \text{ or } 5.732$$
$$r \doteq \sqrt{5.732}, \text{ or } 2.4$$

The radius of the cross section is about 2.4 cm.

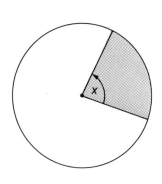

The part of a circle between two radii, shown shaded, is called a **sector**, and the angle, x, between the radii, is called the **sector angle**. The ratio of the sector area to the total area is the same as the ratio of the sector angle to 360°.

$$\frac{\text{area of sector}}{\text{area of circle}} = \frac{x}{360}$$

We use this proportion to determine the sector angles when we construct a circle graph.

Example 3. The table shows the budget of a typical Canadian family according to a recent survey. Show this information on a circle graph.

Budget of a Typical Canadian Family	
Housing	40%
Food	24%
Transportation	15%
Clothing	9%
Personal Care & Savings	7%
Recreation & Education	5%

Solution. Let x be the sector angle for Housing.

Then, $\dfrac{x}{360} = \dfrac{\text{area of sector for Housing}}{\text{area of circle}}$

$$= \frac{40}{100}, \text{ or } 0.4$$

$$x = 0.4 \times 360$$

$$= 144$$

The sector angle for Housing is 144°.

The sector angles for the other items of the budget are calculated in a similar manner. The circle graph is then constructed with the aid of a protractor and the sectors labelled.

Budget	Sector Angle
Housing	144°
Food	86.4°
Transportation	54°
Clothing	32.4°
Personal Care & Savings	25.2°
Recreation & Education	18°

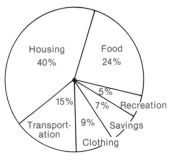

Budget of a Typical Canadian Family

Exercises   6 - 7

A   1. What is the diameter of a circle that has a circumference of

a) 5 cm?   b) 3π cm?   c) π^2 cm?   d) $\dfrac{1}{\pi}$ cm?

2. What is the radius of a circle that has an area of

   a) 78.5 mm²?   b) 300 mm²?   c) 144π mm²?   d) $\dfrac{64}{\pi}$ mm²?

3. What is the sector angle for a sector that is

   a) $\dfrac{1}{6}$ of a circle?   b) $\dfrac{2}{3}$ of a circle?   c) $\dfrac{5}{8}$ of a circle?

4. Write the area of each shaded sector as a percent of the area of the circle:

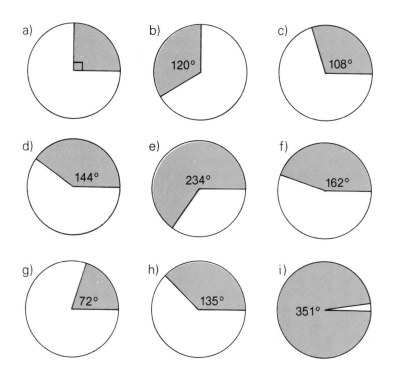

a)   b) 120°   c) 108°

d) 144°   e) 234°   f) 162°

g) 72°   h) 135°   i) 351°

Composition of Air by Volume

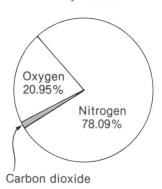

Oxygen 20.95%

Nitrogen 78.09%

Carbon dioxide

5. The circle graph shows the composition of air by volume. What is the angle of the sector representing carbon dioxide?

6. The student population of a high school is distributed as follows:

Grade 9	Grade 10	Grade 11	Grade 12
35%	30%	24%	11%

Show this data on a circle graph.

7. A 2.4 kg sample of nuts is found to contain 384 g of pecans, 864 g of peanuts, 768 g of hazel nuts, and the rest are cashews. Draw a circle graph to display this information.

8. The diameter of a weather balloon increases to 1.5 times its initial size by the time it ascends to the cloud base. By what factors have the following increased?

   a) the circumference   b) the cross-sectional area

9. Which would drain a tank faster—one drain 4 cm in diameter or two drains each 2 cm in diameter? How many times faster?

10. The formula for the volumes, V, of a sphere, cylinder, and cone are given below.

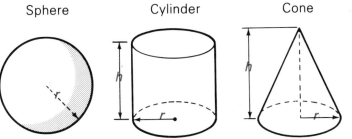

Sphere	Cylinder	Cone

weather balloon

$$V = \frac{4}{3}\pi r^3 \qquad V = \pi r^2 h \qquad V = \frac{1}{3}\pi r^2 h$$

Calculate the ratio

   a) $\dfrac{V(\text{cylinder})}{V(\text{sphere})}$ when the radius of both cylinder and sphere is r and the height of the cylinder is $2r$;

   b) $\dfrac{V(\text{cylinder})}{V(\text{cone})}$ when cylinder and cone have the same base and height;

   c) $\dfrac{V(\text{sphere})}{V(\text{cone})}$ when the radius of both sphere and cone is r and the height of the cone is $2r$.

11. What is the angle between the hands of a clock at each of these times:

   a) 01:00?   b) 02:20?   c) 03:30?   d) 04:45?

12. A ball just fits into a can, touching bottom, sides, and lid. If it is inserted into the can when the can is full of water, what fraction of the water is displaced?

13. One of Jupiter's moons is three times the diameter of Earth's moon. How many times greater is the mass of Jupiter's moon than Earth's? What assumption are you making?

14. Imagine that a steel ring circles Earth, just touching it, at the equator. What length would need to be added to the ring to raise it 1 m from the surface all the way round?

Earth

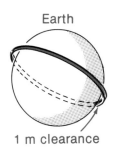

1 m clearance

The Golden Ratio

Twenty-three centuries ago Euclid posed the question:

What are the dimensions of a rectangle that has the property that when it is divided into a square and a rectangle, the smaller rectangle has the same shape as the original?

The answer to Euclid's question can be found by writing the proportion:

$$\frac{\text{Length of original rectangle}}{\text{Width of original rectangle}} = \frac{\text{Length of smaller rectangle}}{\text{Width of smaller rectangle}}$$

Let the length of the original rectangle be x units and its width be 1 unit. The side of the square will then be 1 unit,

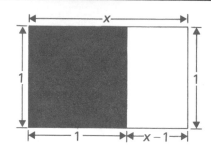

and the proportion can be written: $\frac{x}{1} = \frac{1}{x-1}$

If we multiply both sides of the proportion by $x - 1$, we get: $x^2 - x = 1$

or, $x^2 - x - 1 = 0$

This has the solution:

$$x = \frac{1 + \sqrt{5}}{2}$$

The number, $\frac{1 + \sqrt{5}}{2}$, is called the **golden ratio**.

1. Calculate, to three decimal places, the expression:

    $$\frac{1 + \sqrt{5}}{2}$$

2. Substitute your result in the proportion $\frac{x}{1} = \frac{1}{x-1}$ as a check.

3. Simplify the expression:

    $$1 + \cfrac{1}{1 + \cfrac{1}{1 + \cfrac{1}{1 + \cfrac{1}{1 + 1}}}}$$

How close is this value to the value of the golden ratio?

A rectangle with a length-to-width ratio of the golden ratio is called a **golden rectangle**. Study the diagram. Then, using a pair of compasses and a straight-edge, construct a golden rectangle.

Since Euclid's time, the golden ratio has been found in mathematics, architecture, art, and nature. In fact, the golden ratio is probably sec-

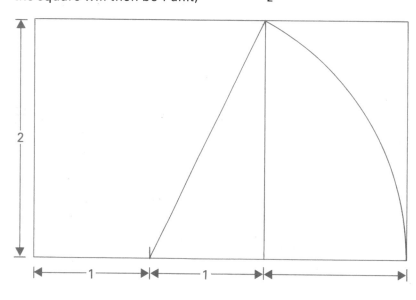

The Parthenon (completed 438 B.C.)

"The Sacrament of the Last Supper"

A Spiral Galaxy

A Nautilus Shell

ond only to π in its frequency of occurrence in the mathematical sciences.

Architecture
The Parthenon, in Athens, was designed having its width-to-height ratio the golden ratio. The photograph has a rectangle superimposed. Measure the sides of this rectangle and calculate their ratio. Is it a golden rectangle?

Art
The golden ratio can be found in the works of many of the great artists. One modern-day artist of international renown is Salvador Dali. This is a reproduction of his painting "The Sacrament of the Last Supper". Measure the sides of the picture and calculate their ratio. Is it a golden rectangle?

Nature
The rectangle obtained by dividing a golden rectangle into a square and a rectangle is itself a golden rectangle. Repeating this process with the smaller rectangle a number of times gives us a set of successively smaller golden rectangles. The result of joining a set of corresponding vertices of these rectangles with a curve is a spiral. Such spirals can be found throughout nature, from the spiral galaxy in the heavens to the nautilus shells in the oceans.

Draw a Diagram

The greatest problem solvers have always known how useful a diagram can be in solving problems. Not only are many problems considerably clarified by a diagram, but often the diagram points the way to the solution.

Example 1. Which will drain water from a pool faster, a circular pipe 8 cm in diameter or two circular pipes each 4 cm in diameter?

Diagram.

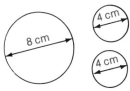

Solution. The rate at which the pool is drained depends on the cross-sectional area of the drain. The diagram clearly shows that the 8 cm diameter pipe has a much greater cross-sectional area than the two 4 cm pipes. Therefore, the 8 cm diameter pipe will drain the pool faster than the two smaller pipes.

Example 2. Two motorboats, *A* and *B*, set out simultaneously but from opposite banks to cross a river. The boats, travelling at constant but different speeds, first pass each other 350 m from the nearer bank. They continue to the opposite bank and return without stopping, passing the second time when they are 250 m from the other bank. How wide is the river?

Diagram.

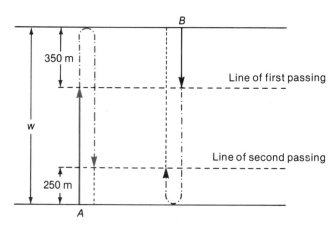

Solution. Let w be the width of the river and A the faster boat. Total distance travelled by both boats

at first passing: w

Total distance travelled at second passing: $3w$

Therefore, each boat has travelled three times as far at time of second passing as at time of first passing.

At first passing, boat B has travelled: 350 m

At second passing, boat B has travelled:

$(w + 250)$ m

$$w + 250 = 3 \times 350$$

The solution of this equation is the width of the river, in metres.

Exercises

Draw a diagram to solve each of these problems.

1. A cupboard, 2.15 m high, is to be fitted with six equally-spaced shelves. Each shelf is 2 cm thick. Determine the exact distance between shelves.

2. A train 1 km long travels at 30 km/h through a tunnel 1 km long. How long does it take the train to clear the tunnel?

3. Two trains, each 1 km long, meet on double tracks travelling at 60 km/h. How long does it take the trains to pass each other?

4. A sheet of 50 stamps is printed in 10 rows of 5 stamps. The edges of the stamps forming the sides of the sheet are straight. How many stamps have

   a) 2 straight edges?

   b) 1 straight edge?

   c) no straight edge?

5. Two swimmers set out simultaneously, but from opposite ends, to swim two lengths of a swimming pool. The swimmers, travelling at constant but different speeds, first pass each other 20 m from the nearer end. They next pass each other 10 m from the other end when they are both on the second lap. What is the length of the pool?

6. Determine how to cut a rectangle 9 cm by 4 cm into two pieces that can be rearranged to form a square with a side length of 6 cm.

Review Exercises

1. In a bracelet of 18 karat gold, the ratio of gold to copper is $18:6$.
   a) Write this ratio in lowest terms.
   b) What mass of gold is in a bracelet with a mass of 72 g?

2. Which is the greater ratio,
   a) $\frac{7}{9}$ or $\frac{10}{13}$?
   b) $5:6$ or $29:34$?
   c) $8:17$ or $11:25$?
   d) $21:8$ or $24:9$?

3. Solve for the missing term:
   a) $\frac{7}{10} = \frac{m}{15}$
   b) $\frac{n}{4} = \frac{13}{8}$
   c) $\frac{39}{a} = \frac{13}{3}$
   d) $\frac{17}{21} = \frac{51}{b}$

4. Solve for x and y:
   a) $x:y:3 = 8:5:9$
   b) $x:y:7 = 9:5:12$

5. Bill contributes $3.50 and Laura $6.50 for the purchase of a $10 lottery ticket. It is drawn for a prize of $125 000. How should the money be divided?

6. Partners in a business agree to share the profits in the ratio $3:8:4$. How much does each receive from a total profit of $45 000?

7. The shadow of a tree is 26.5 m long when that of a metre rule is 58 cm long. How high is the tree?

8. On a map, two towns are 85 mm apart. If the scale of the map is $1:1\,500\,000$, what is the actual distance between the towns?

9. The ferry crossing of the St. Lawrence River at Baie Comeau is a distance of 61 km. What will this distance be on a map drawn to a scale of $1:200\,000$?

10. A person's heart beats 13 times in 10 s.
    a) How many times does it beat in 1 h?
    b) How long does it take to beat 1000 times? to beat 1 000 000 times?

11. A car's rate of fuel consumption is 8.5 L/100 km.
    a) How much fuel is needed to travel 350 km?
    b) If the car's tank holds a maximum of 58 L of fuel, will one tankful be enough to travel the 687 km from Corner Brook to St. John's in Newfoundland?

Corner Brook
St. John's
Newfoundland

12. Hose *A* fills a swimming pool in 12 h. Hose *B* fills the same pool in 18 h. How long will it take to fill the pool with both hoses operating together?

13. Beryl can mow a lawn in 2 h. Joan takes 3 h to mow the same lawn. How long will it take them to mow the lawn working together, each with her own mower?

14. The distance from Winnipeg to Vancouver is approximately 2500 km. If part of the trip is driven at a steady speed of 100 km/h and the rest of the trip is driven at a steady 70 km/h, find the time driven at each speed for the trip to take 30 h.

15. A car's rate of fuel consumption averages 10.5 L/100 km in the city and 8.0 L/100 km in the country. In a week's driving, the car travels 300 km using 27 L of fuel.
    a) What fraction of the 300 km was in the city?
    b) How much fuel was used in the country driving?

16. Good quality seed sells for $9.75/kg and standard quality seed sells for $6.00/kg. What quantities of each should be used to produce 50 kg of a mixture to sell for $7.50/kg?

17. A person's rate of pay increases from $5.00/h to $7.00/h. What is the percent increase?

18. What is the sale price of a $165 bicycle selling at a 15% discount?

19. If the price of oil is $150/m³ and the price is raised 10% every year, what will be its price at the end of three years?

20. Find the interest on the following:
    a) a $215 credit-card bill for 1 month at 18% per annum
    b) $150 in a daily-interest account for 100 days at $12\frac{1}{4}\%$ per annum

21. Part of Marie's capital of $20 000 is invested at 16% and the rest is invested at 12%. If the total interest she receives in a year is $3000, how did she divide her investment?

22. The table gives the composition of a hot dog. Display this information on a circle graph.

23. Tennis balls are sold in cylinders of 3. Is the ratio $\frac{\text{height}}{\text{circumference}}$ of the cylinder greater or less than 1?

Composition of a Hot Dog	
Ingredient	Mass g
Fat	11.0
Protein	4.2
Water	20.4
Other	2.3

A Short History of π

1
From ancient times, people have known that the circumference of a circle...

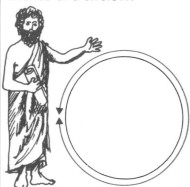

2
...is more than 3 times its diameter.

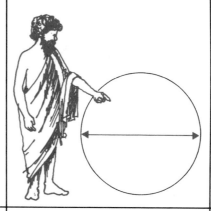

3
Through the centuries different civilizations have tried to find a fraction which would be exactly equal to the ratio:

$$\frac{\text{circumference}}{\text{diameter}}$$

About 1700 B.C., the Egyptians used the fraction $\frac{256}{81}$ to approximate this ratio.

4
About 220 B.C. the Greeks used the Greek letter π to represent this ratio. And they used the fraction $\frac{22}{7}$ as its approximate value.

$$\pi = \frac{22}{7}$$

5
In their search for the elusive rational number equal to π, various other civilizations used these approximations.

Civilization	Date	Value for π
Chinese	470	$\frac{355}{113}$
Hindu	530	$\frac{3927}{1250}$
European	1220	$\frac{864}{275}$

6
Finally, in 1761, Johann Lambert proved that there is no rational number equal to π. That is, the decimal form of π does not repeat.

$$\pi = 3.141\ 592\ 653\ 589\ 793\ 238\ldots$$

In 1973, two mathematicians computed π to one million decimal places using a computer. As expected, the decimal did not repeat.

Use your calculator to determine which civilization up to the year 1220 A.D. had the best approximation for π.

Cumulative Review (Chapters 4 - 6)

1. Solve:

   a) $7 + y = -6$ b) $9 - t = -13$ c) $-18 = -6p$

   d) $\frac{1}{7}x = -2$ e) $\frac{r}{3} = \frac{1}{5}$ f) $-2.75x = 16.5$

   g) $5x - 2 = 13$ h) $3 - 3y = -3$ i) $3x - \frac{1}{7} = 8$

2. Solve:

   a) $5x - 3 = 2x + 6$ b) $-3y + 5 = 2y - 10$

   c) $2(-x - 3) = -2$ d) $-2(3x + 4) = -7$

   e) $-2(t + 1) = 3(t - 2)$ f) $5(7x - 3) = 17x - (2 - 5x)$

   g) $\frac{4}{3} - \frac{1}{2}x = \frac{2}{3}x$ h) $1.25 - 0.8x = 0.4x - 0.19$

3. A father is three times as old as his daughter. In 12 years' time he will be only twice as old. What are their ages now?

4. Two numbers differ by 3. The sum of the larger and one-fourth the smaller is 13. Find the numbers.

5. Simplify:

   a) $2^3 + 2^4$ b) $2^3 \times 2^4$ c) $(-3)^2 + (-3)^3$

   d) $\left(-\frac{2}{3}\right)^3$ e) $(-2)^7 \div (-2)^4$ f) $2^{-3} \div 2^{-4}$

   g) 3^{-2} h) $\left(-\frac{2}{3}\right)^{-2}$ i) $\left(\frac{3}{4}\right)^{-1} + \left(\frac{3}{4}\right)^0$

   j) $3y^2 \times 4y^3$ k) $(3y^3)^2$ l) $25x^6 \div (-5x^{-3})$

6. Solve:

   a) $3\sqrt{a} = 21$ b) $5x^2 = 125$ c) $2^x = 32$

   d) $3^x = \frac{1}{9}$ e) $x^{-3} = 0.001$ f) $4^x = 1$

7. Write in scientific notation:

   a) $15\,000$ b) $2\,700\,000$ c) 21

   d) $0.000\,016$ e) $0.000\,37$ f) 0.19

8. Solve:

   a) $\frac{5}{8} = \frac{n}{16}$ b) $\frac{r}{3} = \frac{8}{15}$ c) $\frac{16}{25} = \frac{48}{x}$

   d) $\frac{7}{15} = \frac{49}{y}$ e) $\frac{12}{39} = \frac{s}{13}$ f) $\frac{19}{t} = \frac{57}{42}$

9. Solve for x and y:

   a) $x:y:3 = 8:5:15$ b) $x:y:2 = 9:6:12$

10. Jerry contributes $4.00 and Penny $6.00 for the purchase of a lottery ticket. It is drawn for a prize of $25 000. How should the money be divided?

11. On a map, two cities are 120 mm apart. If the scale of the map is $1:1\,250\,000$, what is the actual distance between the cities?

12. Jill does a job in 3 h. John takes 4 h to do the same job. How long will it take to do the job if they work together?

13. a) Find 16% of a number.

   b) 24% of a number is 18. What is the number?

   c) What percent of 1210 is 484?

14. A calculator regularly priced at $15.75 is on sale for $12.60. What is the percent of discount?

15. A television set that lists for $480 is on sale at a 15% discount. What is the sale price?

16. Three partners in a business agree to share the profits in the ratio $2:4:5$. How much will each receive from a total profit of $88 000?

17. Part of a journey of 380 km was travelled at a steady speed of 120 km/h and the rest at a steady 90 km/h. If the journey took 4 h, find the time driven at each speed.

18. A car's rate of fuel consumption averages 13.2 L/100 km in the city and 8.4 L/100 km on the highway. In a week's driving the car travels 500 km using 62 L of fuel.

   a) What fraction of the 500 km was in the city?

   b) How much fuel was used in the country driving?

19. Good quality coffee sells for $18.50/kg and standard quality coffee sells for $11.50/kg. What quantities of each should be used to produce 20 kg of a blend to sell for $15.70/kg?

20. Bruce wins $12 000. He invests part of it at 12% and the rest at 10.5%. If these investments yield a total interest of $1365 in a year, how did Bruce divide his investments?

7 Statistics and Probability

Statistics show that few car accidents occur in the early morning, only very few of these occur in fog, and fewer still involve vehicles travelling at speeds greater than 150 km/h. Does this mean that it would be safe to drive at more than 150 km/h in the early hours of a foggy morning? Why? (See *Example 1* of Section 7-4.)

7 - 1   Graphing Data

In this fast-moving computer age, we encounter vast quantities of data. **Statistics** is that branch of mathematics which deals with the collection, organization, and interpretation of data. These data are usually organized into tables and presented as graphs. Some common types of graphs are shown below. Try to answer the question that accompanies each graph.

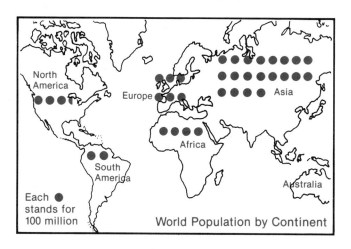

World Population by Continent

Pictographs. The graph shown here uses a symbol, ●, to represent 100 million people. A graph that uses a symbol to represent a certain amount is called a **pictograph**.

• What are the populations of North America and Europe?

Circle Graphs. In a **circle graph**, a complete set of data is represented by the area of a circle. Various parts of the data are represented by the areas of sectors of the circle.

• What percent of Ontario's forest fires are caused by lightning?

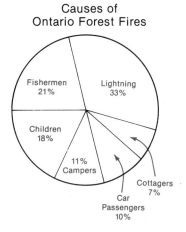

Causes of Ontario Forest Fires

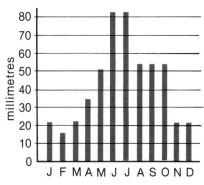

Precipitation — Winnipeg

Bar graphs. The graph shown here uses vertical bars to represent the amount of precipitation each month. Graphs of this type are called **bar graphs**. Bar graphs may have horizontal or vertical bars.

• What appears to be the "rainy season" in Winnipeg?

Histograms. A graph that records the number of times a variable, such as height or mass, has a value in a particular interval is called a **histogram**. This graph shows the number of students whose heights fall in each 5 cm interval from 140 cm to 189 cm.

• Examine the graph and tell how many students are at least 170 cm tall.

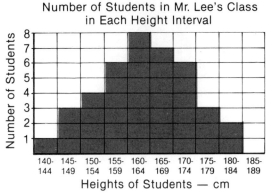

Number of Students in Mr. Lee's Class in Each Height Interval

Heights of Students — cm

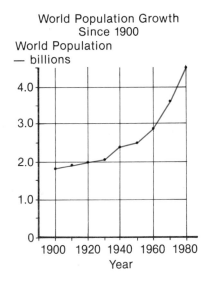

World Population Growth Since 1900

Broken-line Graphs. The graph shown gives the population of the world at the end of each decade from 1900. Since the exact population during the decades is not known, the points plotted are joined by line segments. Such a graph is called a **broken-line graph**.

• About how many years did it take the world's population to grow from 3 billion to 4 billion?

Continuous-line Graphs. A graph that displays the value of one variable, such as speed, corresponding to the value of another variable, such as stopping distance, for all values over a given interval is called a **continuous-line graph**. The graph shown gives the distance required to bring a car to rest from the moment the brakes are applied when travelling at speeds up to 100 km/h.

• What is the car's stopping distance when travelling at 60 km/h?

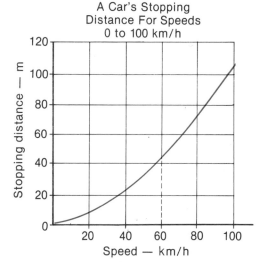

A Car's Stopping Distance For Speeds 0 to 100 km/h

Speed — km/h

, The only points on a broken-line graph that represent actual data are endpoints of the segments. However, *all* points on a continuous-line graph correspond to actual data, unless a statement to the contrary is made.

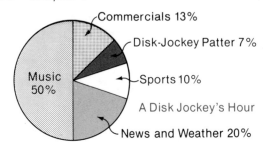

Commercials 13%

Disk-Jockey Patter 7%

Sports 10%

Music 50%

A Disk Jockey's Hour

News and Weather 20%

Exercises 7 - 1

A 1. a) In the circle graph shown, how many minutes each hour does the disk jockey devote to news and weather reports?

   b) About how many minutes each hour are devoted to commercials?

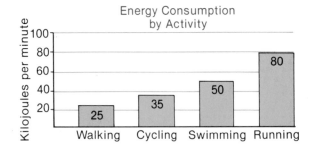

Energy Consumption by Activity

Kilojoules per minute

100
80
60
40
20

Walking 25 Cycling 35 Swimming 50 Running 80

2. Use the "Energy Consumption by Activity" graph to answer the following:

   a) What activity burns up energy twice as fast as walking?

   b) If you cycled for 30 min, how many kilojoules of energy would you use?

   c) How long would you have to run to burn up the 2200 kJ in a chocolate milkshake?

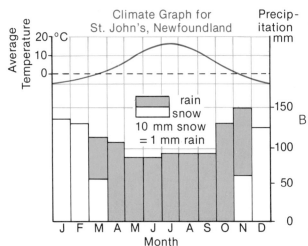

Monthly Record of Sales of "Blue Jeans Blues"

Sales in 100's

8
6
4
2

Jan Feb Mar Apr May Jun July

Month

3. The broken-line graph shows the sales, in one record shop of the hit song, "Blue Jeans Blues", over a six-month period.

   a) About how many records of this hit were sold in March?

   b) What was the peak month for sales?

   c) If you were the shop owner, how many records of this song would you keep in stock for July: 600, 400, 200, or 0?

Climate Graph for St. John's, Newfoundland

Average Temperature

Precipitation mm

20 °C
10
0

rain
snow
10 mm snow = 1 mm rain

150
100
50
0

J F M A M J J A S O N D

Month

B 4. Use the climate graph for St. John's, Newfoundland, to answer the following:

   a) What is the average temperature in St. John's in March?   in July?

   b) For how many months of the year is the average temperature below 0°C?

   c) What is the average precipitation that falls as snow in a year?

5. The annual oil consumption per person for seven industrial nations was reported to be as follows:

Country	Canada	U.S.	U.K.	West Germany	France	Japan	Italy
Oil consumption per person—m³	10.2	9.7	4.5	4.9	4.0	3.5	2.9

Show this information on a bar graph.

6. Illustrate this weekly budget plan by a circle graph.

Weekly Budget

Payroll deduction	$40	Savings	$22
Apartment rent	50	Transportation	10
Food	25	Clothing	10
Recreation	16	Miscellaneous	27

C 7. Display each set of data in a suitable graph.

a)

Where Each Dollar of Your Federated Appeal Contribution Goes

33¢	Health Care and Rehabilitation Services
31¢	Child and Family Care Services
26¢	Youth Group Services
7¢	Campaign expenses
3¢	Administration costs

b)

Fuel Consumption of Recent-Model Cars

Model	$\frac{L}{100\ km}$
Pontiac Trans-Am	18.4
Porsche Turbo	14.1
Saab 900 GL	11.4
Triumph TR7	10.6
Ford Fiesta	8.9
Honda Accord	9.0
Lada 1200	9.9
VW Diesel Rabbit	6.1
VW Scirocco	9.1
Fiat Strada	10.2
Mini 1000	8.2
Toyota Corolla	9.0

c)

The effect of wind speed at a still-air temperature of −12°C

Wind speed km/h	Wind-chill temperature °C
0	−12
10	−17
20	−26
30	−31
40	−35
50	−37
60	−38

7-2 Organizing and Presenting Data

The heights of the players on a school basketball team were recorded by the coach.

John	175 cm	Terry	178 cm	Dick	177 cm
Bruce	180 cm	Kevin	178 cm	Paul	178 cm
Scott	177 cm	Gordon	180 cm	Jason	177 cm
Joe	176 cm	Larry	179 cm	Art	178 cm
Brian	181 cm	Neil	175 cm		

To get a better idea of the distribution of the heights of his players, the coach made a **frequency table**. The **frequency** of a measurement, such as height, is the number of times it occurs. Using the table, the coach then made a **frequency graph**. Each measurement is represented on the graph by a bar whose height corresponds to a frequency. The bars are the same width and have no gaps between them. The vertical scale starts at zero.

Height cm	Number of Players	Frequency
175	//	2
176	/	1
177	///	3
178	////	4
179	/	1
180	//	2
181	/	1

Frequency of Heights of Players

Often, the number of different results is too great for each to be represented by a bar on a frequency graph. Consider the following set of marks obtained by a class on a mathematics test.

72	53	73	59	68	83	71
67	77	78	70	67	63	65
56	86	47	78	72	79	67
74	62	84	92	88	71	74
81	70	66	64	75	65	46

Of the 35 marks listed, 27 are different. This could mean a frequency graph with 27 bars. The number of bars can be reduced substantially be grouping the marks into convenient intervals. The marks that fall into each interval can then be tallied in a frequency table.

Interval	Number of Students	Fre-quency
40–49	//	2
50–59	///	3
60–69	ⅬⅎⅬⅎ	10
70–79	ⅬⅎⅬⅎ ////	14
80–89	Ⅼⅎ	5
90–99	/	1

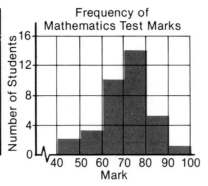

Frequency of Mathematics Test Marks

Intervals of 5 points can be used if a wider spread of the distribution of marks is desired.

Interval	Number of Students	Fre-quency
45–49	//	2
50–54	/	1
55–59	//	2
60–64	///	3
65–69	Ⅼⅎ //	7
70–74	Ⅼⅎ ////	9
75–79	Ⅼⅎ	5
80–84	///	3
85–89	//	2
90–94	/	1

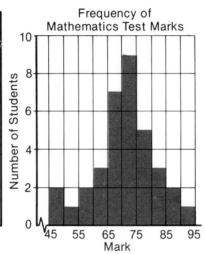

Frequency of Mathematics Test Marks

Exercises 7 - 2

A 1. A car dealer hired students to determine the ages of the cars owned by the residents of the area. They listed their findings in a table. Display the data by a frequency graph.

Age of car—years	0	1	2	3	4	5	6	7
Number of cars	25	40	50	65	45	30	20	15

2. The maximum daily temperatures in Vancouver during July, over a five-year period, are shown in the table. Make a frequency graph for this data.

Maximum Temperature	Fre-quency
13–15°C	4
16–18	14
19–21	22
22–24	32
25–27	40
28–30	30
31–33	13

Mass kg	Frequency
1.91–1.94	15
1.95–1.98	85
1.99–2.02	150
2.03–2.06	75
2.07–2.10	20

3. The actual mass of a 2 kg box of chocolates was checked by weighing a selection of 345 boxes. Display the results, listed in the table, on a frequency graph.

4. The scores obtained by the 25 players in a high-school golf tournament held at the Glen View golf course were as follows:

87	96	94	102	84	92	104	79	97
93	84	98	85	89	68	78	75	99
95	71	79	69	103	72	84		

   a) Make a frequency table with 10-stroke intervals.
   b) Draw a frequency graph.

B 5. It is important that structural materials have the right qualities of strength and hardness. Hardness tests were performed on 30 samples of aluminum used in aircraft construction and the following hardness numbers obtained:

29	38	26	29	30	41	35	35	24	28
33	29	37	25	29	32	33	35	30	35
32	28	30	31	34	38	28	32	39	34

   Draw a frequency graph using intervals of 2.

6. A school's grade 9 students obtained the following marks in an English examination:

55	66	64	98	56	69	68	62	52	69	65	63	51
90	69	68	32	66	72	44	80	61	84	74	66	79
61	89	78	63	66	59	75	53	69	23	92	78	73
67	38	65	67	41	75	63	71	57	77	66	56	63
73	100	56	76	71	61	51	46	84	55	63	68	86
65	69	66	60	62	68	82	73	65	76	79	88	44

   Display these results on a frequency graph using intervals of 10.

C 7. Two groups of students wrote the same mathematics test. Group A was given notice of the test and were able to prepare for it. Group B had the test sprung upon them. The groups obtained the following marks out of 20.

A						B				
16	17	18	20	11		12	11	15	18	12
18	20	19	15	20		6	9	11	5	11
15	15	17	12	19		11	14	11	16	17
8	13	16	17	14		12	13	9	8	19
19	14	20	12	15		10	10	7	11	18

   Compare the results of the two groups of students by drawing the frequency graphs. Use intervals of 5.

7 - 3   Measures of Central Tendency

Typical salaries of five professions are given below.

Doctor: $50 000

Lawyer: $46 000

Police officer: $22 000

Trucker: $30 000

Airline pilot: $42 000

Who earn more, doctors or lawyers?

It appears from the data given that doctors have greater incomes than lawyers. This can be misleading. Many lawyers earn more than $50 000 per year, and many doctors earn less than $46 000.

We want a single amount that best represents the income of all doctors or all lawyers. We are looking for an "average" income. We may define an average in different ways depending on the information we are seeking. Three of the most commonly used averages are the **mean**, the **median**, and the **mode**.

The **mean** is the arithmetical average of the numbers in a set — the sum of all the numbers divided by the number of numbers. The mean income for all doctors is the sum of the incomes of all doctors divided by the number of doctors.

The **median** is the middle number when the numbers are arranged in sequence. If there is an even number of numbers, the median is the average of the two middle numbers. The median income for all doctors is the middle amount when all incomes are arranged from greatest to least.

The **mode** of a set of numbers is the most frequently occuring number. There may be more than one mode, or there may be no mode. If more doctors report an income of $48 000 than any other amount, then $48 000 is their mode income.

Since the mean, median, and mode of a set of numbers are single values around which the numbers cluster, they are referred to as **measures of central tendency**.

*rainfall measuring
apparatus*

Example 1. The recorded rainfall, in millimetres, for seven consecutive days in Kitimat, B.C., was: 12, 14, 8, 8, 8, 12, and 15. What are the mean, median, and mode of this set of data?

Solution. Mean: $\dfrac{12+14+8+8+8+12+15}{7} = \dfrac{77}{7}$, or 11

Median: Arranged in order, the numbers are: 8, 8, 8, 12, 12, 14, 15.
The median is 12.

Mode: Since 8 occurs the most, the mode is 8.

Sometimes one of the measures of central tendency is more representative of a set of data than the other two.

Example 2. The following is the annual payroll of the Beta Metal Works:

1 Manager	$60 000
1 Foreman	$40 000
3 Craftsmen	$20 000
5 Laborers	$11 500

a) Determine the mean, median, and mode for the payroll.

b) Which measure could be used to make salaries *look* high?   low?

c) Which measure most fairly represents the pay scale in the factory?

Solution. a) Mean salary:

$$\dfrac{\$60\,000 + \$40\,000 + (3 \times \$20\,000) + (5 \times \$11\,500)}{10}$$

$$= \$21\,750$$

median: $\dfrac{\$11\,500 + \$20\,000}{2}$, or $15 750

mode: $11 500

b) The mean of $21 750 is achieved by only 2 of the 10 employees. Therefore, it is an artificially high representation of the typical employee's salary. The mode is artificially low since every employee earns at least that amount.

c) The median is probably the most descriptive of the incomes.

Exercises 7 - 3

A 1. Find the mean, median, and mode of each of the following sets of data:

   a) 10, 12, 8, 9, 12, 14, 11, 15, 9, 12

   b) 1, 2, 2, 3, 3, 3, 4, 4, 4, 4, 4, 4, 5

   c) 2.3, 4.1, 3.7, 3.2, 2.8, 3.6

   d) 15, 18, 16, 21, 18, 14, 12, 19, 11

   e) 9, 12, 7, 5, 18, 15, 5, 11

   f) 7, 9, 18, 8, 5, 6, 3

   g) $\dfrac{1}{2}, \dfrac{1}{4}, \dfrac{2}{3}, \dfrac{5}{12}, \dfrac{3}{4}, \dfrac{2}{3}, \dfrac{1}{2}, \dfrac{7}{12}, \dfrac{1}{6}, \dfrac{1}{2}$

   2. Calculate the mean, median, and mode of this set of hardness figures for aluminum sheet:

29	38	26	29	30	41	35	35	24	28
33	29	37	25	29	32	33	35	30	35
32	28	30	31	24	38	28	32	39	34

   3. Over a period of time, shares of stock were purchased as follows: 10 shares at $8 per share; 20 shares at $9.50 per share; and 15 shares at $8.50 per share. What was the mean price per share?

   4. For the numbers: 5, 6, 7, 8, 9, find the effect on the mean

   a) if each number is increased by 2;

   b) if each number is doubled.

   5. If the mean of the numbers 8, 12, 13, 14, x is 13, what is the value of x?

B 6. The mean of seven marks on a mathematics test is 68. However, the correction of an error in grading raises one student's mark by 14. Calculate the new mean.

   7. The Cabot Manufacturing Company has the following employees at the rates of pay shown:

Position	Weekly Pay	Position	Weekly Pay
1 President	$950	1 Typist	$200
1 Secretary	350	3 Assemblers	225
1 Designer	350	3 Packers	195
1 Foreman	300	3 Apprentices	150

   Which measure of central tendency most fairly represents the pay structure of the company?

8. Two groups of students write the same mathematics test and obtained the following marks out of 20.

A	16	17	18	20	11
	18	20	19	15	20
	15	15	17	12	19
	8	13	16	17	14
	19	14	20	12	15

B	12	11	15	18	12
	6	9	11	5	11
	11	14	11	16	17
	12	13	9	8	19
	10	10	7	11	18

Calculate the three measures of central tendency for both groups.

C 9. Find five numbers that have
   a) a mean of 9 and a median of 8;
   b) a mean of 14 and a median of 8.

10. The number of accidents at a ski resort for each of five months was:

Dec.	Jan.	Feb.	Mar.	Apr.
25	35	40	35	5

Which measure of central tendency best describes this data?

11. Which measure—mean, median, or mode—is most suitable for the following?
   a) the average number of children in a Canadian family
   b) a person's average weekly salary
   c) a class's average mark in a test
   d) the average rainfall in Quebec City
   e) the average time you spend on homework each night
   f) the average number of hours a ten-year-old child spends watching TV shows each week
   g) the average price of gasoline in a given area
   h) the average size of shoe sold by a store

12. In a set of data, the smallest number is increased by 5 and the largest number is decreased by 5. What changes occur in the measures of central tendency?

7 - 4 Misusing Statistics

Graphs of the same kind can give quite different impressions of a set of data depending on the scales used.

The table shows the number of cars produced by one plant of the Standard Automobile Corporation during a five-year period. This data was used in constructing each of the broken-line graphs below. Do they each convey the same impression?

Year	Cars Produced
1978	20 150
1979	22 300
1980	23 850
1981	25 600
1982	27 100

Graph A is an honest graph which shows that the company is making moderate increases in production.

Graph B is misleading because it uses a scale that really begins well above zero. In this graph there appears to be a spectacular increase in production over the five-year period. Broken scales are used to make graphs look steeper than they really are.

In **graph C**, an expanded vertical scale is used together with a compressed horizontal scale to give the appearance of a greater increase in production than actually took place. As a general rule, the horizontal and vertical axes should be approximately the same length.

A. Honest graph

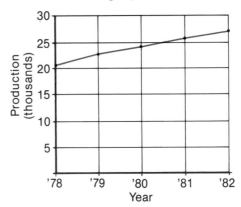

C. Misleading graph

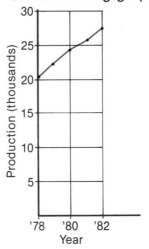

B. Misleading graph

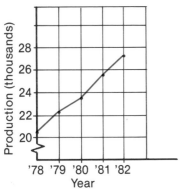

Not all misuses of statistics involve graphs. The following examples contain statements that are typical of many that can be found in newspapers and magazines.

Example 1. Comment on the reasoning:
"As few accidents happen in the early morning, very few of these as a result of fog, and fewer still as a result of travelling faster than 150 km/h, it would be best to travel at high speed on a foggy morning."

Solution. There are few accidents as a result of high speed on a foggy morning because:
  i) there is little traffic on the roads in the early morning;
  ii) many motorists would wait for the fog to clear, and those who do drive would reduce their speeds;
  iii) few motorists drive at more than 150 km/h at any time. Therefore, it would be wrong to conclude from the statistics that it would be best to travel at high speed on a foggy morning.

Example 2. Comment on this newspaper report:

The Gazette

Two Out of Three Have Heart Trouble

In an interview with a Gazette reporter, Dr. Cole of Eastern Hospital stated that he was consulted by 30 patients last week. Of those 30, he found that 20 had had heart trouble. This is an alarmingly high

Solution. The headline is quite inaccurate because it is based only on the people who visited Dr. Cole. They very likely went to see him because they were not in good health. The 30 patients were not a representative portion of the population. Any representative portion of the population would include many more people than 30, most of whom would be perfectly healthy.

Exercises 7 - 4

A 1. You are the Public Relations Manager for an insurance company. A case is being made for cheaper insurance rates for
persons under 25 years of age based on these statistics. What
points would you make in reply?

Age Group	Number of Drivers in Fatal Accidents	Percent
under 25	98	39.2
over 25	152	60.8
	Total 250	100.0

2. Say whether or not the following statements are correct in-
terpretations of the statistics. Give reasons.

a) A cure for the common cold has been found. In a recent test,
300 cold victims took the new wonder drug, Coldgone. After
four days, only 7 persons still had colds.

b) Last year, 50 motorists and 8 bicyclists were killed in traffic
accidents. This proves that it is safer to ride a bicycle than
drive a car.

c) The data in the table shows that teena-
gers are more likely to have a skiing acci-
dent than persons in other age groups.
Persons over 50 years of age are the best
skiers.

d) Almost 48% of injuries in downhill skiing
are to skiers who have never had lessons.
This means that skiers who take lessons
are involved in more skiing accidents than
those who do not. It follows that it is better not to take lessons.

Age of Skier Years	Percent of Skiing Accidents
under 10	10
10-19	61
20-29	19
30-39	5
40-49	3
over 50	2

e) Last year, there were 45 000 job vacancies in Canada. There
were also 500 000 unemployed workers. This shows that
the unemployed are not interested in working.

f) In 15 consecutive tosses a coin shows heads. It will therefore show heads on the next toss.

B 3. Comment on the statistics displayed in these graphs:

a) Government Spending

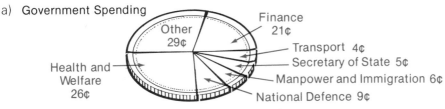

b)

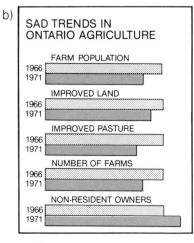

SAD TRENDS IN
ONTARIO AGRICULTURE

FARM POPULATION
1966
1971

IMPROVED LAND
1966
1971

IMPROVED PASTURE
1966
1971

NUMBER OF FARMS
1966
1971

NON-RESIDENT OWNERS
1966
1971

From an election pamphlet

c)

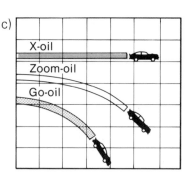

X-oil

Zoom-oil

Go-oil

To prove that X-oil is better suited for smaller, hotter, higher-revving engines, we tested X-oil against Zoom-oil and Go-oil. As the graph above plainly shows, only X-oil didn't break down.

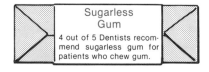

Sugarless Gum
4 out of 5 Dentists recommend sugarless gum for patients who chew gum.

Year	Profit $ millions
1977	5.0
1978	5.2
1979	5.4
1980	5.8
1981	6.5
1982	7.0
1983	7.4
1984	8.5

4. Printed on the wrapper of this stick of gum is the statement: "4 out of 5 Dentists recommend Sugarless Gum for their patients who chew gum." Comment on this statement.

5. Give three examples of statistical data from newspapers and magazines, or from radio and television commercials, and comment on the statements made.

6. The table records a company's annual profits for the years 1977 to 1984.
   a) Draw an honest graph to represent the data.
   b) Draw a graph on which the annual profit seems to be the same.
   c) Draw a graph that exaggerates the increase in profits.

7. The following table gives a company's monthly sales for a year.

Month	Sales $1000s	Month	Sales $1000s
Jan.	95	Jul.	80
Feb.	98	Aug.	78
Mar.	98	Sep.	75
Apr.	95	Oct.	72
May	90	Nov.	68
Jun.	85	Dec.	64

   a) Draw an honest graph to represent the data.
   b) Draw a graph on which the monthly sales seem to be the same.
   c) Draw a graph that exaggerates the decrease in sales.

7 - 5 Sampling

How is it determined what television program is the most popular of those on the air at
a particular time?

It is clearly impossible to poll the entire
population to find out who is watching what.
Instead, a representative portion of the population, called a **sample**, is polled. If the sample is sufficiently large, and carefully chosen,
the viewing preferences of the sample will
accurately reflect the preferences of the entire population. In fact, it has been found that
a carefully chosen sample of about 1500
people will reflect the viewing preferences of
the entire population of North America.

This principle is applied every time a chef tastes a spoonful
of soup, or a mother tests the baby's bath water with her
elbow. The chef assumes that the spoonful of soup tastes like
the whole pot. The mother assumes that the temperature of
the water is the same throughout the tub.

In statistics, **population** is taken to mean the whole of
anything of which a sample is being taken. The pot of soup and
the tub of bath water are referred to as populations.

A sample that is chosen in such a way that it may be expected to be typical of the population it represents is called a
random sample. Only a sufficiently large random sample can
be expected to be representative of a population.

To obtain information about a population, follow these
steps:
- decide on an appropriate sample size
- choose a device for selecting a random sample of that size
- collect the data from the sample
- organize and interpret the data
- make inferences about the characteristics of the
  population

There are several ways of collecting data:
- Personal Interviews — door-to-door, at shopping centres
  or other appropriate locations, by telephone
- Questionnaires — by mail, with a purchased article, in
  newspapers
- Tests and Measurements — recording instruments, quality control, time study

Examples:

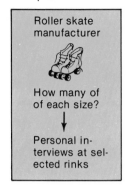

Roller skate manufacturer

How many of of each size?

↓

Personal interviews at selected rinks

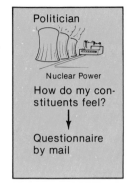

Politician

Nuclear Power

How do my constituents feel?

↓

Questionnaire by mail

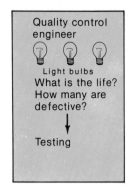

Quality control engineer

Light bulbs
What is the life? How many are defective?

↓

Testing

Example 1. A Vancouver company is hired by a television station to conduct a poll to predict the outcome of an upcoming national election. To gather this information, the company considers sampling Canadian voters in one of the following ways:

a) Interview 100 people at random.

b) Poll a random sample of 1000 people in British Columbia.

c) Put an advertisement in all major newspapers asking people to tell their political preference.

d) Send 10 questionnaires to all major businesses to be completed by anyone selected at random.

Describe the main weaknesses of each of these methods.

Solution. a) The sample is much too small to be reliable.

b) Political preferences are often regional in nature. A sample of voters in British Columbia is not likely to be representative of the political opinions of all Canadians.

c) Generally, only people with very strong political views will take the trouble to respond to an advertisement. The sample will not be random.

d) This sample tends to exclude such voting groups as farmers, homemakers, and senior citizens, and is therefore not a random sample.

Exercises 7 - 5

A 1. How would you collect data to find the following? Give reasons.
  a) The popularity of a TV program
  b) The most popular breakfast cereal
  c) The average number of children in a family
  d) The number of occupants in cars in rush hours
  e) The most popular recording artist
  f) An average family's food budget
  g) The color preference of kindergarten children

  2. How would you collect data to find the following? What kind of people or items would be in your sample?
  a) The player most likely to be voted "outstanding rookie"
  b) The top 10 movies of the year
  c) The life span of flashlight bulbs
  d) The amount spent by car owners on repairs each month
  e) The time required to eat lunch in a cafeteria
  f) The most popular soft drink
  g) The breaking strength of fishing lines
  h) The percent of foreign-made cars in a town

  3. Explain why data is collected from a sample and not a population:
  a) Quality control in the manufacture of flash cubes
  b) The number of pets per family
  c) The purity of processed food
  d) The strength of aluminum extension ladders
  e) The cost of ski equipment
  f) Percent of the population with the various blood types
  g) The effectiveness of a new pain-killing drug
  h) The weekly income of students

Number of persons in household	Number of households
1	3
2	7
3	12
4	21
5	14
6	8
7	5

B 4. To collect data to find the average number of persons in a household, a student visits every household within three blocks of her home.
  a) How many households did she visit?
  b) How many persons live within three blocks of the student?

Make of car	Number of each make
Ford	47
General Motors	64
Chrysler	29
American Motors	25
Foreign	82

5. The make of every third car in a large parking lot is noted and the numbers of each make is recorded.
   a) Which is the most popular make of car?
   b) How many cars were in the lot?
   c) If the parking fee is $3.50 per car, what are the total receipts?

6. Decide what kind of a sample you need, then work singly, in pairs, or in groups to collect the following data:
   a) The age, height, and mass of the students in the class
   b) The number of persons in cars in the rush hour
   c) The amount spent on lunch in the cafeteria
   d) The most popular musical group
   e) The time spent waiting in line in the cafeteria
   f) The weekly earnings of students
   g) The percent of times a thumbtack lands point up when dropped from a height of 25 cm
   h) The number of letters in English words

7 - 6 Predicting

One of the principal uses of statistics is in prediction. By studying samples of voter opinions, we can forecast election outcomes with a high degree of accuracy. By studying mortality tables, we can predict with great precision the number of deaths in Canada in a given year.

Outcomes of elections cannot be controlled. We cannot say for certain what will happen. Nevertheless, sampling enables us to assess the likelihood that a particular outcome will occur.

Suppose you select a letter, at random, from any word on any page of a novel. What is the likelihood that the letter you select is an *e*? You might select 1000 letters, at random, and count the number of times *e* occurs. If there are 100 *e*'s out of 1000 letters, we say that the **relative frequency** of *e* is $\frac{100}{1000}$, or $\frac{1}{10}$. The chance that *e* would be selected is about 1 in 10.

> If an outcome, *A*, occurs *r* times in *n* repetitions of an experiment then the relative frequency of *A* is $\frac{r}{n}$.

The examples that follow show the use of sampling techniques to determine the relative frequency of an outcome. The relative frequency is then used to predict the number of times an outcome will occur in future experiments.

Example 1. The owner of an art shop decides to make and sell sheets of adhesive letters used for notices and posters. What percent of the letters should be: *e, t, i, s, a, o, n, r?*

Solution. It is necessary to find the frequencies of occurrence of the letters in the English language by examining a sufficiently large sample of prose or poetry. This sample is the first three verses of "The Tiger" by William Blake (1757–1827).

> Tiger, tiger, burning bright
> In the forests of the night,
> What immortal hand or eye
> Could frame thy fearful symmetry?
>
> In what distant deeps or skies
> Burnt the fire of thine eyes?
> On what wings dare he aspire?
> What the hand dare seize the fire?
>
> And what shoulder and what art
> Could twist the sinews of thy heart?
> And, when thy heart began to beat,
> What dread hand and what dread feet?

The frequencies with which these letters occur in the sample are shown in the table.

Letter	*e*	*t*	*i*	*s*	*a*	*o*	*n*	*r*
Tally	⦀⦀⦀⦀⦀⦀⦀⦀	⦀⦀⦀⦀⦀⦀⦀⦀	⦀⦀⦀⦀⦀	⦀⦀⦀	⦀⦀⦀⦀⦀	⦀⦀	⦀⦀⦀⦀	⦀⦀⦀⦀
Frequency	38	36	18	15	29	12	20	23
Relative Frequency	0.13	0.12	0.06	0.05	0.10	0.04	0.07	0.08

The relative frequency of each letter is obtained by dividing its frequency by the number of letters, 301. The relative frequency multiplied by 100 is the percent of all the letters the owner of the art shop makes that should be *e, t, i,* and so on. For example, 13% of the letters should be *e*'s.

Example 2. When a paper cup is tossed, there are three possible outcomes:

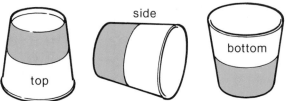

side

bottom

top

a) Find the relative frequency of each outcome after 400 tosses.

b) In 1000 tosses, approximately how many times would the cup land on the bottom?

Solution. a) Tossing the cup 400 times produced these results:

Outcome	Frequency	Relative Frequency
top	106	$\frac{106}{400} = 0.265$
side	246	$\frac{246}{400} = 0.615$
bottom	48	$\frac{48}{400} = 0.120$

b) The results of the experiment suggest that the cup lands on the bottom about 12% of the time. Therefore, we can predict that the cup will land on the bottom about 0.12×1000, or 120, times in 1000 tosses.

Exercises 7 - 6

A 1. A class of mathematics students tossed pennies a total of 319 020 times. Heads occured 160 136 times. What was the relative frequency of heads?

2. A die with faces marked:

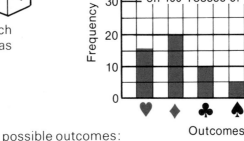

was rolled 100 times. The frequency of each outcome is shown on the graph. What was the relative frequency of

a) ♥ ?   b) ● ?   c) ♣ ?

Do you think it is a "fair" die? Explain.

3. When a thumbtack is tossed, there are two possible outcomes:

point up   point down

a) Toss 10 thumbtacks onto a desk. Record the number that land "point up", and calculate the relative frequency.

b) Repeat the experiment 9 more times, recording the number that land "point up".

c) What is the relative frequency for "point up" after tossing 100 thumbtacks?

4. a) Toss a coin the number of times indicated and record the frequency of heads.
   i) 10 times   ii) 20 times   iii) 30 times

   b) Calculate the relative frequency of heads in each case.

   c) Combine your results with those of other students to obtain the relative frequency of heads for a greater number of tosses.

   d) How does the relative frequency of heads compare with 0.5 for a greater number of tosses?

5. If, in a coin-tossing experiment, you calculated the relative frequency of heads to be 0.47, what should be the relative frequency of tails?

B 6. Choose 200 lines from a magazine story or newspaper article. Count the number of complete sentences and the number of words in each sentence. What is the relative frequency of sentences containing
   a) fewer than 9 words?   b) more than 12 words?

7. a) Toss two coins 30 times and record the number of times they show:
   i) two heads;   ii) two tails;   iii) one head, one tail.

   b) Calculate the relative frequency in each case.

c) Combine your results with those of other students to find the relative frequencies for a greater number of tosses.

d) If two coins were tossed 5000 times, about how many times should they show:
   i) two heads? ii) two tails? iii) one head, one tail?

8. A thumbtack is tossed 400 times and lands "point up" 250 times. About how many times should it land "point up" if it is tossed 5000 times?

9. a) Toss a paper cup 30 times and record the frequency of each of the three possible outcomes.

   b) Calculate the relative frequency of each outcome.

   c) Combine your results with those of other students to obtain the relative frequencies for a greater number of tosses.

   d) If a paper cup it tossed 5000 times, about how many times should each outcome occur?

C 10. In the second week of the baseball season, a player has had 9 hits out of 20 official times at bat.

   a) Calculate the player's batting average.

   b) In the next game, the player gets 0 hits out of 3 times at bat. Calculate his batting average after this game.

   c) In the final month of the baseball season, the player has had 106 hits out of 425 official times at bat. Calculate his batting average now.

   d) The player gets 0 hits out of his next 3 times at bat. What does this make his average?

   e) Why did a game with 0 hits out of 3 times at bat make less difference to the player's batting average at the end of the season than at the beginning?

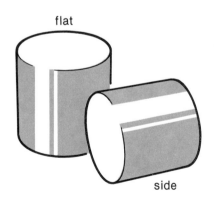
flat

side

11. When a cylinder is tossed there are two possible outcomes, From a broom handle, cut cylinders 1 cm, 2 cm, 3 cm, and 4 cm long. Record the outcome of each of 50 tosses, flat or side, for the three cylinders.

   a) What is the relative frequency of the cylinder landing flat?

   b) Make a statement of the effect of the ratio of length to diameter on the way a cylinder lands.

12. You intend to toss a coin 100 times to determine the relative frequency of heads. Investigate whether it makes a significant difference if you toss

   a) 1 coin, 100 times;   b) 2 coins, 50 times;

   c) 4 coins, 25 times;   d) 10 coins, 10 times.

7 - 7   Probability

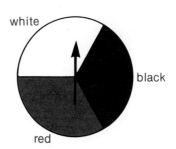

white

black

red

When spun, the pointer on this wheel may stop on white, black, or red. What is the likelihood that it will come to rest on black?

Since all three portions are the same size, there is one chance in three that the pointer will stop on black. We say that the **probability** that the pointer will stop on black is $\frac{1}{3}$, and we write:

$$P(\text{black}) = \frac{1}{3}.$$

If a die is rolled, there are six equally likely outcomes. The chances that the die will show  are 1 in 6. We say that the probability that the die will show a 2 is $\frac{1}{6}$, and we write:

$$P(2) = \frac{1}{6}.$$

Any outcome, or set of outcomes, of an experiment is called an **event.**

> If an experiment has n equally likely outcomes of which r are favorable to event A, then the probability of event A is: $P(A) = \frac{r}{n}.$

Example 1.   A lottery issues 1000 tickets. What is the probability of your winning if you hold:

   a) 1 ticket? b) 17 tickets? c) 100 tickets?

Solution.   Each ticket has an equal chance of being drawn, so

   a) $P(\text{win}) = \frac{1}{1000}$, or 0.001

   b) $P(\text{win}) = \frac{17}{1000}$, or 0.017

   c) $P(\text{win}) = \frac{100}{1000}$, or 0.1

Example 2.   A jar contains 3 black balls, 4 white balls, and 5 striped balls. If a ball is picked, at random, what is the probability that it is

   a) black?   b) white?   c) striped?

Solution. There are 12 balls and each one has an equal chance of being picked.

a) $P(\text{black}) = \frac{3}{12}$, or 0.25

b) $P(\text{white}) = \frac{4}{12}$, or approximately 0.333

c) $P(\text{striped}) = \frac{5}{12}$, or approximately 0.417

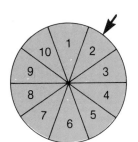

Example 3. For the wheel shown, determine the probability for each of these events:

a) Event *A:* Wheel stops on a number equal to or less than 3.

b) Event *B:* Wheel stops on a number greater than 6.

c) Event *C:* Wheel stops on an even number.

Solution. a) Event *A* has three favorable outcomes:

$$1, 2, 3. \qquad P(A) = \frac{3}{10}, \text{ or } 0.3$$

b) Event *B* has four favorable outcomes:

$$7, 8, 9, 10. \qquad P(B) = \frac{4}{10}, \text{ or } 0.4$$

c) Event *C* has five favorable outcomes:

$$2, 4, 6, 8, 10. \qquad P(C) = \frac{5}{10}, \text{ or } 0.5$$

Example 4. a) For the wheel in *Example 3,* if event *D* is that the wheel stops on a number from 1 to 10, what is $P(D)$?

b) Event *E* is that the wheel stops on a number greater than 10. What is $P(E)$?

Solution. a) Every possible outcome is favorable to event *D*. Therefore, $P(D) = \frac{10}{10}$, or 1.

b) No outcome is favorable to event *E*. Therefore, $P(E) = \frac{0}{10}$, or 0.

WIZARD OF ID by permission of Johnny Hart and Field Enterprises, Inc.

Probability and relative frequency are closely linked. The probability of an event indicates what the relative frequency should be as the experiment is performed.

For the wheel shown, we know that $P(\text{black}) = \frac{1}{3}$. This does not mean that in 30 spins the pointer will stop on black exactly 10 times. It means that as the number of spins increases, the relative frequency of the pointer stopping on black gets closer and closer to $\frac{1}{3}$.

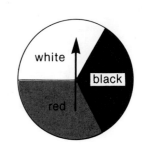

Example 5. About how many time should the pointer stop on black in 2000 spins?

Solution. Since $P(\text{black}) = \frac{1}{3}$, the spinner should stop on black about $\frac{1}{3} \times 2000$ times, or about 667 times.

Exercises 7 - 7

A 1. When a pointer is spun, what is the probability that it will stop in the colored sector?

a)

b)

c)

2. Five hundred tickets are printed for a lottery. Carla bought 7 tickets. What is the probability of her winning if

a) all the tickets were sold?

b) 370 tickets were sold and the rest destroyed?

3. A traffic light is red for 30 s, green for 25 s, and orange for 5 s in every minute. What is the probability that the light is orange when you first see it?

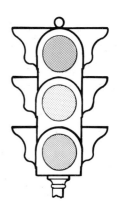

4. A pair of opposite faces of a white die are colored red. If the die is tossed, what is the probability that the top is

a) white?   b) red?

5. Calculate the probability of tossing a regular tetrahedron so that it lands with the 4 face down if the number on the faces are:

a) 2, 4, 6, 8   b) 1, 4, 4, 7   c) 1, 3, 5, 7

6. What is the probability of receiving a $3 bill in change when groceries are purchased at a store?

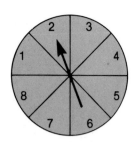

7. When the pointer on this wheel is spun, what is the probability that it will stop on
   a) an odd number?   b) an even number?
   c) a one-digit number?   d) a two-digit number?

8. What is the probability of a die, when tossed, showing
   a) 5?   b) an odd number?   c) a prime number?
   d) a number less than 3?   e) a single-digit number?

9. About how many times should a coin show heads if it is tossed
   a) 25 times?   b) 100 times?   c) 1000 times?

10. About how many times should a die show 5 if it is tossed
    a) 25 times?   b) 100 times?   c) 1000 times?

11. If each wheel is spun 50 times, about how many times should it come to rest with the arrow pointing to 2?

    a)   b)   c)   d)

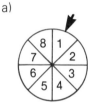 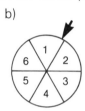  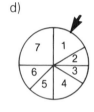

12. What is the probability that all the students in your class are older than 10 years of age?

13. Kara says that the probability that a person can cross a street safely is $\frac{1}{2}$ because there are two possible outcomes, crossing safely or not crossing safely. Do you agree? Why?

B 14. You can pick one marble from any one of the three bags and you win a prize if you pick a red one. Which bag should you choose in order to have the best chance of winning?

   Bag A   Bag B   Bag C

    contains 3 red 7 white   contains 2 red 3 white   contains 4 red 11 white

Distribution of Tiles		
A–9	J–1	S–4
B–2	K–1	T–6
C–2	L–4	U–4
D–4	M–2	V–2
E–12	N–6	W–2
F–2	O–8	X–1
G–3	P–2	Y–2
H–2	Q–1	Z–1
I–9	R–6	Blank–2

15. In the SCRABBLE® Brand Crossword Game, the letters of the alphabet are distributed over the 100 tiles as shown in the table. From a full bag of tiles, what is the probability of randomly selecting
    a) a "B"?   b) an "E"?   c) an "S"?

16. The words "Statistics and Probability" are spelled out with SCRABBLE® Brand Crossword tiles and the tiles put in a bag. What is the probability that a tile drawn from the bag at random will be

a) a vowel?
b) a consonant?
c) one of the first 10 letters of the alphabet?

Car's age years	Number
0	25
1	40
2	50
3	70
4	45
5	35
6	20
7	15

17. The table lists the number of cars in a parking lot by their ages. Calculate the probability that the age of a car selected at random will be

a) 2 years;
b) greater than 4 years;
c) less than 3 years;
d) 3 to 5 years.

18. What is the probability that a card drawn at random from a deck of 52 cards will be

a) red?
b) a spade?
c) a black 7?
d) a face card (Jack, Queen, or King)?

C 19. The probability of getting 4 heads on the toss of 4 coins is $\frac{1}{16}$.

a) What is the probability of not getting 4 heads?
b) What is the probability of getting 4 tails?

20. In a card game known as "In Between", a player is dealt two cards. To win, the player must be dealt a third card with a value in between the first two. Calculate the probability of winning if the two cards already dealt are:

a) a three and a seven;
b) a nine and a Queen,
c) a four and a ten;
d) a six and a seven.

Province	Population
Nfld.	577 400
P.E.I.	123 200
N.S.	848 500
N.B.	703 500
Que.	6 302 300
Ont.	8 517 700
Man.	1 126 400
Sask.	960 000
Alta.	2 030 800
B.C.	2 587 200
N.W.T.	43 200
Yukon	21 800
Canada	23 842 000

21. The table gives the estimated population of Canada, by province, in October 1979. What is the probability that a Canadian selected at random then lived in

a) Quebec? b) Alberta? c) the Atlantic provinces?

22. A die is loaded so that the outcomes have the relative frequencies shown in this table:

Outcome	1	2	3	4	5	6
Relative Frequency	0.12	0.17	0.17	0.08	0.35	0.11

What is the probability of throwing

a) a number less than 3?
b) an even number?

23. What is the probability that, in any year chosen at random, the month of February has
    a) exactly 28 days?   b) 29 days?
    c) 30 days?   d) more than 27 days?

24. Life insurance companies use birth and death statistics in calculating the premiums for their policies. The table shows how many of 100 000 people at age 10 are still living at ages 30, 50, 70, and 90.

Age	Number People Living
10	100 000
30	95 144
50	83 443
70	46 774
90	2 220

   a) What is the probability that a 10-year-old child will live
      i) to age 50?   ii) to age 70?   iii) to age 50 but not 70?
   b) What is the probability that a 30-year-old person will live to age 90?

7 - 8 The Probability of Successive Events

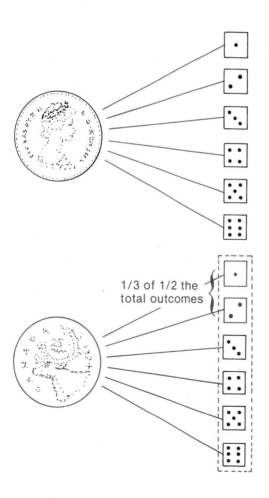

1/3 of 1/2 the total outcomes

Suppose a coin and die are tossed. What is the probability of the event:
> the coin shows tails *and*
> the die shows a number less than 3?

The **tree diagram** at the left shows all the possible outcomes of the tosses.

Event *A* — tails and a number less than 3 — has two favorable outcomes:

Tail, $\boxed{\,\cdot\,}$; and Tail, $\boxed{\cdot\,\cdot}$

These outcomes are shown in color.

$$P(A) = \frac{2}{12} \begin{array}{l}\text{favorable outcomes}\\ \text{total outcomes}\end{array}$$

$$= \frac{1}{6}$$

Another way of finding $P(A)$ is to notice that the number of outcomes involving "tails" is $\frac{1}{2}$ the total number of outcomes. The outcomes for tails *and* the die showing a number less than 3 are $\frac{1}{3}$ of $\frac{1}{2}$ the total outcomes.

$$P(A) = \frac{1}{3} \times \frac{1}{2}, \text{ or } \frac{1}{6}$$

The probability of two or more events happening in succession is the product of the probabilities of each event.

Example 1.   A coin is tossed three times. What is the probability that it shows a head each time?

Solution.   For each toss, $P(\text{head}) = \frac{1}{2}$

$$P(3 \text{ heads}) = \frac{1}{2} \times \frac{1}{2} \times \frac{1}{2}$$

$$= \frac{1}{8}$$

The rule for successive events must be used carefully, because the probability of the second event may depend on whether the first event occurred.

Example 2.   A bag contains 2 black balls and 2 red balls. Find the probability of drawing 2 red balls in succession if

a)   the first ball is replaced before drawing the second;

b)   the first ball is not replaced.

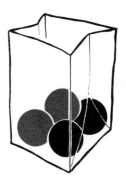

Solution.   a)   First draw: There are 4 balls, of which 2 are red.

$$P(\text{red, on first draw}) = \frac{2}{4}, \text{ or } \frac{1}{2}$$

Second draw: There are 4 balls, of which 2 are red.

$$P(\text{red, on second draw}) = \frac{1}{2}$$

Then, $P(2 \text{ reds in succession}) = \frac{1}{2} \times \frac{1}{2}$

$$= \frac{1}{4}$$

b) $P(\text{red, on first draw}) = \frac{1}{2}$

Second draw: There are 3 balls, of which 1 is red.

$$P(\text{red, on second draw}) = \frac{1}{3}$$

Then, $P(2 \text{ reds in succession}) = \frac{1}{2} \times \frac{1}{3}$

$$= \frac{1}{6}$$

Exercises 7 - 8

A 1. A coin and a regular tetrahedron (with faces marked 1, 2, 3, 4) are tossed. Draw a tree diagram to find the probability of getting
   a) a head and a 1;
   b) a tail and an even number.

2. If it is equally likely that a child be born a girl or a boy, what is the probability that
   a) a family of two children will be both girls?
   b) a family of 5 children will be all boys?

3. What is the probability of tossing a coin five times and getting tails each time?

4. A True-False test has 6 questions. If all the questions are attempted by guessing, what is the probability of getting all 6 right?

5. What is the probability of rolling three consecutive 6's with one die?

B 6. If a thumbtack is tossed, the probability that it lands with the point up is 0.6. What is the probability of tossing a thumbtack four times and having it land with the point up each time?

7. Two people are selected at random. What is the probability that they both have birthdays in September?

8. A box of 100 flash cubes contains 3 defective ones. If 2 cubes are taken from the box, what is the probability that both are defective?

9. Three bags contain black balls and red balls in the numbers shown.

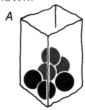
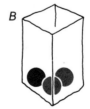

For each bag, what is the probability of drawing 2 red balls in succession if
   a) the first ball is replaced before the second is drawn?
   b) the first ball is not replaced?

10. What is the probability of drawing 3 red balls in succession from a bag containing 3 red balls and 3 black balls if
    a) each ball is replaced before the next is drawn?
    b) the balls are not replaced after drawing?

11. A card is drawn from each of two well-shuffled decks. What is the probability of drawing two cards that are both
    a) spades?   b) red?
    c) aces?   d) the ace of spades?

12. What is the probability of drawing 4 aces from a deck of cards
    a) if there is replacement and shuffling after each draw?
    b) if there is no replacement of the cards drawn?

13. Mrs. Hansoti makes equal numbers of white and brown sandwiches. Half of each kind she makes with margarine and the other half she makes with butter. Each of the combinations she fills with cucumber, or lettuce and tomato, or egg in equal numbers.
    a) Copy and complete this tree diagram.

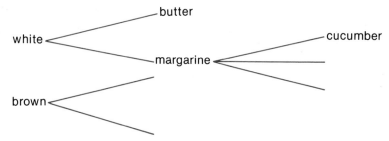

    b) What is the probability that a sandwich, picked at random,
        i)   will be brown?
        ii) will be made with butter?
        iii) will have cucumber?
        iv) will be white with egg?

C 14. A die and two coins are tossed. What is the probability of getting
    a) a 2 and two heads?
    b) a head, a tail, and an odd number?

15. Suppose a die is tossed until a 6 appears. What is the probability that the throw on which it appears is
    a) the second?
    b) the third?
    c) the tenth?

THE MATHEMATICAL MIND
GAMES OF CHANCE

We know from dice found in the tombs of Ancient Greeks and Egyptians that games of chance have been played for thousands of years. However, it was not until the sixteenth and seventeenth centuries that a serious attempt was made to study games of chance using mathematics.

Chevalier de Méré, a professional gambler and amateur mathematician, had

Blaise Pascal 1623-1662

many questions about dice probabilities. He turned to the great mathematician, Blaise Pascal, for the answers. In order to answer, Pascal turned to his friend, Pierre de Fermat, and together they began a systematic study of games of chance. The theory of probability was founded.

One of de Méré's questions was: "What is the probability of throwing two dice and *not* getting a 1 or 6?"

Pascal answered: "For each die, the probability is $\frac{4}{6}$, or $\frac{2}{3}$. For both, the probability is $\frac{2}{3} \times \frac{2}{3}$, or $\frac{4}{9}$ — about 0.44."

With this information, de Méré offered the equivalent of this gamble:

Bet $1.
Throw 2 dice.
If ⚀ or ⚅
do NOT show,
you win $2.

He now knew that for every 100 people who played the game, about 44 would win. That meant he would take in $100 and pay out $88. He could expect to win about $12 every time 100 people played. Now that you know this, would you spend $1 to play this game?

Although the theory of probability began with the study of gambling, many more-important applications have since been found. Weather forecasting, genetics, insurance premiums all involve probability. The theory of probability is also needed in statistics to estimate the chance of error when samples are taken.

For each of the following games of chance.

a) determine whether you can expect to win, or lose, money if you play the game a great number of times;

b) decide whether you are willing to play the game;

c) explain your decision.

1. Bet $1.

   Toss a coin.

   If it shows you win $2.

2. Bet $50.

   Toss a coin.

   If it shows you win $100.

3. Bet $1.

   Draw a card from a well-shuffled deck.
   If it shows a spade, you win $5.

4. Bet $1.

   Draw a card from a well-shuffled deck.

   If it shows , or , or , or 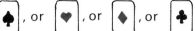 you win $10.

5. Bet $1.

   Toss two coins.

   If they show you win $3.

6. Bet $3.

   Throw 1 die.
   Win the amount shown, in dollars.

7. Bet $1.

   Throw 2 dice.

   If they show you win $25.

8. Bet $1.

   Throw 2 dice.

   If does *not* show, win $2.

In a lottery when you buy a ticket, you are betting that the number on your ticket will be drawn. In each of the following, assume that the ticket costs $5, and,

a) determine whether you can expect to win, or lose, money if you bought great numbers of tickets;

b) decide whether you are willing to buy a ticket;

c) explain your decision.

1. Tickets with a number ending in 6 win $1.

2. Tickets with a number ending in 37 win $10.

3. Tickets with a number ending in 008 win $100.

4. Ticket number 9 480 913 wins $1 000 000.

7 - 9 Strategy Games

Games such as tic-tac-toe, checkers, and backgammon are called **strategy games**. A player's strategy depends on his knowledge of the game, and his strategy may change during the game depending on what his opponent does.

Show-A-Card

Alan and Brenda are playing the game Show-A-Card. Alan has the cards: 2♣ and 3♥ . Brenda has: 3♠ and 4♦ .

Each player shows one card at the same time. If the colors match, Alan wins; if the colors do not match, Brenda wins. The amount won is the difference in the numbers on the cards, in cents.

The possible plays and payoffs can be summarized in a **payoff table**:

Signs indicate payments from Brenda's point of view, that is:

−1 means that Brenda loses 1¢.

+2 means that Brenda wins 2¢;

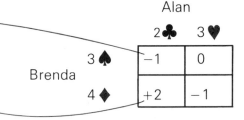

	Alan	
	2♣	3♥
Brenda 3♠	−1	0
4♦	+2	−1

A reasonable strategy for each player is to play each card about half the time, but not according to any pattern. A coin could be tossed, without the other player seeing the result, and one card could be played for heads and the other card played for tails. Thus, each card is played with probability $\frac{1}{2}$.

This strategy means that the probability of each outcome is $\frac{1}{2} \times \frac{1}{2}$, or $\frac{1}{4}$. If, say, 100 games were played, 2♣ 3♥ and 3♠ 4♦ should each come up about $\frac{1}{4} \times 100$, or 25, times, giving Alan a total payoff of 50¢.

But 2♣ 4♦ should also come up about 25 times, giving Brenda 2¢ each time for a total of 50¢, as well. This appears to be a fair game because each player can expect to win the same amount in 100 games.

However, a player wants to win, not just break even. If Alan thinks Brenda is playing with the probability $\frac{1}{2}$ strategy, he should play 2♣ less often and 3♥ more often. The next example shows how and why.

Example 1. In 100 games of Show-A-Card, Alan plays 2♣ with probability $\frac{1}{4}$ and 3♥ with probability $\frac{3}{4}$, while Brenda plays each card with probability $\frac{1}{2}$.

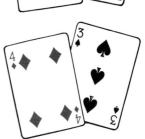

a) How does Alan play with these probabilities?

b) What are the expected payoffs?

Solution. a) Alan tosses two coins. If they both show heads, he plays 2♣. Otherwise, he plays 3♥

b) In each part of the table below,

   ① is the probability of that combination of cards,

   ② is the approximate number of times that the combination should occur in the 100 games,

   ③ is the expected payoff.

Alan

	plays 2♣ with probability $\frac{1}{4}$,	plays 3♥ with probability $\frac{3}{4}$.
plays 3♠ with probability $\frac{1}{2}$	① $\frac{1}{2} \times \frac{1}{4} = \frac{1}{8}$ ② $\frac{1}{8} \times 100 = 12.5$ ③ $12.5(-1) = -12.5$	① $\frac{1}{2} \times \frac{3}{4} = \frac{3}{8}$ ② $\frac{3}{8} \times 100 = 37.5$ ③ $37.5(0) = 0$
plays 4♦ with probability $\frac{1}{2}$	① $\frac{1}{2} \times \frac{1}{4} = \frac{1}{8}$ ② $\frac{1}{8} \times 100 = 12.5$ ③ $12.5(+2) = +25$	① $\frac{1}{2} \times \frac{3}{4} = \frac{3}{8}$ ② $\frac{3}{8} \times 100 = 37.5$ ③ $37.5(-1) = -37.5$

Brenda (label at left of the two rows)

Brenda can expect to win about 25¢, but Alan can expect to win about 12.5¢ + 37.5¢, or 50¢. The game is now in Alan's favor.

Example 2. Would Alan win more by playing 3♥ all the time?

Solution. Brenda would notice if Alan did this and would play 3♠ all the time. There would be no payoff. Therefore, Alan should not play 3♥ all the time.

Exercises 7 - 9

A 1. While playing Show-A-Card, Brenda notices that Alan is playing each card, 2♣ and 3♥ , about half the time. What strategy should she use?

B 2. In a game of Show-A-Card, Anne has the cards 4♠ and 3♥ , and Bert has 5♠ and 5♦ . If the colors match, Anne wins; otherwise Bert wins. The amount won is the difference of the numbers on the cards, in cents.

a) Show the payoff table.

b) What are the expected payoffs if the game is played 100 times with these probabilities:

i) Both players play each card with probability $\frac{1}{2}$.

ii) Anne plays 4♠ with probability $\frac{1}{3}$ and 3♥ with probability $\frac{2}{3}$. Bert continues with probability $\frac{1}{2}$.

iii) Both players play their black card with probability $\frac{1}{3}$ and their red card with probability $\frac{2}{3}$.

3. Repeat Exercise 2 changing the amount won to the sum of the numbers on the cards.

C 4. Helen has the cards 3♦ , 5♣ , and 7♥ . Ronald has 2♦ and 8♠ . If the colors match, Helen wins; otherwise Ronald wins. The amount won is the difference of the amounts on the cards, in cents.

a) Show the payoff table.

b) What are the expected payoffs if the game is played 100 times with these probabilities:

i) Helen plays each card with probability $\frac{1}{3}$, Ronald plays each card with probability $\frac{1}{2}$.

ii) Helen plays 5♣ with probability $\frac{1}{2}$, and 3♦ and 7♥ each with probability $\frac{1}{4}$. Ronald plays 2♦ with probability $\frac{1}{3}$ and 8♠ with probability $\frac{2}{3}$.

5. In the game described in *Example 1* in which Alan plays 2♣ with probability $\frac{1}{4}$ and 3♥ with probability $\frac{3}{4}$, is there any strategy Brenda could use to avoid losing? Do you think this game is fair?

Showing Fingers

In this strategy game, two players simultaneously show one or two fingers. It has been agreed beforehand that if the number of fingers shown is the same, player A wins; if the number of fingers is not the same, player B wins.

B 6. Corinne and John play Showing Fingers. Corinne wins 1¢ if they show the same number of fingers; otherwise John wins 1¢.

   a) Show the payoff table.

   b) What are the expected payoffs if the game is played 100 times with these probabilities.

   i) Both players show one or two fingers with probability $\frac{1}{2}$.

   ii) Corinne shows fingers with probability $\frac{1}{2}$. John shows one finger with probability $\frac{1}{4}$ and two fingers with probability $\frac{3}{4}$.

   iii) One finger is shown with probability $\frac{1}{4}$ by Corinne and $\frac{3}{4}$ by John. Two fingers are shown with probability $\frac{3}{4}$ by Corinne and $\frac{1}{4}$ by John.

   7. Repeat Exercise 6 but let the amount won, in cents, be

   a) the sum of the number of fingers shown;

   b) the product of the number of fingers shown.

C 8. If, in 100 games of Showing Fingers, one player always plays with probability $\frac{1}{2}$, is there any strategy by which the other player can probably win? Do you think this game is fair?

Matching Pennies

In this game, each of two players places a penny to show heads or tails. If the pennies match, both heads (*HH*) or both tails (*TT*), player A wins; otherwise player B wins.

B 9. Jeanne and Jacques play Matching Pennies 100 times. If the pennies match, Jeanne wins 1¢; if they don't match, Jacques wins 1¢.

   a) Show the payoff table.

   b) What are the expected payoffs with these probabilities:

   i) Both players play H and T with probability $\frac{1}{2}$.

ii) Jacques plays *H* with probability $\frac{1}{3}$ and *T* with probability $\frac{2}{3}$, while Jeanne continues to play with probability $\frac{1}{2}$.

iii) Both players play *H* with probability $\frac{1}{3}$ and *T* with probability $\frac{2}{3}$.

10. Repeat Exercise 9 but with the payoffs changed to the following:

   If the pennies show *HH*, Jeanne wins 2¢. If they show *HT*, Jacques wins 1¢. No one wins if the pennies show *TT*.

C 11. If, in 100 games of Matching Pennies, one player always plays *H* with probability $\frac{1}{4}$ and *T* with probability $\frac{3}{4}$, is there any strategy by which the other player can probably win? Do you think this game is fair?

Review Exercises

Country	Defence Spending Percent of G.N.P.
Canada	1.8
U.S.A.	5.0
Denmark	2.4
France	3.3
W. Germany	3.4
Norway	3.2
Britain	4.7

1. The table lists the defence spending of each of the nations of the North Atlantic Treaty Organization (NATO) as a percent of the gross national product (g.n.p.) as determined by the International Institute for Strategic Studies. Show this information on a suitable graph.

2. Draw a circle graph to show the information given in the diagram.

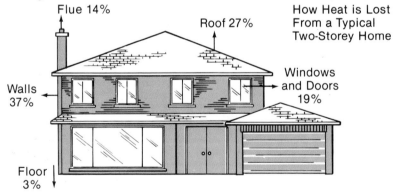

Flue 14% Roof 27% How Heat is Lost From a Typical Two-Storey Home Windows and Doors 19% Walls 37% Floor 3%

3. The workers in a small factory receive these salaries:

$10 000	10 000	10 400	10 800	13 200
10 800	11 200	12 000	12 000	12 400
12 400	12 400	12 400	12 800	10 800
14 000	14 000	14 400	14 800	14 800
15 200	16 000	15 600	15 200	16 000

Display the data on a frequency graph.

4. Calculate the measures of central tendency for the salaries given in Exercise 3.

5. If the mean of the numbers 9, 10, 21, 27, 29, 25, 19, 13, x is 21, what is x?

6. Write nine natural numbers that have a median of 25 and a mean of 21.

7. Comment on the following:

   a) A saleswoman asked for an increase in salary and used this graph to show she deserved one.

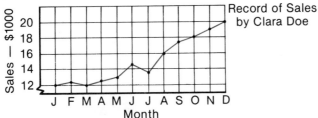

   Record of Sales by Clara Doe

   b)

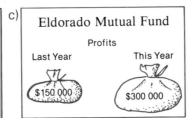

   Supreme TV Repairs
   - Our service personnel have 100 years experience.
   - All parts guaranteed
   - No one has lower rates
   - Same day service

   c) Eldorado Mutual Fund

   Profits

   Last Year   This Year

   $150 000 $300 000

   d) **Gilead Balm**   100 g

   for relief of: headache, toothache, heartache neuralgia, rheumatism, athlete's foot, and other ailments
   Recognized by all doctors

8. How would you collect data to determine the following:

   a) the extent of mercury poisoning in fish in the Great Lakes

   b) the political party most likely to win the next provincial election

   c) the food-purchasing habits of single males

   d) the force required to break a certain gauge of fishing line

9. Shake 5 coins in a paper cup and empty them onto your desk. Record the frequency of heads. Repeat this procedure 24 more times. From your results, if you did this a total of 300 times, with what frequency would you expect 5 coins to show

   a) 3 heads?   b) 4 heads?   c) no heads?

10. A manufacturer of widgets has maintained a minimum standard of 95% dependability over the years.

a) 3 widgets in a batch of 75 are found to be defective. Does the batch meet the minimum standard?

b) How many defective widgets are permissible in a batch of 250?

c) If workmanship and materials are improved so that only 4 defective widgets are being found in every 250, what is the probability that a widget selected at random is not defective?

11. What is the probability that the number on a ball selected at random from 15 balls numbered from 1 to 15 is

a) even?   b) prime?   c) a multiple of 5?

d) a two-digit number?

Heights of Students at Montcalm Secondary School

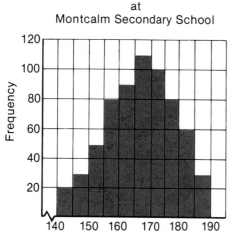

12. The graph shows the frequency distribution of the heights of students at Montcalm Secondary School. What is the probability that a student, selected at random, will be

a) between 150 cm and 165 cm tall?

b) taller than 175? c) shorter than 155 cm?

13. A box of coins contains 36 quarters, 45 dimes, 25 nickels, and 62 pennies. What is the probability that a coin drawn at random will be

a) a quarter?   b) a nickel or a dime?

c) a quarter or a nickel?   d) other than a penny?

14. The faces of two regular tetrahedrons are numbered 1 to 4. If they are tossed, what is the probability of getting

a) a pair?   b) a total of 5?   c) a difference of 1?

15. An aviary has parakeets of four different colors. There are 10 green parakeets, 7 blue, 2 yellow, and 1 white. If two birds escape, what is the probability that they will both be

a) green?   b) blue?   c) yellow?   d) white?

16. A cafeteria offers a number of choices for lunch:
    3 appetizers: soup, juice, or salad
    4 main courses: beef, chicken, pork, or fish
    2 desserts: pie or ice cream
If a three-course meal is selected at random, calculate the probability of getting:

a) soup;   b) soup and beef;   c) juice, fish, and pie.

Mathematics Around Us

Coin-Tossing Computers?

Well, of course, there are no coins inside a computer to be tossed. What a computer can do is *simulate* the tossing of a coin.

A computer can be programmed to generate a very large set of single-digit numbers in which each number has an equal chance of occuring each time. Such a set of numbers is called a set of **random numbers**. If we let "even" numbers correspond to heads and "odd" numbers correspond to tails, the computer has fairly tossed a coin for us.

How can a computer be programmed to generate a set of random numbers? One way is to program the computer to square a 6- or 8- digit number, print out the middle six or eight digits, square them, and again print out the middle digits, and keep on doing this. If we start with the number 123 456, the first twenty-four random numbers would be generated as follows:

	Result	Print Out
123 456² =	15 [241 383] 936	241 383
241 383² =	58 [265 752] 689	265 752
265 752² =	70 [624 125] 504	624 125
624 125² =	389 [532 015] 625	532 015

Number of Random Numbers	Frequency		
	"even" Coin shows *H*	6 Die shows	1 2 coins show *HH*
10	3	1	3
100	49	19	31
1 000	515	164	256
10 000	5 047	1 694	2 480
100 000	50 067	16 858	25 045

The second column in this table shows how often a computer tosses a head (an even number) in the first 10 numbers, in the first 100 numbers, 1000 numbers, 10 000 numbers, 100 000 numbers of a typical set of random numbers.

By programming the computer to disregard 0, 7, 8, 9 whenever they occur in the generation of a set of random numbers, the throw of a die can be simulated. The third column of the table shows how often a 6 occurs in the throw of a die by this method.

The result of tossing 2 coins can be simulated by programming the computer to disregard 0, 5, 6, 7, 8, 9 in its generation of a set of random numbers. The fourth column of the table shows how often 1 occurs in such a program and this can be taken as being equivalent to both coins showing heads.

Random Numbers From a Calculator

This same method of generating random numbers can be used with a calculator if the size of the number is restricted to 4 digits.

Random numbers can also be obtained from a telephone book or by means of playing cards.

Start with 4176:

$$4176^2 = 17\boxed{438\ 9}76$$
$$4389^2 = 19\boxed{263\ 3}21$$
$$2633^2 = 6\boxed{932\ 6}89$$
$$9326^2 = 86\boxed{974\ 2}76$$

Continue many times.

200 Random Numbers

4389	8533	6553	5534	4770
2633	8120	9418	6251	7529
9326	9344	6987	0750	6858
9742	3103	8181	5625	0321
9065	6286	9287	6406	1030
1742	5137	2483	0368	0609
0345	3887	1652	1354	3708
1190	1087	7291	8333	7492
4161	1815	1586	4388	1300
3139	2942	5153	2545	6900

Questions

1. Determine the relative frequencies of 0, 1, 2, 3, ... 9 in the table of 200 random numbers above.

2. By starting anywhere in the table of random numbers, simulate

   a) 20 tosses of a coin;

   b) 20 throws of a die;

   c) 20 tosses of two coins;

   d) 20 tosses of three coins.

3. a) Choose any 4-digit number and use the method described above to obtain your own table of 200 random numbers.

   b) Repeat Exercise 2 using your own table.

c) Combine your results with those of other students to simulate a larger number of tosses.

4. Certain patterns may appear when the squaring method is used to generate random numbers. To see how this can happen, consider the third last entry in the table of 200 random numbers, 7492.

   a) Simplify 7492^2 and confirm that the next two entries are 1300 and 6900.

   b) Attempt to continue this process to obtain more random numbers, and observe the result.

8 Polynomials

The cost, in dollars, of making n tennis rackets is $6n + 2000$. The income, in dollars, from sales is $15n$. If all rackets produced are sold, how many must be made and sold to earn a profit of $34\,000$. (See *Example 6* of Section 8-3.)

8 - 1 Simplifying Polynomials

In Chapter 1, we learned that the simplest algebraic expressions are called **terms.** A term is either a number, called a **constant term,** or the product of a number and one or more variables.

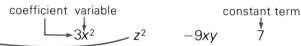

coefficient variable   constant term

$3x^2$ z^2 $-9xy$ 7

An expression with just one term is called a **monomial.** A **polynomial** is the sum or difference of two or more terms in which the variable appears to positive integral powers only. A polynomial of two terms is called a **binomial.** Examples are:

Monomials: x^2, $-3y^3$, $7xy^2$

Binomials: $3x + 7$, $2y^2 - y$, $\frac{1}{2}z^3 - 4zy$

Polynomials: $x^2 - 7x + 12$,
$a^2 + ab + b^2 + 3a + 4b - 12$

These are not polynomials: $5 + \frac{2}{x}$, $\sqrt{x} + 3$,
$3a^{-3} + 2a^{-2} + a^{-1}$

Terms having the same variable to the same power can be combined.

$3x + 7x = 10x$ $7y^2 - 4y^2 = 3y^2$ $6x^2y - 2x^2y = 4x^2y$

Polynomials can be simplified by combining like terms.

Example 1. Simplify: $3x^2 - 2x + 4 + 3x + 1$

Solution. To show the procedure, we align like terms vertically and then add.

$$\begin{array}{r} 3x^2 - 2x + 4 \\ + 3x + 1 \\ \hline 3x^2 + x + 5 \end{array}$$

When the terms of a polynomial are arranged in order from the highest to the lowest powers of the variable, the polynomial is said to be in descending powers of the variable.

These polynomials are in descending powers.

$x^3 - 3x^2 + 4x - 5$

$4x^2 - 9$

$5x^4 - 2x^2 + 11$

These are not in descending powers.

$x^2 + 4 + 2x$

$1 - x^2$

$x^4 + 7x + 2x^3$

coefficient of
1 understood

Example 2. Simplify, and write in descending powers of y:
$$3y - 2y^3 + 5y - 7 + y^2 + y^3$$

Solution. Arrange like terms vertically and add:
$$3y - 2y^3 - 7 + y^2$$
$$\underline{5y + y^3}$$
$$8y - y^3 - 7 + y^2$$
Rearrange in descending powers of y:
$$-y^3 + y^2 + 8y - 7$$

The term with the greatest exponent, or exponent sum, determines the **degree** of a polynomial.

$x^2 + 3x^3 + 2x - 7$ is a third-degree polynomial.

$y^4 + 2y^3x^2 - 5$ is a fifth-degree polynomial. $(3 + 2 = 5)$

$xy + 4x$ is a second-degree polynomial. $(1 + 1 = 2)$

$x + y - 3$ is a first-degree polynomial.

Example 3. Simplify, and state the degree of each polynomial:

a) $-3x^2 + x^3 - 2x + 7 + 5x^2 + 2x^3 - 8x + 3$

b) $7x - 5x^4 + 2y^2 + 3xy^4 + 5x^3 - 6y^2 + 8$

Solution. a) $-3x^2 + x^3 - 2x + 7 + 5x^2 + 2x^3 - 8x + 3$
$$= 3x^3 + 2x^2 - 10x + 10$$
It is a third-degree polynomial.

b) $7x - 5x^4 + 2y^2 + 3xy^4 + 5x^3 - 6y^2 + 8$
$$= 3xy^4 - 5x^4 + 5x^3 - 4y^2 + 7x + 8$$
It is a fifth-degree polynomial.

Exercises 8 - 1

A   1. State the coefficient in each of these terms.
   a) $14x$ b) $7y^2$ c) a d) $-b^2$ e) $-4c^3$

2. What is the variable in each of these terms?
   a) $5t$ b) $3x^2$ c) $2w^2$ d) $6z^5$ e) 6

3. State whether the terms are like or unlike.
   a) $2a, 3a$ b) $5x, x$ c) $4m, n$
   d) $9c, -6c$ e) $7x^2, 3x^2, -x^2$ f) $8a^2, 8a$
   g) $-2a^2, -4b^2$ h) $4t^3, 2t^3, 3t$ i) $2.3b, 6.9b$
   j) $5, \pi, \sqrt{2}$ k) $\frac{3}{4}x, -\frac{2}{3}x, \frac{1}{2}x$ l) $\sqrt{2}n, \frac{3}{2}n$

4. Which are the like terms in each of the following sets?

   a) $5a$, $3b$, $5c$, a^2, $-a$, $3d$, $3e$

   b) $4x$, $3y^2$, $4z$, $2y$, y^2, $4w$

   c) $9g$, $6h$, $9g^2$, $\frac{1}{9}g$, $\frac{1}{6}h^2$, g^2

   d) 16, d^2, d, f, -8, $0.5d$, $7d^3$

   e) $3q^3$, $17q^2$, $-3t^3$, $6q$, $3t^2$, $-15q^2$

5. Simplify:

   a) $(+3m) + (+11m)$

   b) $(-5k) + (+3k)$

   c) $(-2x) + (+7x)$

   d) $(+2n) + (-6n) + (+n)$

   e) $(-3a) + (+4a) + (-5a)$

   f) $(-2x^2) + (+9x^2) + (-x^2)$

   g) $(-y^3) + (-5y^3) + (+2y^3)$

   h) $(-3c) + (-5c)$

   i) $(2.7m) + (-6.9m) + (-5.2m)$

   j) $(\frac{1}{2}t^2) + (-t^2) + (\frac{1}{2}t^2)$

   k) $(\frac{3}{4}x) + (\frac{2}{3}x) + (\frac{1}{2}x)$

6. Simplify:

   a) $3t + 5t$

   b) $4a - 3a + 2a$

   c) $-14m + 5m - m$

   d) $-5s^2 + s^2 + 9s^2$

   e) $12k^3 - 15k^3 + 2k^3$

   f) $-5c^4 - 3c^4 + 7c^4$

   g) $\frac{1}{2}x + \frac{1}{3}x$

   h) $0.5r + 0.7r$

   i) $\frac{3}{4}a^2 - \frac{2}{5}a^2$

   j) $8.2h - 11.2h$

   k) $\frac{3}{2}c^2 - \frac{11}{3}c^2$

   l) $1.3n - 1.7n + 0.8n$

7. Simplify:

   a) $(+6x) - (+2x)$

   b) $(+3a) - (+7a)$

   c) $(+5p) - (-3p)$

   d) $(-n^2) - (+5n^2)$

   e) $(-7x^3) - (-2x^3)$

   f) $(-y) - (-8y)$

   g) $(+4.5c) - (+5.3c)$

   h) $(+2.9x) - (+9.7x)$

   i) $(+2.5b^2) - (-3.5b^2)$

   j) $(+\frac{1}{3}a) - (-\frac{1}{2}a)$

   k) $(-\frac{1}{4}m) - (+\frac{3}{4}m)$

   l) $(+\frac{2}{5}y^2) - (+\frac{2}{5}y^2)$

8. Simplify:

   a) $8x + 2x - 3x$

   b) $-5a - a - 2a$

   c) $-3p^2 + p^2 + 5p^2$

   d) $-2y^2 - 3y + y$

   e) $5m^2 - 2m - 3m$

   f) $7x^3 + 5x^3 - 4x^2$

   g) $-3a^2 + 8a^2 + a^2$

   h) $-x^2 - 2x - 3x^2$

   i) $2.5c - 3.2c^2 + 1.7c$

   j) $\frac{1}{5}m + \frac{1}{2}m + m^2$

9. Simplify:

 a) $32m^2 - 15m - 7m$ b) $-65x^2 + 37x - 27x$

 c) $38c^2 + 45c - 20c$ d) $-18n - 24n + 20n^2$

 e) $6.3x^2 - 9.7x^2 + 2.5x^2$ f) $-4.7x^3 - 3.9x^3 + 11.7x^2$

 g) $-\frac{1}{4}c + \frac{1}{3}c - \frac{1}{2}c$ h) $\frac{1}{5}a^2 - \frac{1}{2}a - \frac{1}{3}a$

10. Simplify, write in descending powers of the variable, and state the degree of these expressions:

 a) $3a + 2 + 5a + 7$ b) $2x - 6 + 8x + 4$

 c) $5n + 1 - 9n + 2$ d) $-6c + 3 - c - 5$

 e) $7x - 2 - 3x - 1$ f) $3x^2 - 5x + x^2 - 2x$

 g) $-4a^2 - 3a - a^2 + 2a$ h) $-m - 2 + 3m - 1$

 i) $x^2 - 3x - 4x + 12$

 j) $5a^3 - 2a + 3a^2 - a^3 + a^2 + 3a$

11. Simplify where possible:

 a) $3a^2 - 2a - a^2 + a$ b) $5x + 7$

 c) $5x^2 + 3x$ d) $8m^2 - 3 - 2m^2 + 4$

 e) $-c - 3 + 2c - 5$ f) $3a^2 - 3a - 3$

 g) $-2 + 5x - x + 4$ h) $-\frac{1}{2}y^2 + \frac{2}{3}y + \frac{1}{4}y^2 - \frac{1}{6}y$

 i) $-n^2 - 2n + 4$ j) $2.5x^2 - 5.2x^2 + x$

12. Which are like terms in the following sets:

 a) $2x, \quad x^2, \quad xy, \quad 2y, \quad 3x^2y, \quad 2xy, \quad -5x^2y$

 b) $3a, \quad 4ab, \quad 5ab^2, \quad -7a, \quad 3b, \quad -5ab, \quad 8$

 c) $5ax, \quad 4a, \quad 7x, \quad 3ax, \quad -7a, \quad 6xa, \quad 7ax^2$

 d) $3a^2b, \quad -5ab, \quad 9ab^2, \quad 6ba, \quad -3a^2b$

13. Simplify:

 a) $2xy - 3x^2y + 2xy^2 - 3yx + 5xy^2 - 2yx^2$

 b) $4xy^2 - 3xy + 2yx^2 - 4y^2x^2 + 2yx - 4y^2x + 3$

 c) $9y^2a - 2a^2 + a^2y - 7ay + 3a^2y - 8ay^2$

 d) $9w + 4z - 3wz + 2w^2z - 8zw^2 + 2w - 3z$

14. Simplify:

 a) $5a^2b + 10ab^2 + 12a^2b^2 + 6ab^2 - 12a^2b - 8a^2b^2 + 7a^2b$

 b) $9xy^2 + 6xy - 8xy^2 - 8x^2y + 6x^2y^2 - 6xy$

 c) $3x + 4y - 5xy + 3y + 5yx - 4x$

 d) $8x^2y - 6xy^2 - 7yx^2 + 6y^2x - x^2y^2$

8 - 2 Evaluating Polynomials

Many formulas in mathematics, science, and engineering involve polynomials. To apply these formulas, we often evaluate the polynomial for any value of the variable.

Example 1. The sum, S, of the first n even numbers is given by the formula: $S = n^2 + n$. What is the sum of the first 30 even numbers?

Solution. When $n = 30$, $n^2 + n = (30)^2 + 30$
$$= 900 + 30, \text{ or } 930$$
The sum of the first 30 even numbers is 930.

Example 2. The number of diagonals, d, in a polygon of n sides is given by the formula: $d = \frac{1}{2}n^2 - \frac{3}{2}n$. How many diagonals has an octagon?

Solution. For an octagon, $n = 8$.
When $n = 8$, $d = \frac{1}{2}(8)^2 - \frac{3}{2}(8)$
$$= 32 - 12, \text{ or } 20$$

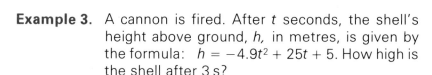

Example 3. A cannon is fired. After t seconds, the shell's height above ground, h, in metres, is given by the formula: $h = -4.9t^2 + 25t + 5$. How high is the shell after 3 s?

Solution. When $t = 3$, $h = -4.9(3)^2 + 25(3) + 5$
$$= -44.1 + 75 + 5, \text{ or } 35.9$$
The shell is 35.9 m above the ground after 3 s.

Exercises 8 - 2

A

1. Evaluate $n^2 - 3n + 4$ for the following values of n:

   a) 1   b) 3   c) 4   d) −1   e) $\frac{1}{2}$ f) 1.5

2. Evaluate $2x^2 - 3x - 5$ for the following values of x:

   a) 0   b) 1   c) −1   d) 7   e) −4   f) −2.5

3. Find the value of each polynomial when $a = 8$.

   a) $a^2 + 5a + 6$ b) $2a^2 - \frac{1}{2}a + 3$

   c) $a^2 - 25$ d) $1.5a^2 - 0.4a + 3.6$

4. Find the value of each polynomial when $x = -5$:

   a) $4x^2 - 7x + 3$ b) $-3x^2 + 2x + 5$

   c) $6 - 3x - x^2$ d) $2.9 - 0.5x + 3.6x^2$

5. For each polynomial:
   i) Write it in descending powers of k;
   ii) State the coefficients and constant term;
   iii) Evaluate the polynomial when $k = 1$.

   a) $-3 + 7k^2 + 2k$ b) $3k^2 + 6 - 4k^3$

   c) $2.5k - 8.4k^2$ d) $\frac{1}{4}k + \frac{1}{2}k^2 + \frac{1}{3}$

6. The temperature, T, in an oven, calibrated in degrees Celsius, n minutes after it has been turned on is given by the formula: $T = -n^2 + 30n + 20$ (for $n < 14$). Find the temperature in the oven, in degrees Celsius,

   a) 7 min after being turned on;

   b) 10 min after being turned on.

7. The number of accidents in one month, n, involving drivers x years of age is given by the formula: $n = 3x^2 - 200x + 5000$. Find the number of accidents in one month involving drivers

   a) 18 years old;   b) 25 years old.

8. Use the formula in *Example 1* to find the sum of

   a) the first 50 even numbers; b) the first 100 even numbers.

9. Use the formula in *Example 2* to find how many diagonals

   a) a pentagon has;   b) a hexagon has.

10. The stopping distance, d, in metres, for a car travelling at v km/h is given by the formula: $d = 0.20v + 0.015v^2$. Find the stopping distance for a speed of 50 km/h; of 100 km/h.

11. On July 20, 1969, Neil Armstrong became the first man to walk on the moon. The speed, v, in metres per second, of his space craft, Apollo XI, t seconds before touchdown was given by: $v = 0.45 + 3.2t$. Its height, h, in metres, above the moon's surface was given by: $h = 0.45t + 1.6t^2$. What was the speed and height of Apollo XI 1 min before touchdown?

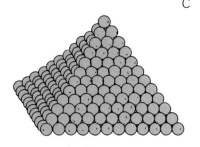

Special

$1.89/dozen

C 12. When there are n oranges in the bottom row of a square pyramid of oranges, the total number of oranges, N, is given by the formula: $N = \frac{1}{3}n^3 + \frac{1}{2}n^2 + \frac{1}{6}n$. Find the value of all the oranges in the display shown in the drawing.

13. The cost, C, in dollars, of producing x bicycles is given by the formula: $C = -0.04x^2 + 50x + 1000$, $x \leqslant 600$.
Find the cost of producing:
a) 50 bicycles;
b) 100 bicycles;
c) 250 bicycles;
d) 450 bicycles;

8 - 3 Adding and Subtracting Polynomials

Addition of polynomials resembles addition of numbers, as the following table shows.

Arithmetic		Algebra
7258	$7(10^3) + 2(10^2) + 5(10) + 8$	$7x^3 + 2x^2 + 5x + 8$
+ 2431	$2(10^3) + 4(10^2) + 3(10) + 1$	$+ 2x^3 + 4x^2 + 3x + 1$
9689	$9(10^3) + 6(10^2) + 8(10) + 9$	$9x^3 + 6x^2 + 8x + 9$

Polynomials, like numbers, can also be added horizontally. We simply combine like terms.

Example 1. Simplify: a) $(3n + 5) + (2n + 3)$;
b) $(-2x^2 + 6x - 7) + (3x^2 - x - 2)$

Solution. a) $(3n + 5) + (2n + 3) = 3n + 5 + 2n + 3$
$= 3n + 2n + 5 + 3$
$= 5n + 8$

b) $(-2x^2 + 6x - 7) + (3x^2 - x - 2)$
$= -2x^2 + 6x - 7 + 3x^2 - x - 2$
$= -2x^2 + 3x^2 + 6x - x - 7 - 2$
$= x^2 + 5x - 9$

Example 2. Simplify: a) $(2a - 3) + (-2a + 3)$
b) $(5x^2 + 2x - 1) + (-5x^2 - 2x + 1)$

Solution. a) $(2a - 3) + (-2a + 3) = 2a - 3 - 2a + 3$
$= 2a - 2a - 3 + 3$
$= 0$

b) $(5x^2 + 2x - 1) + (-5x^2 - 2x - 1)$
$$= 5x^2 + 2x - 1 - 5x^2 - 2x + 1$$
$$= 5x^2 - 5x^2 + 2x - 2x - 1 + 1$$
$$= 0$$

Two polynomials having a sum of 0 are called **opposites.** To find the opposite of a polynomial, multiply by -1.

Example 3. Find the opposite of: a) $7x - 4$;
b) $-x^2 + 2x - 6$.

Solution. a) $(-1)(7x - 4) = -7x + 4$
The opposite of $7x - 4$ is $-7x + 4$.
b) $(-1)(-x^2 + 2x - 6) = x^2 - 2x + 6$
The opposite of $-x^2 + 2x - 6$ is $x^2 - 2x + 6$.

Like addition, subtraction of polynomials may be performed either vertically or horizontally.

> To subtract a polynomial, add its opposite.

Example 4. Simplify: a) $(2x + 7) - (6x - 2)$
b) $(8 - n + n^2) - (-2 + 3n - n^2)$

Solution. a) $(2x + 7) - (6x - 2) = (2x + 7) + (-6x + 2)$
$$= 2x + 7 - 6x + 2$$
$$= 2x - 6x + 7 + 2$$
$$= -4x + 9$$
b) $(8 - n + n^2) - (-2 + 3n - n^2)$
$$= (8 - n + n^2) + (+2 - 3n + n^2)$$
$$= 8 - n + n^2 + 2 - 3n + n^2$$
$$= 8 + 2 - n - 3n + n^2 + n^2$$
$$= 10 - 4n + 2n^2$$

Sometimes addition and subtraction occur in the same problem.

Example 5. Simplify: $(2a - 3) - (a^2 - 5a) + (4 - a)$

Solution. $(2a - 3) - (a^2 - 5a) + (4 - a)$
$$= (2a - 3) + (-a^2 + 5a) + (4 - a)$$
$$= 2a - 3 - a^2 + 5a + 4 - a$$
$$= -a^2 + 2a + 5a - a - 3 + 4$$
$$= -a^2 + 6a + 1$$

Example 6. The cost, in dollars, of making *n* tennis rackets is $6n + 2000$. The income, in dollars, from sales is $15n$.

a) Write a formula for the profit from making and selling *n* rackets.

b) Calculate the profit from the production and sale of
   i) 500 rackets;   ii) 1500 rackets.

c) If all rackets produced are sold, how many must be made and sold to earn a profit of $34 000?

Solution.   a)   Profit, *P,* is income less cost.
$$P = 15n - (6n + 2000)$$
$$= 9n - 2000$$

b)   i)   When $n = 500$, $P = 9(500) - 2000$
$$= 4500 - 2000,$$
or 2500
The profit on the production and sale of 500 rackets is $2500.

ii)   When $n = 1500$, $P = 9(1500) - 2000$
$$= 13\,500 - 2000,$$
or 11 500
The profit on the production and sale of 1500 rackets is $11 500.

c)   Substitute 34 000 for *P:* $34\,000 = 9n - 2000$
$$36\,000 = 9n$$
$$4\,000 = n$$
To earn a profit of $34 000, 4000 rackets must be made and sold.

Exercises 8 - 3

A   1. Simplify:

a) $(6x + 2) + (3x + 4)$ b) $(5a - 3) + (2a + 7)$

c) $(8 - 4m) + (-3 - 2m)$ d) $(-x + 4) + (7x - 2)$

e) $(-1 - 3t) + (4 - 5t)$ f) $(9c - 2) + (-5c - 3)$

g) $(4n^2 - 3n - 1) + (2n^2 - 5n - 3)$

h) $(3x^2 + 6x - 8) + (-5x^2 - x + 4)$

i) $(2 - 3c + c^2) + (5 - 4c - 4c^2)$

j) $(8 - 2n - n^2) + (-3 - n + 4n^2)$

2. Which of the following are opposites?

a) $-2x + 3$ and $2x - 3$

b) $4 - 5n$ and $-5n + 4$

c) $3a - 5$ and $3a + 5$

d) $1 - 3t$ and $-1 + 3t$

e) $n - 3$ and $3 - n$

f) $7x - 5$ and $5 - 7x$

g) $-2x + 7$ and $2x + 7$

h) $8a^2 + 2a - 3$ and $-8a^2 - 2a + 3$

i) $5x^2 - 6x + 1$ and $6x - 5x^2 + 1$

j) $\frac{1}{2}k - 3 + k^2$ and $-\frac{1}{2}k + 3 - k^2$

3. Write the opposite of:

a) $2 - 3x;$

b) $5a + 4;$

c) $\frac{1}{2}x - 5;$

d) $4n^2 - 3n + 1;$

e) $-3 - 2t + t^2;$

f) $0.2a^2 + 0.4a - 0.6.$

4. Simplify:

a) $(3x - 2) - (x - 1) + (4x - 3)$

b) $(2a + 3) + (6a - 1) - (a - 5)$

c) $(7c - 5) - (-c + 3) - (2c - 1)$

d) $(4x^2 - 3x) - (x^2 + 2x) + (3x^2 - x)$

e) $(2m^2 - 5) + (3m - 2) - (m^2 + 1)$

f) $(5t - 4) + (3t^2 - t) - (-2t + t^2)$

g) $(2 - 3n) - (1 - n) + (5 - 2n)$

h) $(5 - 2s) - (3 - s) + (7s - 2)$

i) $(3 - 4x + x^2) - (2x - x^2) + (4 - x + 5x^2)$

j) $(3n^2 - 6n + 5) - (3n^2 - 2n - 1) + (n^2 + 4n - 3)$

5. The cost, in dollars, of producing n records is $1.9n + 20\,000$. The income from selling them, in dollars, is $4.4n$.

a) Write a formula for the profit from producing and selling n records.

b) Calculate the profit from the production and sale of
   i) 10 000 records;   ii) 20 000 records.

c) If all records made are sold, how many must be made and sold
   i) to earn a profit of $10 000?
   ii) to earn a profit of $20 000?
   iii) to break even?

B   6. Simplify:
   a) $(17x - 25) + (34x + 19) - (23x - 11)$
   b) $(45 - 10x) - (-15 - 25x) - (35x + 10)$
   c) $(25n^2 - 6) - (30n^2 - 2n) + (5n^2 + 3n)$
   d) $(37 - 42t) - (61 + 23t) + (21 - 17t)$
   e) $(16n^2 - 10n - 4) + (3n^2 + 25n - 21) - (n^2 - 15n + 19)$
   f) $(2.5x - 3.7) - (1.4x + 4.2) + (-0.8x - 1.3)$
   g) $(6.9 - 1.2t) - (8.4 - 3.5t) + (-4.5 - 2.3t)$
   h) $\left(\frac{1}{2}x - \frac{1}{3}\right) + \left(\frac{1}{4}x + \frac{1}{2}\right) - \left(\frac{3}{4}x - \frac{1}{4}\right)$

   7. If $y = (8x - 5) - (x - 4) + (3x + 1)$,
   a) find the value of y for the following values of x:
      i) 4   ii) -2 iii) 1   iv) 10
   b) find the value of x when $y = 30$.

   8. Solve:
   a) $(3x - 2) - (x + 7) = -5$
   b) $(2 - 5x) + (3 - x) = 8$
   c) $(7 - x) - (2x - 1) = x - 3(x - 5)$
   d) $2 + (4x - 4) = (7 - x) - (x + 3)$
   e) $-9 = (3 - 9x) - (2x + 1)$

C   9. The profit earned, P, in millions of dollars, from constructing an office building having x storeys is $P = 3.5x - (0.5x + 0.1x^2) - 1$.
   a) Express P as a polynomial in descending powers of x.
   b) Find the profit earned from constructing an office building with: i) 5 storeys; ii) 15 storeys.

   10. Simplify:
   a) $(3x^2 + 5xy + 7y^2) - (2x^2 - 4xy + 9y^2)$
   b) $(5x^2 - 3y^2) + (x^2 + 4y^2)$
   c) $(x^2 + 5x^2y - 7) - (3x^2 - 5x^2y + y^2) - (x^2 - 7y^2)$
   d) $(5x^2 - 5x + 7) + (2y^2 - 3x + 7) - (2x^2 - 2x + 7y)$
   e) $(3x^2 - 5x + 7) - [4 - (2x^2 + 3x) - 2]$
   f) $(7y^2 - 3y + 6) - [(2y^2 + 3) - (y^2 + 6y - 8)]$
   g) $(8x^2y^2 + 3xy - 12) - x^2y^2 - (5xy - 6 - 2x^2y^2)$

   11. Arrange the following expressions in descending powers of x, and evaluate for $x = -2$:
   a) $7 - (3x^2 + 2x) - (5x + x^2 - 6) - (3x + 3x^2 - 12)$
   b) $(x^2 + x - 6) - 5x - [(6x - 2x^2 + 3) - (4 - x^2)]$

8 - 4 Multiplying and Dividing Polynomials by Monomials

In earlier work, we found products of monomials as follows:

Multiply the
coefficients.

$$(3x^5)(4x^2) = 12x^7$$

Add the exponents—
the rule for multiplying powers.

Example 1. Find the product: a) $(-6b^3)(3b^2)$;

b) $(-4x^2)(-5x)(3x^3)$.

Solution. a) $(-6b^3)(3b^2)$
$= -18b^5$

b) $(-4x^2)(-5x)(3x^3)$
$= (+20x^3)(3x^3)$
$= 60x^6$

When a polynomial is multiplied by a monomial, the distributive law is used.

$$(5x + 3)(2x) = 10x^2 + 6x$$

Example 2. Find the product: a) $3x(5x - 4)$;

b) $-6x^2(-3 + x + 2x^2)$.

Solution. a) $3x(5x - 4) = (3x)(5x) + (3x)(-4)$
$= 15x^2 - 12x$

b) $-6x^2(-3 + x + 2x^2)$
$= (-6x^2)(-3) + (-6x^2)(x) + (-6x^2)(2x^2)$
$= 18x^2 - 6x^3 - 12x^4$

Example 3. The value of a wheat crop per hectare, V, in dollars, x days after planting is given by the formula:

$$V = 0.05x(140 - x), \quad x \leqslant 70.$$

a) Express V as a polynomial in x;

b) Find the value of the crop, per hectare, 60 days after planting.

Solution. a) $V = 0.05x(140 - x)$
$= (0.05x)(140) + (0.05x)(-x)$
$= 7x - 0.05x^2$

b) When $x = 60$, $V = 7(60) - 0.05(60)^2$
$$= 420 - 180$$
$$= 240$$

The value of the crop 60 days after planting is $240/ha.

To divide a polynomial by a monomial, perform the division with each term.

Example 4. Simplify: $\dfrac{4x^3 - 6x^2 + 2x}{2x}$

Solution. $\dfrac{4x^3 - 6x^2 + 2x}{2x} = \dfrac{4x^3}{2x} - \dfrac{6x^2}{2x} + \dfrac{2x}{2x}$
$$= 2x^2 - 3x + 1$$

Exercises 8 - 4

A   1. Find the product:

   a) $(6n)(5n)$ b) $(-2a)(3a)$ c) $(-5x)^2$
   d) $(3n^2)(-6n)$ e) $(-5x^2)(-2x^2)$ f) $(3a^3)(2a)$
   g) $(5x^4)(2x)$ h) $(8y)(-7y^2)$ i) $(-x)(-5x^3)$
   j) $(\frac{1}{2}n)(\frac{1}{4}n)$ k) $(2.5m)(1.2m^2)$ l) $(0.5x^3)(3x^2)$

   2. Find the product:

   a) $(12x)^2$ b) $(-10a)(17a^2)$ c) $(-25n^2)(8n^2)$
   d) $(-35c^3)(-4c^2)$ e) $(17x^2)(5x^3)$ f) $(-28n)(5n^3)$
   g) $(3x)(5x)(2x)$ h) $(-4n)(-2n)(-3n)$ i) $(3a)^2$
   j) $(-2x^2)^2$ k) $(-2x^2)(6x^2)(-3x)$ l) $(-10m)(-8m)(-5m^2)$

   3. Find the product:

   a) $5(x - 3)$ b) $7(a + 1)$ c) $-3(2 + n)$
   d) $-4(x - 2)$ e) $-1(2x - 5)$ f) $3(6x - 4)$
   g) $-6(5 + 2t)$ h) $5(x^2 - 6x + 3)$ i) $-2(-3 + 5n - 3n^2)$
   j) $7(x^2 - 3x + 9)$ k) $0.4(1.5x - 2.5)$ l) $\frac{1}{2}(\frac{1}{3} - \frac{1}{2}a)$

   4. Find the product:

   a) $x(3x + 2)$ b) $a(5a - 1)$ c) $n(3 - 7n)$
   d) $-x(x - 2)$ e) $-c(3c + 5)$ f) $x^2(3x - 1)$
   g) $y^3(y - 5)$ h) $r^2(2 - 7r)$ i) $n^2(3n^2 - 5n + 1)$
   j) $-x^3(5x^2 - x)$ k) $a^2(3a^2 - 2a + 1)$ l) $-s(7 - 2s + s^2)$

5. Find the product:

   a) $5x(2x + 3)$ b) $2a(3a - 4)$ c) $3c(5 - 2c)$

   d) $-4n(2n - 1)$ e) $-7y(2y^2 - 5)$ f) $6k(3 - k + 2k^2)$

   g) $2x^2(3x - 5)$ h) $-4a^2(3a^2 - 2a)$ i) $5s(3s^2 - 2s - 7)$

   j) $3p^2(2 - 3p - p^2)$ k) $-7a^2(3a^2 - 2a - 4)$

   l) $2n(3 - 6n + 9n^2)$ m) $-1.5x^2(4 - 1.5x - 12x^2)$

6. Simplify:

   a) $\dfrac{6a + 15}{3}$ b) $\dfrac{24x - 4}{4}$ c) $\dfrac{-10 + 4m}{-2}$

   d) $\dfrac{5x^2 - 10x}{5x}$ e) $\dfrac{18a - 21a^2}{3a}$ f) $\dfrac{-28n^2 - 7n}{7n}$

   g) $\dfrac{18x^2 - 6x + 30}{6}$ h) $\dfrac{-5 - 15d + 10d^2}{-5}$

   i) $\dfrac{4x^3 - 12x^2 - 8x}{4x}$ j) $\dfrac{8a + 2a^2 - 2a^3}{2a}$

7. The number of baskets of apples, N, that can be produced by x trees in a small orchard $(x \leqslant 125)$ is given by the formula: $N = x(25 - 0.1x)$. How many baskets of apples can be produced by: a)   60 trees?   b)   80 trees?   c)   125 trees?

8. A field is x metres wide and $(2x + 3)$ metres long.

   a)   Write expressions for the area and perimeter of the field.

   b)   Find the area and the perimeter if $x = 250$ m.

9. Simplify:

   a) $\dfrac{x^3 + x^2}{x^2}$ b) $\dfrac{6x^3 - 4x^2}{2x^2}$ c) $\dfrac{15a^3 - 12a}{3a}$

   d) $\dfrac{-7c^2 + 6c^3}{-c^2}$ e) $\dfrac{12x^5 - 6x^3}{6x^3}$ f) $\dfrac{32a^3 + 40a^4}{8a^3}$

   g) $\dfrac{15x^4 - 30x^3 + 5x^2}{5x^2}$ h) $\dfrac{18a^4 + 6a^3 - 12a^2}{6a^2}$

   i) $\dfrac{-12y^3 + 8y^4 - 4y^5}{-4y^3}$ j) $\dfrac{28x^6 - 35x^4 + 7x^2}{7x^2}$

10. A person of mass x kg on the end of a diving board causes it to dip d cm. The relation between d and x is $d = 0.000\,01x^2(x + 50)$. Find how much the board dips under a person whose mass is

    a)   50 kg;   b)   100 kg.

11. The surface area, S, and volume, V, of a right circular cylinder of radius r and height h are given by the formulas: $S = 2\pi r^2 + 2\pi rh$ and $V = \pi r^2 h$. What is the surface area of a can of height 10 cm and volume 160π cm^3?

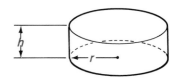

8 - 5 Common Factors

To factor a number or a polynomial means to write it as a product.

factors   factors

$15 = 3 \times 5$ $6x + 3 = 3(2x + 1)$

If two or more terms have the same factor, it is called a **common factor**.

Example 1. a) Find the common factors of 24, 42, and 60.

b) Express $(24 + 42 + 60)$ as a product.

Solution. a) $24 = (2)(2)(2)(3)$
$42 = (2)(3)(7)$
$60 = (2)(2)(3)(5)$
2, 3, and (2)(3), or 6, are common factors of 24, 42, and 60. 6 is the **greatest common factor**.

b) $24 + 42 + 60 = 6(4) + 6(7) + 6(10)$
$= 6(4 + 7 + 10)$, or $6(21)$

Notice the use of the distributive law.

Example 2. a) Find the common factors of $2x^3$, $4x^2$, $-6x$.

b) Express $2x^3 + 4x^2 - 6x$ as a product.

Solution. a) $2x^3 = (2)x \cdot x \cdot x$
$4x^2 = (2)(2)x \cdot x$
$-6x = (-3)(2)x$
Since 2 and x are factors common to all three terms, then $2x$ is the greatest common factor.

b) $\quad 2x^3 + 4x^2 - 6x$
$= (2)x \cdot x \cdot x + (2)(2)x \cdot x + (-3)(2)x$
$= 2x(x^2 + 2x - 3)$

Example 3. Factor and check: a) $6x - 15$;
b) $a^3 - 4a^2 + 2a$.

Solution. a) $6x - 15 = 3(2x) - 3(5)$
$= 3(2x - 5)$
Check by multiplying: $3(2x - 5) = 6x - 15$

b) $a^3 - 4a^2 + 2a = a(a^2) - a(4a) + a(2)$
$$= a(a^2 - 4a + 2)$$

Check: $a(a^2 - 4a + 2) = a^3 - 4a^2 + 2a$

When factoring, always look for the *greatest* common factor.

Example 4.   Factor, where possible:

a) $18n^2 - 12n$
b) $36x^3 - 27x^2 + 54x$
c) $-20x^4y + 10x^2y$
d) $2x + 5$

Solution.   a) $18n^2 - 12n = 6n(3n - 2)$

b) $36x^3 - 27x^2 + 54x = 9x(4x^2 - 3x + 6)$

c) $-20x^4y + 10x^2y = -10x^2y(2x^2 - 1)$

d) $2x + 5$ cannot be factored using integers. If
we use fractions, many forms are possible,
some of which are:

$$2(x + 2.5); \qquad \frac{1}{2}(4x + 10); \qquad 10(\frac{1}{5}x + \frac{1}{2})$$

Exercises 8 - 5

A

1. Identify the greatest common factor in each set of terms:

a) 9,   6,   12
b) 16,   28, -44
c) $2x$, $4x^2$, $6x^3$
d) $-14x$, $21x^2$, $28x^3$
e) $8y^4$, $2y^3$, $-3y^2$
f) $16z^2$, $12z^3$, $32z^5$

2. Express as products:

a) $5y - 10$
b) $12a + 18$
c) $3x^2 + 6x - 12$
d) $2a^2 - 10a + 2$
e) $4w + 3w^2 - 7w^3$
f) $8y^3 - 4y^2 + 2y$
g) $6s + 2s^2 + 4s^3$
h) $-7k^3 - 35k^4 + 49k^5$
i) $6m^2 - 36m^3 - 54m^4$
j) $13w^5 - 65w^3 + 13w$

3. Factor and check:

a) $14x^2 + 35x - 7$
b) $10 - 25a + 30a^2$
c) $20n^2 - 30n + 80$
d) $5x + 10x^2 + 15x^3$
e) $9c^3 + 15c$
f) $x^3 + x^2$
g) $4x - 8x^2 + 12x^3$
h) $6y^2 - 3y^3 + 12y^4$
i) $12m + 16m^2 - 4m^3$
j) $17k - 85k^2 - 51k^3$

4. Factor, where possible:

   a) $16x + 40$ b) $5n - 2$

   c) $2a^2 - 6a$ d) $a^3 - 9a^2 + 3a$

   e) $7x^2 - 10x + 3$ f) $5x^3 + 3x^2 - x$

   g) $9a^3 + 7a^2 + 18a$ h) $8d - 24d^2 + 8d^3$

B   5. Factor:

   a) $b^4 - 3b^3$ b) $3b^4 - 6b^5$

   c) $a^4 + 3a^3 - 2a^2$ d) $21x + 42x^2 + 63x^3$

   e) $-d^4 + 5d^5$ f) $24m^3 + 6m^2 - 12m$

   g) $12x + 9x^2 - 3x^3 - 6x^4$ h) $2x^3 + 3x^4 - 4x^5 + 5x^6$

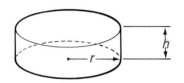

6. The surface area, S, of a right circular cylinder of radius r and height h is given by the formula: $S = 2\pi rh + 2\pi r^2$.   Express S as a product.

7. Identify the greatest common factor in each set of terms:

   a) $3x^2y,\ 6xy,\ -9xy^2$ b) $5ab^3,\ -10ab^2,\ 20a^2b^2$

   c) $6mn,\ 12m^2n,\ -3m$ d) $7m^2n,\ -14m^4,\ -21n^2m^2$

   e) $8x^2y,\ 16x^2y^2,\ 22xy^2$ f) $21x^2y^2z,\ 42xy^2z^2,\ 14x^2yz^2$

C   8. Factor, where possible:

   a) $18a - 6ab^2$ b) $12x^2y + 16xy$

   c) $8ab^2c - 12a^2bc^2$ d) $5m^2 + 15mn + 7n^2$

   e) $24x^2y^2z - 16xy^2z^2$ f) $3x^2 + 6y^2 - 12x^2y^2$

   g) $5xy + 6y + 3x$ h) $9a^3b^2 + 7a^2b^2 + 18a^2b^3$

   i) $7a^3b^2 + 14a^2b^3 - 21a^2b^2$ j) $8x^3y + 16x^3y^2 - 32x^3y^3$

9. Express as a product:

   a) $a(a + 6) + 7(a + 6)$ b) $x(x - 9) - 2(x - 9)$

   c) $8(1 + y) - 3y(1 + y)$ d) $5(2 - x) + x(2 - x)$

   e) $2x(x + 3) + 4(x + 3)$ f) $3a(2a - 1) - 6(2a - 1)$

10. Factor:

   a) $6t(3t - 1) + 9(3t - 1)$ b) $a^2(a + 2) - a(a + 2)$

   c) $5x^2(x + 7) - 10x(x + 7)$ d) $(2n + 3)n - (2n + 3)4$

   e) $(r - 3)2r + (r - 3)8$ f) $(x + 4)x^2 + (x + 4)x^3$

   g) $3a^3(a - 1) - 6a^2(a - 1)$ h) $12a^2(2a - 5) + 16a(2a - 5)$

11. Factor using fractions:

   a) $3m + 5$ b) $2m - 7$ c) $4x + 4y$

   d) $6x^2 - x$ e) $5a^2 + 2a - 10$ f) $2x^2 + 3x + 5$

8 - 6   Product of Two Binomials

To find the product of two binomials (polynomials with two terms), we use the distributive law twice and combine any like terms.

Example 1.   Find the product: $(x + 7)(x + 2)$

   Solution. $(x + 7)(x + 2) = (x + 7)(x) + (x + 7)(2)$
$$= x^2 + 7x + 2x + 14$$
$$= x^2 + 9x + 14$$

The product in *Example 1* can be illustrated geometrically. The length of the large rectangle is $x + 7$ and its width is $x + 2$. Therefore, its area is:
$$(x + 7)(x + 2).$$
The area is also equal to the sum of the areas of the four small rectangles:
$$x^2 + 7x + 2x + 14.$$
Therefore, $(x + 7)(x + 2) = x^2 + 9x + 14$

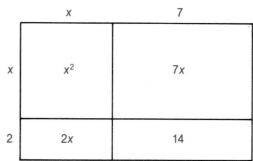

Example 2.   Find the product: $(3a - 2)(a + 4)$

   Solution.   The diagram shows how the products are obtained.
$$(3a - 2)(a + 4) = 3a^2 + 12a - 2a - 8$$
$$= 3a^2 + 10a - 8$$

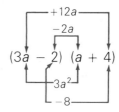

Example 3.   Find the product: a) $(2x + 1)(7x - 3)$

                                        b) $(n - \frac{1}{2})(n - \frac{1}{4})$

                                        c) $5(3 + x)(2 - x^2)$

   Solution.   a) $(2x + 1)(7x - 3) = 14x^2 - 6x + 7x - 3$
$$= 14x^2 + x - 3$$

             b) $(n - \frac{1}{2})(n - \frac{1}{4}) = n^2 - \frac{1}{4}n - \frac{1}{2}n + \frac{1}{8}$

$$= n^2 - \frac{3}{4}n + \frac{1}{8}$$

             c) $5(3 + x)(2 - x^2) = 5(6 - 3x^2 + 2x - x^3)$
$$= 30 - 15x^2 + 10x - 5x^3$$
$$= 30 + 10x - 15x^2 - 5x^3$$

Sometimes, as in *Example 3(c)*, a product has no like terms.

Exercises 8 - 6

A 1. Find the product:

    a) $(x + 3)(x + 4)$ b) $(n + 2)(n + 6)$ c) $(a - 5)(a - 3)$

    d) $(t - 1)(t - 4)$ e) $(x - 2)(x + 5)$ f) $(n + 3)(n - 4)$

    g) $(a + 6)(a - 8)$ h) $(x + 9)(x - 7)$ i) $(x + 12)(x - 5)$

    j) $(s - 11)(s - 3)$ k) $(n + \frac{1}{2})(n + \frac{1}{4})$ l) $(a - \frac{2}{3})(a + \frac{1}{2})$

2. Find the product:

    a) $(3x + 2)(x - 1)$ b) $(2a - 5)(a - 3)$ c) $(4n - 7)(n + 5)$

    d) $(x + 3)(6x - 5)$ e) $(2x + 1)(3x - 1)$ f) $(5n - 1)(2n - 2)$

    g) $(7c - 5)(2c + 1)$ h) $(6x - 2)(3x + 1)$ i) $3(x - 1)(x + 2)$

    j) $(3a + 1)(2a - 5)2$ k) $-4(8y - 3)(5y - 1)$

3. Find the product:

    a) $(x + 3)(x + 3)$ b) $(n - 5)(n - 5)$ c) $(2a - 1)(2a - 1)$

    d) $(4 - 3x)(4 - 3x)$ e) $(x + 7)(x - 7)$ f) $(a - 5)(a + 5)$

    g) $(3n - 2)(3n + 2)$ h) $(1 - 2x)(1 + 2x)$ i) $5(x - 1)(x - 1)$

    j) $(1 + 3x)(1 + 3x)x$ k) $-3y(y - 3)(y + 3)$

B 4. Find the product:

    a) $(4 - a)(3 + a)$ b) $(6 - 3n)(2 + n)$ c) $(x + 3)(5 + x)$

    d) $(x - 4)(2 + x)$ e) $(2a - 5)(3 + a)$ f) $(c + 3)(5 - 2c)$

    g) $(5 - a)(3 - a)$ h) $4(x - 2)(x - 5)$ i) $3(t - 2)(2 - t)$

    j) $5(4x - 3)(5x + 1)$ k) $(3 - x)(2 + 3x)7$

5. Find the product:

    a) $(x^2 + 3)(x + 5)$ b) $(a^2 + 2)(a - 3)$ c) $(n^2 + 2)(n^2 + 3)$

    d) $(x - 5)(x^2 + 2)$ e) $(x^2 - 1)(x^2 - 4)$ f) $(a^2 - 4)(a^2 - 9)$

    g) $(3x + 2)(x^2 - 5)$ h) $n(n - 2)(2n^2 + 1)$

    i) $(3a^2 - 1)(2 - a^2)a^2$ j) $8y^2(4y^2 - 3y)(4y^2 + 3y)$

6. Find the missing term:

    a) $(x + 3)(x - 2) = x^2 + \blacksquare - 6$

    b) $(a + 5)(a + 7) = a^2 + \blacksquare + 35$

    c) $(n - 4)(n - 6) = n^2 + \blacksquare + 24$

    d) $(x - 7)(x + 1) = x^2 + \blacksquare - 7$

    e) $(a - 10)(a - 3) = a^2 + \blacksquare + 30$

    f) $(x + \blacksquare)(x - 7) = x^2 - 3x - 28$

    g) $(n - 3)(n + \blacksquare) = n^2 - 8n + 15$

    h) $(x + \blacksquare)(x + \bullet) = x^2 + 7x + 12$

7. The revenue earned, R, in dollars, from a bus tour carrying x people is given by the formula:

$$R = (100 - 0.5x)(x - 10), \quad \text{where } 10 < x \leqslant 105.$$

a) Write the polynomial for R in descending powers of x.

b) Find the revenue earned when the tour group numbers

i) 60 people;   ii) 90 people.

C   8. Find the product of $(n^2 + 3n + 2)(n^2 + 2n + 1)$ by applying the distributive law three times and combining like terms. Write the result in descending powers of n.

9. Find the product:

a) $(x + 5)(x^2 + 2x + 1)$

b) $(a + 3)(a^2 - 4a + 2)$

c) $(t - 4)(t^2 + 3t - 5)$

d) $(2x - 3)(x^2 - 6x + 4)$

e) $(x + 1)(x + 2)(x + 3)$

f) $(2x - 1)(3x - 1)(x + 1)$

g) $3(x - 4)(x^2 - 7x + 5)$

h) $2x(x + 1)(x - 1)(3x + 2)$

10. Find the missing binomials:

a) $(n + 2) ($ $) = n^2 + 7n + 10$

b) $(x - 3) ($ $) = x^2 - 7x + 12$

c) $($ $)(x + 6) = x^2 + 4x - 12$

d) $($ $)(a - 5) = a^2 - 3a - 10$

e) $(x + 2)($ $) = x^2 + 5x + 6$

f) $(t - 4)($ $) = t^2 + t - 20$

g) $($ $)(s - 7) = s^2 - 12s + 35$

h) $($ $)($ $) = x^2 + 9x + 20$

i) $($ $)($ $) = a^2 - 9a + 14$

j) $($ $)($ $) = n^2 - 10n + 25$

11. Find the product:

a) $(x + 2y)(x + 5y)$

b) $(a - 3b)(a + 2b)$

c) $(3m - n)(2m - n)$

d) $(5x + 3y)(4x - y)$

e) $(6r + s)(r - 3s)$

f) $(8a + 7b)(7a + 8b)$

g) $(p - 3q)(2p + 5q)$

h) $(3x - 8y)(2x + 5y)$

i) $(6a + 7b)(7a - 8b)$

j) $(5m + n)(2m - n)$

12. a) Illustrate these products geometrically;

i) $(x + 3)(x + 5)$ ii) $(x - 3)(x + 5)$

iii) $(x + 3)(x - 5)$ iv) $(x - 3)(x - 5)$

b) Explain how the diagrams for (a) can be used to show that: $(-3)(+5) = -15$ and $(-3)(-5) = +15$.

8 - 7 Factoring Trinomials of the Form: $x^2 + bx + c$

A polynomial with three terms is called a trinomial. In Section 8-6, we saw that the product of two binomials is often a trinomial.

$$(x + 3)(x + 4) = x^2 + 7x + 12$$

The reverse process is to factor the trinomial. The integers in the trinomial provide the help needed.

This integer — is the sum of these two integers.

$$x^2 + 7x + 12 = (x + 3)(x + 4)$$

This integer — is the product of these two integers.

Example 1. Factor: $x^2 + 5x + 6$, and check.

Solution. $x^2 + 5x + 6$

Sum of integers is $+5$. Product of integers is $+6$.

$$x^2 + 5x + 6 = (x + 3)(x + 2)$$

Check. $(x + 3)(x + 2) = x^2 + 2x + 3x + 6$
$$= x^2 + 5x + 6$$

The factors of $x^2 + 5x + 6$ are $(x + 3)$ and $(x + 2)$.

Factors of $+6$	
$+6,$	$+1$
$+3,$	$+2$
$-3,$	-2
$-6,$	-1

Example 2. Factor: $a^2 - 8a + 12$, and check.

Solution. $a^2 - 8a + 12$

Sum of integers is -8. Product of integers is $+12$.

$$a^2 - 8a + 12 = (a - 6)(a - 2)$$

Check. $(a - 6)(a - 2) = a^2 - 2a - 6a + 12$
$$= a^2 - 8a + 12$$

The factors of $a^2 - 8a + 12$ are $(a - 6)$ and $(a - 2)$.

Factors of $+12$	
$+12,$	$+1$
$+6,$	$+2$
$+4,$	$+3$
$-4,$	-3
$-6,$	-2
$-12,$	-1

Example 3. Factor: $m^2 - 5m - 14$, and check.

Solution. $m^2 - 5m - 14$

Sum of integers is -5. Product of integers is -14.

$$m^2 - 5m - 14 = (m - 7)(m + 2)$$

Check. $(m - 7)(m + 2) = m^2 + 2m - 7m - 14$
$$= m^2 - 5m - 14$$

The factors of $m^2 - 5m - 14$ are $(m - 7)$ and $(m + 2)$.

Factors of -14	
$+14,$	-1
$-14,$	$+1$
$+7,$	-2
$-7,$	$+2$

An important reason for factoring a trinomial is that the factored form is simpler. For any value of x:

$x^2 + 7x + 12$ is the sum of *three* numbers.

$(x + 3)(x + 4)$ is the product of *two* numbers.

To evaluate $x^2 + 7x + 12$ for several values of x, it is easier to use the factored form: $(x + 3)(x + 4)$.

Example 4. Evaluate: $x^2 + 7x + 12$ and $(x + 3)(x + 4)$ for $x = 7, x = 16,$ and $x = -3$.

Solution.

Value of x	Value of $x^2 + 7x + 12$	Value of $(x + 3)(x + 4)$
7	$7^2 + 7(7) + 12$ $= 49 + 49 + 12$ $= 110$	$(7 + 3)(7 + 4)$ $= 10(11)$ $= 110$
16	$16^2 + 7(16) + 12$ $= 256 + 112 + 12$ $= 380$	$(16 + 3)(16 + 4)$ $= 19(20)$ $= 380$
-3	$(-3)^2 + 7(-3) + 12$ $= 9 - 21 + 12$ $= 0$	$(-3 + 3)(-3 + 4)$ $= 0(1)$ $= 0$

Not all trinomials can be factored using integers. To determine whether a trinomial can be factored with integers, consider the possible factors of the constant term.

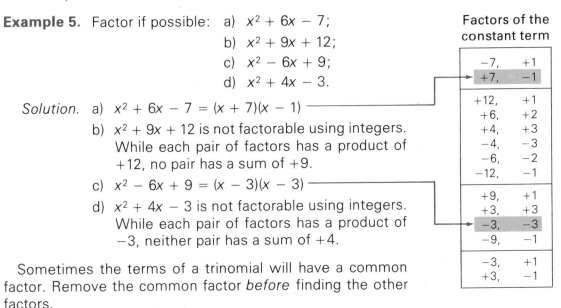

Example 5. Factor if possible: a) $x^2 + 6x - 7$;

b) $x^2 + 9x + 12$;

c) $x^2 - 6x + 9$;

d) $x^2 + 4x - 3$.

Factors of the constant term

$-7,$	$+1$
$+7,$	-1
$+12,$	$+1$
$+6,$	$+2$
$+4,$	$+3$
$-4,$	-3
$-6,$	-2
$-12,$	-1
$+9,$	$+1$
$+3,$	$+3$
$-3,$	-3
$-9,$	-1
$-3,$	$+1$
$+3,$	-1

Solution. a) $x^2 + 6x - 7 = (x + 7)(x - 1)$

b) $x^2 + 9x + 12$ is not factorable using integers. While each pair of factors has a product of $+12$, no pair has a sum of $+9$.

c) $x^2 - 6x + 9 = (x - 3)(x - 3)$

d) $x^2 + 4x - 3$ is not factorable using integers. While each pair of factors has a product of -3, neither pair has a sum of $+4$.

Sometimes the terms of a trinomial will have a common factor. Remove the common factor *before* finding the other factors.

Example 6. Factor completely: a) $3x^2 - 18x + 24$

b) $x^4 + 4x^3 - 5x^2$

Solution. a) $3x^2 - 18x + 24 = 3(x^2 - 6x + 8)$

$= 3(x - 4)(x - 2)$

b) $x^4 + 4x^3 - 5x^2 = x^2(x^2 + 4x - 5)$

$= x^2(x + 5)(x - 1)$

When factoring, always look for a common factor first.

Exercises 8 - 7

A   1. Factor:

a) $x^2 + 7x + 10$ b) $a^2 + 5a + 6$ c) $m^2 + 10m + 24$

d) $a^2 + 8a + 7$ e) $x^2 + 6x + 9$ f) $20 + 9x + x^2$

g) $n^2 + 14n + 49$ h) $t^2 + 13t + 40$ i) $1 + 2x + x^2$

2. Factor:

a) $x^2 - 8x + 15$ b) $c^2 - 12c + 32$ c) $a^2 - 10a + 21$

d) $x^2 - 6x + 5$ e) $x^2 - 12x + 36$ f) $14 - 9n + n^2$

g) $y^2 - 15y + 54$ h) $s^2 - 18s + 81$ i) $30 - 11k + k^2$

3. Factor and check:

a) $x^2 - 6x + 8$ b) $x^2 + 9x + 18$ c) $a^2 - 11a + 18$

d) $m^2 + 11m + 28$ e) $n^2 - 10n + 25$ f) $30 - 13n + n^2$

g) $p^2 + 16p + 64$ h) $y^2 - 13y + 42$ i) $56 + 15x + x^2$

4. Factor:

a) $x^2 + 6x - 16$ b) $a^2 + 4a - 12$ c) $x^2 + 6x - 27$

d) $c^2 + 2c - 35$ e) $x^2 + x - 12$ f) $1 + a - 30a^2$

g) $y^2 + y - 56$ h) $t^2 - 2t - 24$ i) $15 - 2x - x^2$

5. Factor:

a) $r^2 - 5r - 36$ b) $a^2 - 4a - 45$ c) $n^2 - 3n - 54$

d) $m^2 - 2m - 48$ e) $k^2 - 2k - 63$ f) $1 - 7x - 30x^2$

6. Factor and check:

a) $x^2 - 5x - 24$ b) $x^2 + 5x - 50$ c) $a^2 + a - 72$

d) $n^2 - 3n - 40$ e) $m^2 + m - 42$ f) $8 - 7x - x^2$

g) $y^2 + 3y - 4$ h) $s^2 - 7s - 18$ i) $2 + t - t^2$

7. Factor:
   a) $x^2 - 7x - 8$
   b) $a^2 + 5a - 14$
   c) $t^2 - 2t - 3$
   d) $n^2 + 13n + 42$
   e) $x^2 - 17x + 72$
   f) $c^2 - 11c + 30$
   g) $m^2 + 6m - 55$
   h) $a^2 + 10a + 9$
   i) $s^2 + s - 20$
   j) $c^2 + 9c - 36$
   k) $12 + 4m - m^2$
   l) $15 - 8y + y^2$

8. Consider the trinomial $x^2 - 2x - 15$.
   a) Find its value when $x = 17$.
   b) Factor the trinomial.
   c) Evaluate the factored expression when $x = 17$.

9. Consider the trinomial $x^2 - 13x + 36$.
   a) Find its value when $x = 59$.
   b) Factor the trinomial.
   c) Evaluate the factored expression when $x = 59$.

10. Factor if possible:
    a) $x^2 + 16x + 63$
    b) $a^2 + 12a + 30$
    c) $x^2 - 4x + 32$
    d) $t^2 + 11t + 24$
    e) $n^2 - 12n + 35$
    f) $k^2 - 5k - 21$
    g) $x^2 + 7x - 60$
    h) $n^2 + 7n - 7$
    i) $x^2 - 6x + 24$
    j) $a^2 - a - 45$
    k) $56 + t - t^2$
    l) $6 - x + x^2$

11. Factor completely:
    a) $2x^2 + 12x + 10$
    b) $5a^2 - 10a - 40$
    c) $10n^2 + 10n - 20$
    d) $4a^2 - 16a - 20$
    e) $3x^2 + 15x + 6$
    f) $7a^2 - 35a + 42$
    g) $x^3 - 2x^2 - 3x$
    h) $a^3 - 2a^2 - 9a$
    i) $2y^3 + 14y^2 + 24y$
    j) $3x^3 + 6x^2 - 24x$
    k) $60n + 50n^2 + 10n^3$
    l) $14s + 7s^2 - 7s^3$

12. Find integers to replace ▌ so that each trinomial can be factored:
    a) $x^2 + ▌x + 12$
    b) $x^2 - ▌x + 20$
    c) $x^2 + ▌x - 18$
    d) $x^2 + 5x + ▌$
    e) $x^2 + 4x + ▌$
    f) $x^2 - 2x + ▌$

13. If a hockey arena increases its ticket prices by x dollars, the predicted revenue, R, in thousands of dollars, from all ticket sales is given by the formula: $R = 35 + 2x - x^2$, $x < 7$.
    a) Factor the trinomial.
    b) Find the predicted revenue if ticket prices are increased by:
       i) \$1;
       ii) \$2;
       iii) \$3.

14. If a transit company increases its fares by x cents, the total daily revenue, R, in thousands of dollars, is given by the formula:
$$R = 55 + 6x - x^2, \quad x < 11.$$
   a) Factor the trinomial.
   b) Find the total daily revenue if fares are increased by:
      i) 1¢;   ii) 2¢;   iii) 3¢;   iv) 5¢;   v) 6¢.

15. Small businesses that cannot afford a computer can rent computer time. One rental company charges $120/min. A computer takes 8 s to evaluate a trinomial expression but takes only 5 s to evaluate it in its factored form.
   a) How much is saved by using the factored form?
   b) Express the amount saved as a percent.

C 16. Consider the equation $y = x^2 + 12x + 11$.
   a) What are the factors of the trinomial?
   b) Evaluate y for $x = 1, 2, 3, \ldots, 10$.
   c) For which values of x is y a perfect square?

17. Find values of x which make these trinomials perfect squares:
   a) $x^2 + 7x + 6$ b) $x^2 + 8x - 9$ c) $x^2 - 11x + 24$

18. Factor:
   a) $x^2 + 8xy + 15y^2$
   b) $x^2 - 9xy + 14y^2$
   c) $x^2 - 4xy - 5y^2$
   d) $x^2 + 7xy - 18y^2$
   e) $c^2 - 4cd - 21d^2$
   f) $m^2 + 14mn + 45n^2$
   g) $6x^2 + xy - y^2$
   h) $8a^2 - 2ab - b^2$
   i) $28a^2 + 3ab - b^2$
   j) $12x^2 + 8xy + y^2$

8 - 8 Trinomials That Are Perfect Squares

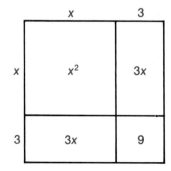

Squaring a binomial means to multiply it by itself. The diagram shows how this can be represented geometrically.

Example 1. Expand: $(x + 3)^2$

Solution.
$$\begin{aligned}(x + 3)^2 &= (x + 3)(x + 3) \\ &= x^2 + 3x + 3x + 9 \\ &= x^2 + 6x + 9\end{aligned}$$

There is a pattern to the terms of the trinomial obtained from squaring a binomial.

Example 2. Expand: a) $(x - 4)^2$;
      b) $(3x - 2)^2$;
      c) $(5 + 6n)^2$.

Solution. a) $(x - 4)^2 = x^2 - 8x + 16$
      b) $(3x - 2)^2 = 9x^2 - 12x + 4$
      c) $(5 + 6n)^2 = 25 + 60n + 36n^2$

In each of the three cases in the last example:
- the first term of the expansion is the square of the first term of the binomial;
- the second term of the expansion is twice the product of the terms of the binomial;
- the third term of the expansion is the square of the second term of the binomial.

With practice, the pattern that results from squaring a binomial can be used to square numbers mentally.

Example 3. Calculate mentally: 41^2

Solution. Think: $41^2 = (40 + 1)^2$
      Think: $= 40^2 + 2(40)(1) + 1^2$
            $= 1600 + 80 + 1$
            $= 1681$

Before factoring a trinomial, test to see if it fulfils the requirements of a perfect square.

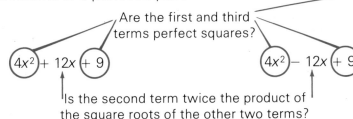

Are the first and third terms perfect squares?

Their signs must be positive.

Is the second term twice the product of the square roots of the other two terms?

For the above examples, the requirements are met.
$$4x^2 + 12x + 9 = (2x + 3)^2 \quad \text{and} \quad 4x^2 - 12x + 9 = (2x - 3)^2.$$

Example 4. Factor: a) $n^2 - 14n + 49$;
      b) $9k^2 - 24k + 16$.

Solution.   a) $n^2 - 14n + 49$
      n^2 and 49 are perfect squares, and
      $-14n = 2(n)(-7)$.
      Therefore, $n^2 - 14n + 49 = (n - 7)^2$

b) $9k^2 - 24k + 16$
$9k^2$ and 16 are perfect squares, and
$-24k = 2(3k)(-4)$.
Therefore, $9k^2 - 24k + 16 = (3k - 4)^2$

Exercises 8 - 8

A   1. Expand:

a) $(x + 5)^2$ b) $(x + 1)^2$ c) $(a - 3)^2$ d) $(n - 7)^2$
e) $(c + 4)^2$ f) $(x - 1)^2$ g) $(2a + 3)^2$ h) $(3a - 1)^2$
i) $(2 + a)^2$ j) $(5 - x)^2$ k) $(t + 9)^2$ l) $(10 - m)^2$

2. Expand:

a) $(7x - 5)^2$ b) $(6a + 4)^2$ c) $(3n - 8)^2$ d) $(5 - 9x)^2$
e) $(10 - 3c)^2$ f) $(12x + 5)^2$ g) $(n + 8)^2$ h) $(7s + 5)^2$
i) $(n + 0.2)^2$ j) $(x - \frac{1}{2})^2$ k) $(1.2x - 5)^2$ l) $(x + 3)^2$

3. Find the square:

a) $x + 7$ b) $a - 2$ c) $n + 4$ d) $2x - 1$
e) $5c + 3$ f) $2 - 3m$ g) $4 + 5x$ h) $8 - 3x$
i) $5 + 5s$ j) $2(x + 3)$ k) $5(a - 1)$ l) $7(2x - 3)$

4. Expand:

a) $3(x - 2)^2$ b) $2(x + 5)^2$ c) $5(a - 1)^2$
d) $4(2x - 3)^2$ e) $-3(5 - n)^2$ f) $x(3x - 4)^2$
g) $2x(x - 9)^2$ h) $-a(5 - 2a)^2$ i) $-3x(2 - 3x)^2$

5. Expand:

a) i) $(x - 3)^2$ ii) $(-x + 3)^2$
b) i) $(3 - x)^2$ ii) $(-3 + x)^2$

6. Use the pattern for squaring a binomial to calculate, mentally:
a) 31^2;   b) 13^2;   c) 29^2;   d) 103^2;   e) 61^2;   f) 62^2.

7. Factor and check:

a) $x^2 - 8x + 16$ b) $x^2 - 12x + 36$ c) $a^2 + 4a + 4$
d) $p^2 - 2p + 1$ e) $x^2 + 6x + 9$ f) $t^2 - 10t + 25$
g) $x^2 + 14x + 49$ h) $25y^2 - 10y + 1$ i) $b^2 - 12b + 36$
j) $49 - 14d + d^2$ k) $x^2 + 20x + 100$ l) $y^2 - 40y + 400$

8. Find the missing integers or binomials:
   a) $(x + \blacksquare)^2 = x^2 + 8x + 16$
   b) $(x + \blacksquare)^2 = x^2 - 12x + 36$
   c) $(a + \blacksquare)^2 = a^2 + \bullet a + 49$
   d) $(2a + \blacksquare)^2 = 4a^2 + \bullet a + 9$
   e) $(\blacksquare - 2x)^2 = 49 - \bullet x + 4x^2$
   f) $(\qquad)^2 = x^2 + 6x + 9$
   g) $(\qquad)^2 = x^2 - 20x + 100$

9. Which of the following are the squares of a binomial?
   a) $x^2 - 10x + 25$
   b) $x^2 + 6x + 36$
   c) $a^2 - 8a - 16$
   d) $n^2 - 2n + 1$
   e) $x^2 + 3x + 6$
   f) $t^2 + 10t + 12$
   g) $x^2 + 7x + 49$
   h) $25x^2 - 10x + 1$
   i) $a^2 - 5a + 25$
   j) $49 - 14x + x^2$
   k) $100 + 20x + x^2$
   l) $16t^2 - 24t + 9$

10. Expand:
    a) $(x^2 + 3)^2$
    b) $(x^2 - 5)^2$
    c) $(m^2 - 10)^2$
    d) $(x^3 + 2)^2$
    e) $(c^3 - 1)^2$
    f) $(2x^2 - 3)^2$
    g) $(4a - a^2)^2$
    h) $x(2x^2 + 1)^2$
    i) $3c(2 - c^2)^2$

11. Factor and check:
    a) $9 - 24w + 16w^2$
    b) $64 - 48y + 9y^2$
    c) $4b^2 - 20b + 25$
    d) $9t^2 - 42t + 49$
    e) $0.09x^2 - 0.12x + 0.04$
    f) $p^2 - \frac{4}{3}p + \frac{4}{9}$

12. Expand:
    a) $(x + 5y)^2$
    b) $(a - 3b)^2$
    c) $(2x - 3y)^2$
    d) $(5m + 2n)^2$
    e) $(x + 2)^3$
    f) $(c - 1)^3$
    g) $(x + 2)(x - 2)^2$
    h) $(n - 5)(n + 1)^2$
    i) $(x^2 + 3x + 2)^2$
    j) $(a^2 - a + 3)^2$

13. Factor and check:
    a) $x^2 - 14xy + 49y^2$
    b) $m^2 + 12mn + 36n^2$
    c) $4a^2 + 4ab + b^2$
    d) $25c^2 - 30cd + 9d^2$
    e) $64x^2 - 48xy + 9y^2$
    f) $16x^2 + 56xy + 49y^2$

14. a) Calculate: 15^2, 25^2, 35^2, 45^2.
    b) Make up a rule for squaring two-digit numbers ending in 5.
    c) Verify your rule for other two-digit numbers ending in 5.
    d) Explain why the rule works.

8 - 9 Product of a Sum and a Difference

An interesting case of the product of two binomials occurs when one binomial is the sum of two terms and the other is the difference of the *same* two terms.

Example 1. Find the product: $(x + 5)(x - 5)$.

Solution. $(x + 5)(x - 5) = x^2 - 5x + 5x - 25$
$= x^2 - 25$

Notice that the product, $x^2 - 25$, is the difference of the squares of the two terms.

Example 2. Find the product: a) $(2x - 7)(2x + 7)$:
b) $(4 + n)(4 - n)$.

Solution. a) $(2x - 7)(2x + 7) = 4x^2 - 49$
b) $(4 + n)(4 - n) = 16 - n^2$

This suggests a way to find certain products mentally.

Example 3. Find the product: a) 41×39; b) 18×22.

Solution. a) Think: $41 \times 39 = (40 + 1)(40 - 1)$
$= 1600 - 1$
$= 1599$

b) Think: $18 \times 22 = (20 - 2)(20 + 2)$
$= 400 - 4$
$= 396$

If a binomial is recognized as the difference of two squares, it can be factored immediately.

Example 4. Factor:

a) $49x^2 - 64$; b) $a^2 - 0.16$; c) $\frac{1}{4}x^2 - \frac{1}{9}$.

Solution. a) $49x^2 - 64 = (7x - 8)(7x + 8)$
b) $a^2 - 0.16 = (a + 0.4)(a - 0.4)$
c) $\frac{1}{4}x^2 - \frac{1}{9} = (\frac{1}{2}x - \frac{1}{3})(\frac{1}{2}x + \frac{1}{3})$

Example 5. Factor if possible: a) $5x^2 - 45$; b) $8x^3 - 2x$;

c) $\frac{1}{16}y^2 - \frac{4}{9}$; d) $x^2 + 36$;

e) $x^4 - 16$; f) $x^2 - 5$.

Solution. a) $5x^2 - 45 = 5(x^2 - 9)$
$$= 5(x - 3)(x + 3)$$

b) $8x^2 - 2x = 2x(4x^2 - 1)$
$$= 2x(2x - 1)(2x + 1)$$

Removing a common factor

c) $\frac{1}{16}y^2 - \frac{4}{9} = (\frac{1}{4}y - \frac{2}{3})(\frac{1}{4}y + \frac{2}{3})$

d) $x^2 + 36$ cannot be factored using real numbers.

e) $x^4 - 16 = (x^2 - 4)(x^2 + 4)$
$$= (x - 2)(x + 2)(x^2 + 4)$$
Factoring is in two steps because $x^2 - 4$ is also the difference of squares.

f) $x^2 - 5$ cannot be factored using integers or rational numbers. It can be factored using irrational numbers:
$x^2 - 5 = (x + \sqrt{5})(x - \sqrt{5})$.

Exercises 8 - 9

1. Find the product:

   a) $(x + 3)(x - 3)$ b) $(x + 7)(x - 7)$ c) $(a - 4)(a + 4)$

   d) $(n - 9)(n + 9)$ e) $(t - 1)(t + 1)$ f) $(c + 10)(c - 10)$

   g) $(s + 25)(s - 25)$ h) $(m + 40)(m - 40)$

   i) $(x - 1.5)(x + 1.5)$ j) $(c + 3.1)(c - 3.1)$

   k) $(x - \frac{1}{2})(x + \frac{1}{2})$ l) $(k + \frac{2}{3})(k - \frac{2}{3})$

2. Find the product:

   a) $(3x + 2)(3x - 2)$ b) $(4x - 3)(4x + 3)$

   c) $(7 + a)(7 - a)$ d) $(3 - 5x)(3 + 5x)$

   e) $(2.5 - c)(2.5 + c)$ f) $(6s - 5)(6s + 5)$

   g) $(2 - 3.5y)(2 + 3.5y)$ h) $(0.5m - 1)(0.5m + 1)$

   i) $(\frac{1}{2}x + \frac{1}{3})(\frac{1}{2}x - \frac{1}{3})$ j) $(\frac{2}{3}a - \frac{3}{4})(\frac{2}{3}a + \frac{3}{4})$

3. Find the product:

   a) $(12x - 7)(12x + 7)$ b) $(5y + 13)(5y - 13)$

   c) $(15a + 11)(15a - 11)$ d) $(25s - 20)(25s + 20)$

   e) $(x + \sqrt{2})(x - \sqrt{2})$ f) $(r - \sqrt{3})(r + \sqrt{3})$

   g) $(x^2 + 6)(x^2 - 6)$ h) $(2n^2 - 5)(2n^2 + 5)$

   i) $(3 - 7y^2)(3 + 7y^2)$ j) $(a^3 - 5)(a^3 + 5)$

   k) $(x^2 - 8y)(x^2 + 8y)$ l) $(3a^2 + 2b)(3a^2 - 2b)$

4. Factor:
   a) $x^2 - 25$
   b) $a^2 - 49$
   c) $x^2 - 36$
   d) $y^2 - 100$
   e) $x^2 - 1$
   f) $x^2 - \frac{1}{9}$
   g) $m^2 - \frac{1}{4}$
   h) $25 - 4b^2$
   i) $81 - 49r^2$

5. Factor:
   a) $9a^2 - 4$
   b) $25x^2 - 9$
   c) $16s^2 - 1$
   d) $36 - 100n^2$
   e) $100x^2 - 121$
   f) $144p^2 - 49$
   g) $\frac{1}{4}x^2 - \frac{4}{9}$
   h) $6.25 - n^2$
   i) $2.25 - d^2$

6. Factor if possible, and check:
   a) $x^2 - 4$
   b) $x^2 + 4$
   c) $a^2 + 9$
   d) $x^2 - 9$
   e) $x^2 - 10$
   f) $c^2 - 6$
   g) $3x^2 - 1$
   h) $100t^2 - 81$
   i) $49a^2 + 36$

7. Factor completely:
   a) $2x^2 - 18$
   b) $5x^2 - 5$
   c) $3a^2 - 48$
   d) $3n^2 + 30$
   e) $7 - 28y^2$
   f) $2a^2 + 12$
   g) $x^3 - 25x$
   h) $a^3 - 49a$
   i) $4c - 81c^3$

8. Find these products mentally:
   a) 31×29
   b) 32×28
   c) 59×61
   d) 57×63
   e) 101×99
   f) 105×95

B 9. Find the product:
   a) $(x - 1)(x + 1)(x - 2)(x + 2)$
   b) $(x - 1)(x + 1)(x^2 + 1)$
   c) $(a - 3)(a + 3)(a - 5)(a + 5)$
   d) $(a + 5)(a - 5)(a^2 + 25)$
   e) $(y - 2)(y + 2)(1 - y)(1 + y)$
   f) $(n - 3)(n + 3)(2 + n)(2 - n)$

10. Factor completely, and check:
   a) $x^4 - 1$
   b) $a^4 - 16$
   c) $16c^4 - 1$
   d) $x^4 - 81$
   e) $2 - 2x^4$
   f) $2y - 32y^5$

11. a) Find the product: i) $(x - 1)(x + 1)$; ii) $(x - 1)(x^2 + x + 1)$
    iii) $(x - 1)(x^3 + x^2 + x + 1)$
    b) Predict the product of $(x - 1)(x^4 + x^3 + x^2 + x + 1)$ and check by multiplying.

12. Calculate mentally:
   a) $(1\,000\,000)^2 - (999\,999)^2$
   b) 501×499

8 - 10   Working With Polynomials

The simplification of polynomials may involve several operations. The usual order of performing operations is followed.

Example 1. Simplify: a) $2x(x - 3) - 5(x^2 - 1)$;
                                              b) $(x - 5)(x + 2) - (x - 3)(x + 1)$.

Solution.   a) $2x(x - 3) - 5(x^2 - 1) = 2x^2 - 6x - 5x^2 + 5$
                                                                      $= -3x^2 - 6x + 5$

   b) $(x - 5)(x + 2) - (x - 3)(x + 1)$
          $= (x^2 - 3x - 10) - (x^2 - 2x - 3)$
          $= x^2 - 3x - 10 - x^2 + 2x + 3$
          $= -x - 7$

Polynomials often arise in the solutions of problems.

Example 2.   The dimensions of a cereal box are $(5x - 1)$ cm by $3x$ cm by x cm.

   a) Find expressions for
       i) its volume, V, ii) its surface area, S, and write them in descending powers of x.

   b) What is the volume and surface area when $x = 7$ cm?

Solution.   a)   i) $V = x(3x)(5x - 1)$
                                    $= x(15x^2 - 3x)$
                                    $= 15x^3 - 3x^2$

   ii) $S_{back + front} = 2(3x)(5x - 1)$
         $S_{sides} = 2x(5x - 1)$
         $S_{top + bottom} = 2x(3x)$
         $S_{total} = 2(3x)(5x - 1) + 2x(5x - 1) + 2x(3x)$
                        $= 2(15x^2 - 3x) + 10x^2 - 2x + 6x^2$
                        $= 30x^2 - 6x + 10x^2 - 2x + 6x^2$
                        $= 46x^2 - 8x$

   b) When $x = 7$ cm,
          $V = 15(7)^3 - 3(7)^2$ $S = 46(7)^2 - 8(7)$
             $= 15(343) - 3(49)$ $= 46(49) - 56$
             $= 4998$ $= 2198$
          When $x = 7$ cm, the volume of the box is 4998 cm³, and the surface area is 2198 cm².

Sometimes, in problem solving, it is necessary to solve an equation involving a polynomial.

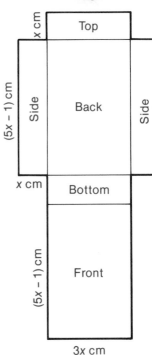

Example 3. If $y = x^2 - 4x + 3$,

    a) find y when $x = 7$;

    b) find x when $y = 8$.

Solution.   a) $y = x^2 - 4x + 3$

      When $x = 7$,

$$y = (7)^2 - 4(7) + 3$$
$$= 49 - 28 + 3$$
$$= 24$$

      When $x = 7$, $y = 24$.

    b) $y = x^2 - 4x + 3$

      When $y = 8$,

$$8 = x^2 - 4x + 3$$
$$0 = x^2 - 4x - 5$$
$$0 = (x - 5)(x + 1)$$

If the product of two factors is 0, then either or both must be 0.

If $x - 5 = 0$ If $x + 1 = 0$

    $x = 5$ $x = -1$

When $y = 8$, $x = 5$ or $x = -1$.

Example 4. The length of a rectangle is 6 cm longer than twice its width. Find the dimensions of the rectangle if its area is 20 cm².

Solution. Let the width of the rectangle be x cm.
Then the length is $(2x + 6)$ cm
and the area is $x(2x + 6)$ cm².

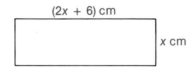

(2x + 6) cm

x cm

$$x(2x + 6) = 20$$
$$2x^2 + 6x = 20$$
$$2x^2 + 6x - 20 = 0$$
$$2(x - 2)(x + 5) = 0$$

Either $x - 2 = 0$ or $x + 5 = 0$
    and $x = 2$,   and $x = -5$.

Since the width cannot be a negative quantity, the width of the rectangle is 2 cm and its length is $(4 + 6)$ cm, or 10 cm.

Check. $2 \times 10 = 20$. The solution is correct.

Exercises   8 - 10

A   1. Simplify:
   a) $3(2x + 1) + 2(x + 4)$ b) $7(3a - 2) + 3(a - 1)$
   c) $6(2x - 3) - 5(x + 4)$ d) $-2(4a + 3) + 3(a - 7)$
   e) $x(5 - 3x) - 2x(x + 1)$ f) $5x(2x - 3) + 4(x^2 + x)$
   g) $4(2 - 3n) + 5(1 + 2n)$ h) $3y(y - 5) - y(y - 1)$
   i) $2x(3x - 1) + 5x(4x + 2)$ j) $a(2 - 7a) - 3a(a - 1)$

   2. Simplify:
   a) $3x(x - 1) + 2(x^2 - x)$ b) $t(2t - 3) - t(3t - 2)$
   c) $5(2 - 3a) - a(2 - a)$ d) $2c(c - 3) + c(c + 5)$
   e) $4(3x^2 - 2) - x(5 - x)$ f) $3y(7y - 1) - y(5y - 2)$
   g) $3(2x^2 + 4x + 3) + 2(x^2 + 5x + 1)$
   h) $7(a^2 - 2a + 4) - 3(a^2 + a - 1)$
   i) $5(2x^2 - 6x - 3) - (x^2 - 2x + 5)$
   j) $-2(x^2 - 3x + 1) - 3x(2 - x)$

   3. Simplify:
   a) $(x - 5)^2 + (x + 3)^2$ b) $(2a - 3)^2 - (a - 1)^2$
   c) $(5x - 2)^2 - (5x + 2)(5x - 2)$ d) $(a + 8)^2 - (a - 8)^2$
   e) $(x + 5)(x + 1) + (x - 2)(x - 3)$
   f) $(a - 3)(a + 7) + (a - 6)(a + 1)$
   g) $(3x - 1)(3x + 1) + (x - 4)(x + 4)$
   h) $(x + 1)^2 + (x + 2)^2 + (x + 3)^2$

   4. Evaluate the expression $2(x - 3) - 5(x + 1) - (x - 4)$ for
   a) $x = 2$;   b) $x = -2$;   c) $x = -5$;   d) $x = 0.1$.

   5. The dimensions of a room are $(3x + 2)$ m by $3x$ m by $(x + 2)$ m.
   a) Find the expressions for
      i) its volume,
      ii) the total area of floor, walls, and ceiling, and write them
          in descending powers of x.
   b) Find the volume and total area of floor, walls, and ceiling
      when: i) $x = 1$; ii) $x = 3$.

   6. If $y = x^2 + 4x - 21$,
   a) find y when $x = 8$;   b) find x when $y = 0$;
   c) find x when $y = -16$;   d) find x when $y = -24$.

7. If $y = x^2 - 2x - 48$,
   a) find y when $x = -3$; b) find x when $y = 0$;
   c) find x when $y = -33$; d) find x when $y = 15$.

B 8. The width of a rectangle is 1 cm shorter than its length. Find the dimensions if the area is 3.75 cm².

9. Simplify:
   a) $1.5(2x - 3) - 0.5(x + 1)$ b) $2.3(x - 7) - 1.4(3x + 2)$
   c) $0.2(a - 5) + 1.2(5 - a)$ d) $3.6(c - 2) + 0.4(c + 5)$
   e) $\frac{1}{2}(a - 3) + \frac{1}{4}(2a - 5)$ f) $\frac{2}{3}(x + \frac{1}{2}) - \frac{1}{2}(x - \frac{1}{3})$
   g) $5(2x - 3) - 4(x - 2) + 7(2x + 1)$
   h) $6(3a - 5) + 4(3 - a) - (a - 1)$
   i) $x(5x - 1) - 3(x + 2) + 2(x + 5)$
   j) $3a(2a - 1) + a(7a + 3) - 4a(a - 2)$

10. Simplify:
    a) $(x + 4)^2 - (x - 4)^2$
    b) $2(x + 3)^2 - 3(x - 2)(x + 3)$
    c) $(2n + 3)(n - 5) - (4n - 1)(n + 3)$
    d) $(6a - 5)(2a + 1) + (5a - 2)(a + 2)$
    e) $(x - 7)(3x + 2) - (2x - 5)(x + 3)$
    f) $3(x - 2)(x + 2) + 4(x - 1)(x + 1)$
    g) $4(a - 3)(a + 5) + 2(a + 1)(a - 1)$
    h) $3(x - 7)(x + 2) - 4(x - 2)^2$
    i) $(x - 3)(x + 5) + (x - 1)(x - 7) - (x + 4)(x + 3)$
    j) $(a - 1)^2 + (a - 2)^2 + (a - 3)^2$

11. Evaluate the following expressions for $n = 3$:
    a) $(n + 1)^2 - 3(n - 1)(n + 1)$
    b) $5(n^2 - 4) - 2(n^2 - 6n + 1)$
    c) $2(n^2 - 6n + 7) + 4(n^2 + 3n - 1)$

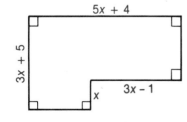

12. For the figure shown,
    a) write polynomials, in descending powers of x, for
       i) the area; ii) the perimeter;
    b) calculate the area and perimeter when $x = 10$.

13. If $y = 2x^2 + 24x + 70$,
    a) find y when $x = -1$; b) find x when $y = 0$;
    c) find x when $y = 30$; d) find x when $y = 16$.

14. One number is 6 more than three times another number. Their product is 189. What are the numbers?

15. The number of diagonals, d, in a polygon of n sides is given by the formula: $d = \frac{1}{2}n^2 - \frac{3}{2}n$.

    a) How many diagonals has a decagon?

    b) How many sides has a polygon having 14 diagonals?

16. If the cost of a football ticket is increased by x dollars, the total revenue, R, in thousands of dollars, from ticket sales and concessions is given by:
$$R = (11 - x)(6 + x) + 1.5(11 - x).$$

    a) Simplify the expression by writing it as a trinomial in descending powers of x.

    b) Find the total revenue if ticket prices are increased by
      i) $1.00;   ii) $2.00;   iii) $3.00.

17. If a hockey arena increases its ticket prices by $x each, the total revenue, R, in thousands of dollars, from ticket sales and concessions, is given by:
$$R = (50 - 2x)(20 + x) + 7(50 - 2x).$$

    a) Write the expression as a trinomial in descending powers of x.

    b) Find the total revenue
      i) if ticket prices are increased by $1.00; by $2.00;
      ii) if ticket prices are decreased by $1.00; by $2.00.

18. a) On a calendar month, select any three-by-three square of numbers, as shown. Multiply the numbers in opposite corners. Subtract the lesser product from the greater. What is the result?

    b) Repeat the procedure with another square of nine numbers from the same or another month. What is the result?

    c) Let the first number in the square be n. Find expressions for the other numbers. Use these expressions to show that the result of the procedure in (a) is always 28.

    d) Repeat the above procedure and find the results for
      i) four-by-four squares;
      ii) five-by-three rectangles.

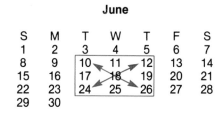

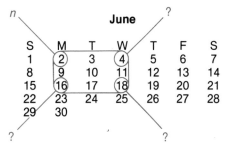

PROBLEM SOLVING STRATEGY

Look for a Pattern

Sometimes a problem can be solved by considering simpler versions of the problem and looking for a pattern to their solutions. If we find one, we then assume the pattern holds true for the solution to the original problem.

Example. Twelve people attend a party. If each person shakes hands with everyone else, how many handshakes are there in all?

Solution. Let each person be represented by a letter. Consider simpler versions of the problem: How many handshakes will there be with 2 people? 3 people? 4 people? 5 people?

		Number of Handshakes
With 2 people there is 1 handshake.	A •———• B	1
A third person shakes hands with the first two. Therefore, with 3 people there are 2 more handshakes.		1 + 2
A fourth person shakes hands with the first three. Therefore, with 4 people there are 3 more handshakes.		1 + 2 + 3
With 5 people there are 4 more handshakes.		1 + 2 + 3 + 4

As each person joins the group, the number of handshakes is increased by the number of people already in the group.

Now that we can see a pattern in the solutions to the simpler versions of our problem, we can extend the pattern to a twelfth person joining a party of 11 people. The number of handshakes will be:

$$1 + 2 + 3 + 4 + \ldots + 11, \text{ or } 66.$$

How many handshakes would there be in a group of 20 people? in a group of n people?

Exercises

Solve each problem by looking for a pattern in the solutions of simpler versions of the problem. Explain why the pattern applies to the original problem.

1. What is the one's digit of 7^{145} when expanded?

2. For which integers, n, do n and n^5 have the same one's digit?

3. a) Given 10 points on a circle, how many chords can be drawn joining them?
   b) How many chords can be drawn joining n points on a circle?

4. Simplify:
   a) $1 - 2 + 3 - 4 \ldots - 10$
   b) $1 - 2 + 3 - 4 \ldots + 19$

5. Consider the expression: $1 - 2 + 3 - 4 \ldots n$.
   a) For what values of n is the value of the expression positive?
   b) For what values of n is the value of the expression negative?
   c) Write a formula for the value of the expression when n is even; when n is odd.

6. a) Use the pattern suggested by the diagram to evaluate:
      $1 + 3 + 5 + 7 + \ldots + 15$.
   b) Find the formula for the expression:
      $1 + 3 + 5 + 7 + \ldots + (2n - 1)$.

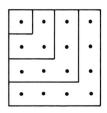

7. a) Use the pattern suggested by the diagram to evaluate:
      $1 + 2 + 3 + \ldots + 9 + 10 + 9 + \ldots + 3 + 2 + 1$.
   b) Find the formula for the expression:
      $1 + 2 + 3 + \ldots + (n - 1) + n + (n - 1) + \ldots + 3 + 2 + 1$.

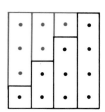

CALCULATOR POWER

Evaluating Polynomials

Suppose we have to evaluate $4x^2 - 44x + 108$ for $x = 3.5$.

Substituting 3.5 for x we get: $4(3.5)^2 - 44(3.5) + 108$.

On the simplest of calculators, each term has to be evaluated separately, noted down, and then reentered to be combined. Even on a calculator with a memory, to evaluate the polynomial as it stands requires a sequence of operations such as:

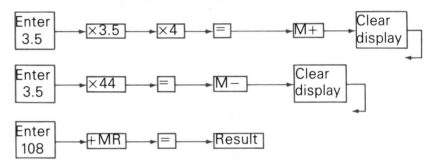

We can evaluate a polynomial more easily on the simplest of calculators if we first factor two or more terms. By factoring the first two terms of $4x^2 - 44x + 108$, we can write:

$$(x - 11)4x + 108.$$

Now, to evaluate $(x - 11)4x + 108$ for $x = 3.5$, only the following sequence of operations is necessary:

Enter 3.5 → -11 → $\times 3.5$ → $\times 4$ → $+108$ → $=$ → Result

The polynomial $(x - 11)4x + 108$ is said to be in a **nested form**. We can express any polynomial in a nested form by successive grouping and factoring. For example:

$$5x^3 + 3x^2 + 2x + 7 = [5x^2 + 3x + 2]x + 7$$
$$= [(5x + 3)x + 2]x + 7$$

Evaluate for $x = 2.5$.

Exercises

1. Express $3x^3 - 7x^2 + 3x - 9$ in a nested form and evaluate for:
   a) $x = 4$;   b) $x = 2.5$;   c) $x = 8.7$.

2. Express $2x^3 - 5x^2 + x + 11$ in a nested form and evaluate for:
   a) $x = 5$;   b) $x = 6.5$;   c) $x = 1.9$.

Review Exercises

1. Simplify:
   a) $(-5n) + (+8n)$ b) $6x - (-2x)$ c) $3y^2 - (-5y^2)$
   d) $-2y^2 - 3y^2 - 5y^2$ e) $-5b^2 + 8b^2 - (-2b^2)$
   f) $-11r^3 - 5r^3 + 6r$ g) $2.7y^3 - 1.8y^3 + 2y$

2. Simplify, arrange in descending powers, and state the degree:
   a) $5x - 2x + 7 - 4x$
   b) $15y^2 - 3y + 6y^3 - 2y^2 + 7y$
   c) $2a - 5a^4 + 6a^2 - 4a^2 + 7a + 8a^4 - a^3$
   d) $4 - 6a + a^2 - 3 + 4a^2 + 12a - a - 2a^2 + 1$

3. Evaluate $3y^2 - 5y + 2$ for
   a) $y = 0;$ ′   b) $y = -1;$ c) $y = -3;$ d) $y = -4.5.$

4. Simplify:
   a) $(5y - 3) + (2y + 4)$ b) $(7x - 2) - (5x + 3)$
   c) $(8r - 5) - (-3r + 2)$ d) $(5y^2 - 3y) + (6y - 2y^2)$
   e) $(9d - 2) - (5d - 3)$ f) $(4a + 7) - (3 - 2a)$
   g) $(x - 3y) + (4y - 3x) - (x - 2y)$
   h) $(4x - 3) - (2x + 1) - (-3x + 4)$
   i) $(1 - 2p + 3p^2) + (-4 + 3p - 7p^2)$
   j) $(3q^2 + 5q - 2) - (8q^2 - 3q + 4)$

5. Find the product:
   a) $(7a)(5a)$ b) $(-3b)(5b)$ c) $(-4y)^2$
   d) $(-9x^3)(7x^2)$ e) $(-2y)^3$ f) $(4xy^2)(-3x^2y)$
   g) $(-3x)(-5x)(-2x^2)$ h) $(-2y^3)(-3y)(4y^2)$

6. Find the product:
   a) $4(y - 2)$ b) $8(a - 3)$ c) $-4(x + 2)$
   d) $3x(5 - x)$ e) $2y(y - 6)$ f) $-5x(3 - x)$
   g) $5y(7 - 2y + 3y^2)$ h) $-6x(3x^2 + 5x - 12)$

7. Simplify:
   a) $\dfrac{15x - 20}{5}$ b) $\dfrac{-12 + 2x}{-2}$ c) $\dfrac{12x^2 + 4x}{4x}$
   d) $\dfrac{10a - 6a^2}{2a}$ e) $\dfrac{3x^3 - 12x^2}{3x^2}$ f) $\dfrac{10a^3 - 6a^2}{2a^2}$
   g) $\dfrac{18c^2 - 6c^3 + 12c^4}{6c^2}$ h) $\dfrac{5x^5 + 20x^4 - 5x^3}{5x^3}$

8. Factor and check:
   a) $6y + 18y^2$
   b) $-3a + 12a^4$
   c) $5a^2 - 25a^3$
   d) $3a^3 + 4a^2 + 7a$
   e) $6x^2y - 3xy + 9xy^2$
   f) $8ab - 4a^2b^2 + 6ab^2$

9. Find the product:
   a) $(x - 3)(x - 4)$
   b) $(y + 7)(y + 3)$
   c) $(a - 2)(a + 5)$
   d) $(n - 6)(n + 7)$
   e) $(n + 4)(n - 7)$
   f) $3(x - 1)(x + 5)$
   g) $4(x - 5)(x - 6)$
   h) $y(y - 2)(y + 9)$
   i) $-3y(1 - y)(7 + y)$

10. Factor:
    a) $x^2 - 7x + 10$
    b) $x^2 - 6x + 9$
    c) $x^2 + 6x + 5$
    d) $a^2 - 4a - 12$
    e) $x^2 - x + 12$
    f) $15 + 2x - x^2$
    g) $y^2 - y - 72$
    h) $n^2 + 3n - 40$
    i) $8 + 7x - x^2$
    j) $2x^2 - 12x + 10$
    k) $5m^2 + 10m - 40$
    l) $y^3 + 2y^2 - 3y$

11. Expand:
    a) $(x + 3)^2$
    b) $(y - 5)^2$
    c) $(5 - q)^2$
    d) $(6 - 5y)^2$

12. Find the product:
    a) $(a - 4)(a + 4)$
    b) $(x + 8)(x - 8)$
    c) $(2 - y)(2 + y)$
    d) $(\frac{1}{2}n - 1)(\frac{1}{2}n + 1)$
    e) $3(x - 6)(x + 6)$
    f) $-5y(3 - y)(3 + y)$

13. Factor:
    a) $b^2 - 25$
    b) $x^2 - 81$
    c) $y^2 - 121$
    d) $m^2 - \frac{1}{4}$
    e) $9x^2 - 16$
    f) $25 - 4y^2$
    g) $2x^2 - 32$
    h) $64m - 4m^3$

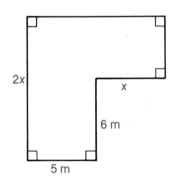

2x

x

6 m

5 m

14. A garden plot has the dimensions shown.
    a) Write an expression for the perimeter of the plot.
    b) Write an expression for the area of the plot.
    c) Determine the perimeter and area when $x = 7$ m.

15. The number of car accidents, N, on a certain street in a week is given by $N = 2c^2 - 6c + 1, c \geq 3$, where c is the number of cars, in thousands, that use the street each week. How many accidents are there in a week when
    a) $c = 4$?
    b) $c = 6$?
    c) $c = 3$?
    d) $c = 5$?

16. The temperature, T, in degrees Celsius, in a sauna n min after being turned on is given by $T = -0.04n^2 + 2.5n + 16$ for the first 30 min. What is the temperature in the sauna when
    a) $n = 0$ min?
    b) $n = 15$ min?
    c) $n = 25$ min?

9 Relations

The height, h, in metres, of a rocket x seconds after blast-off is given by the formula: $h = x^2 + 4x$. What is the height of the rocket 7.5 s after blast off? How many seconds does the rocket take to reach a height of 50 m? (See *Example 2* in Section 9-4.)

9 - 1 The Cartesian Coordinate System

The story goes that the great 17th century French mathematician, René Descartes, in bed because of illness, was watching a spider on the ceiling. As he watched, he thought of how he might describe the position of the spider at any instant. The technique he devised is called the **Cartesian coordinate system**. Although the system is named after him, another French mathematician, Pierre de Fermat, is equally deserving of acclaim. For, quite independently, he developed the same technique at much the same time.

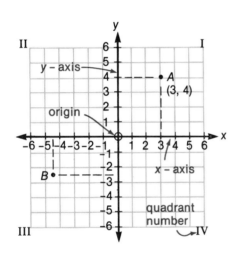

The use of coordinate geometry allows points, lines, circles, and other curves to be represented by numbers and equations. That is, geometrical problems can be solved using the methods of arithmetic and algebra. Descartes, Fermat, and others were able to apply these methods to the solving of practical problems in astronomy, optics, ballistics, and navigation.

Point *A* has **coordinates** (3, 4). The *x*-coordinate is always written first. Therefore, (3, 4) is an **ordered pair** of numbers. The coordinates of point *B* are $(-4.5, -2.5)$.

> For every point in the plane, there corresponds exactly one ordered pair of real numbers. For every ordered pair of real numbers, there corresponds exactly one point in the plane.

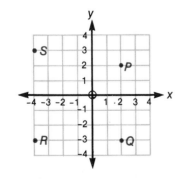

Example 1. a) Give the coordinates of points *P*, *Q*, *R*, and *S*.

b) Plot the points: $A(-3, -1)$, $B(4, -4)$, $C(-2, 4)$, and $D(4, 4)$.

c) Where are the points with
   i) -3 as their first coordinate?
   ii) 2 as their second coordinate?
   iii) 0 as their second coordinate?

Solution. a) The coordinates of the four points are:

$P(2, 2)$, $Q(2, -3)$, $R(-4, -3)$, $S(-4, 3)$.

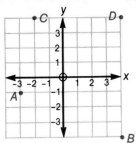

c) i) The points with -3 as their first coordinate lie on a line 3 spaces to the left of the y-axis and parallel to it.

ii) The points with 2 as their second coordinate lie on a line 2 spaces above the x-axis and parallel to it.

iii) The points with 0 as their second coordinate lie on the x-axis.

Example 2. Plot the points: $P(-2.5, 2.5)$, $Q(2.5, 2.5)$, $R(4, -1)$, $S(-1, -1)$.

a) What quadrant is each point in?

b) Draw PQ, QR, RS, and SP.

c) What kind of polygon is determined by the segments?

Solution. a) P—II, Q—I, R—IV, S—III

b)

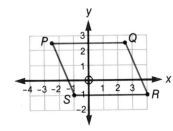

c) The polygon appears to be a parallelogram.

Exercises   9 - 1

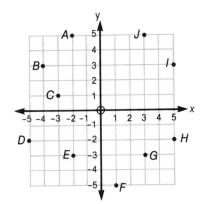

A   1. Give the coordinates of points A to J and the quadrant for each.

2. Plot the points: $M(3, 4)$, $N(-2, -6)$, $P(-4, 5)$, $Q(0, 4)$, $R(-4, 0)$, $S(-5, -2)$, $T(0, -5)$, $U(3, -4)$.

3. Where are the points with

a) first coordinate 0?   b) y-coordinate -4?

c) x-coordinate negative?

d) x-coordinate negative, y-coordinate positive?

e) x-coordinate $= y$-coordinate?

4. Plot these points on the same axes and draw the segments named:

   a) $A(-4, 3)$, $B(-3, -2)$, $C(-2, 1)$, $D(-1, -2)$, $E(0, 3)$
      AB, BC, CD, DE

   b) $F(2, 3)$, $G(2, -2)$, $H(4, 3)$, $J(4, -2)$, $K(2, 0)$, $L(4, 0)$
      FG, HJ, KL

   c) $M(6, 3)$, $N(8, 3)$, $P(7, 0)$, $R(7, -2)$
      MP, NP, RP

   d) What word have you formed?

5. a) Plot: $T(-4, 1)$, $C(-3, -3)$, $E(5, -1)$, $R(4, 3)$

   b) Draw segments: TC, CE, ER, RT.

   c) What familiar polygon have you drawn?

B   6. Plot each set of three points and determine which sets of points do *not* form a triangle.

   a) $(-3, 1)$, $(1, -1)$, $(3, -4)$ b) $(0, 0)$, $(6, 2)$, $(9, 3)$

   c) $(2, -3)$, $(5, -1)$, $(8, 1)$ d) $(4, -6)$, $(2, -2)$, $(-4, 10)$

   e) $(0, 0)$, $(8, 3)$, $(13, 5)$ f) $(-1, 1)$, $(2, 3)$, $(5, 5)$

7. The coordinates of three vertices of a square are given. Find the coordinates of the fourth vertex.

   a) $(3, 5)$, $(-4, 5)$, $(-4, -2)$ b) $(1, 2)$, $(5, -2)$, $(1, -6)$

   c) $(-3, -1)$, $(4, 2)$, $(2, -3)$ d) $(-3, -1)$, $(4, 0)$, $(1, -4)$

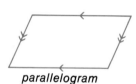

parallelogram

8. The coordinates of three vertices of a parallelogram are given. Find the coordinates of a fourth possible vertex.

   a) $(3, 0)$, $(6, 3)$, $(0, 3)$ b) $(2, 1)$, $(5, 3)$, $(2, 7)$

   c) $(1, 6)$, $(-5, 1)$, $(3, 1)$ d) $(0, -3)$, $(0, 3)$, $(2, 0)$

   e) Can there be more than one correct solution?

9. What polygon has vertices: $(a, 0)$, $(b, 0)$, $(a, b - a)$, $(b, b - a)$?

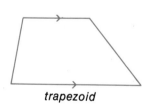

trapezoid

10. For each set of four points, plot the points and join them by line segments to form a quadrilateral. Then indicate whether the quadrilateral is a parallelogram, trapezoid, rhombus, rectangle, or square, or none of these.

   a) $(1, 1)$, $(-4, 1)$, $(-4, -4)$, $(1, -4)$

   b) $(-2, 5)$, $(-5, -3)$, $(-5, -5)$, $(-2, -3)$

   c) $(3, 0)$, $(0, 3)$, $(-3, 0)$, $(0, -3)$

   d) $(0, 4)$, $(1, 0)$, $(-1, 0)$, $(0, -4)$

   e) $(-2, 0)$, $(0, 5)$, $(3, -2)$, $(1, -7)$

   f) $(0, 2)$, $(-3, 0)$, $(0, -4)$, $(6, 0)$

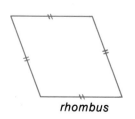

rhombus

9-2 Slope

The *slant* of a ladder against a wall, the *gradient* of a road up a hill, and the *pitch* of a roof are all examples of a mathematical concept called **slope**.

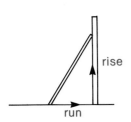
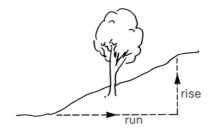
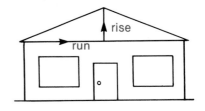

In each case, it is the ratio of the difference in the vertical position of two points (rise) to the difference in horizontal position (run).

$$\text{slope} = \frac{\text{rise}}{\text{run}}$$

In a coordinate system, we can compute the slope of a line or line segment if we know the coordinates of any two points on the line or segment.

$$\text{slope} = \frac{\text{difference in } y\text{-coordinates}}{\text{difference in } x\text{-coordinates}}$$

Example 1. Find the slopes of line JK and line segment PQ.

Solution. Slope of JK: $\dfrac{(5) - (-4)}{(4) - (-2)} = \dfrac{9}{6}$

$= \dfrac{3}{2}$

For every run to the right of 2 there is a rise of 3.

Slope of PQ: $\dfrac{(1) - (7)}{(-2) - (-5)} = \dfrac{-6}{3}$

$= -2$

For every run to the right of 1 there is a rise of -2.

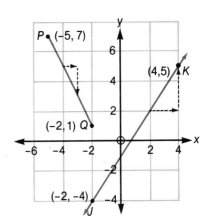

- Lines *rising* to the right have *positive* slope.
- Lines *falling* to the right have *negative* slope.

Example 2. Draw the line passing through

a) the point (0, 0) with slope $\dfrac{2}{3}$;

b) the point (4, 1) with slope $-\dfrac{3}{2}$.

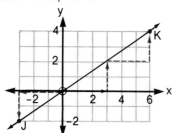

Solution. a) Slope $\frac{2}{3}$ means that for every difference of $+2$ in y there is a corresponding difference of $+3$ in x, and the line rises to the right.

Begin at $(0, 0)$. Move 3 to the *right* and 2 *up*. This is a point on the line. Any number of points can be obtained in this way.

Since $\frac{2}{3} = \frac{-2}{-3}$ we can also move 3 to the *left* and 2 *down* to get another point.

Line *JK* is the required line.

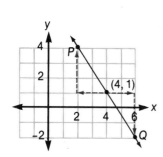

b) Since $-\frac{3}{2} = \frac{-3}{2}$, for every difference of -3 in y there is a corresponding difference of $+2$ in x, and the line *falls* to the *right*.

Begin at $(4, 1)$. Move 2 to the *right* and 3 *down*. This is a point on the line. Other points can be obtained in this way.

Since $-\frac{3}{2} = \frac{3}{-2}$ we can also move 2 to the *left* and 3 *up* to get another point.

Line *PQ* is the required line.

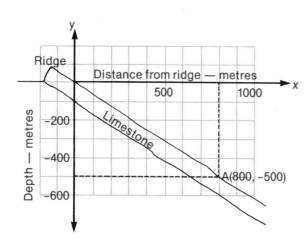

Example 3. Drilling has revealed that a layer of limestone has a uniform slope away from a ridge that the layer has formed. Point *A* is 500 m below a point 800 m from the base of the ridge.

a) Find the slope of the layer.

b) What is the depth of the layer 4 km from the ridge?

Solution. a) Slope of the layer is:

$$\frac{-500 - 0}{800 - 0}, \text{ or } -\frac{5}{8}.$$

b) The layer falls 500 m every 800 m. In 4000 m, the layer will fall $\frac{4000}{800} \times 500$ m, or 2500 m.

The depth of the layer 4 km from the ridge is 2.5 km.

Exercises 9 - 2

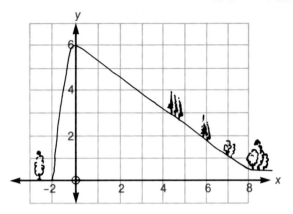

A   1. What is the slope of the ramp?

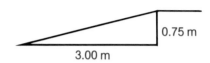

2. The diagram shows a closer view of the ridge in *Example 3*. Find the slope of each side of the ridge.

3. Find the slope of the line passing through:

   a) (−4, 1) and (3, 5);   b) (−3, − 2) and (5, −1);

   c) (2, 3) and (−3, 3);   d) (7, 3) and (−2, −4);

   e) (−3, 6) and (8, −4);   f) (−1, 5) and (6, −3);

   g) (2, 7) and (−5, −5);   h) (−1.5, 6.5) and (8.5, −3.5).

4. Find the slope of each line or line segment:

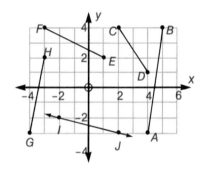

5. On a coordinate system, draw the line through

   a) point (0, 0) with slope 2;

   b) point (1, 4) with slope $\frac{3}{2}$;

   c) point (−1, 2) with slope $-\frac{1}{2}$;

   d) point (−4, 3) with slope $-\frac{2}{3}$;

   e) point (1, 6) with slope −2;

   f) point (−3, 0) with slope $\frac{4}{3}$.

B 6. The diagram is the side view of a ski tow. Each section of the tow approximates a straight line.

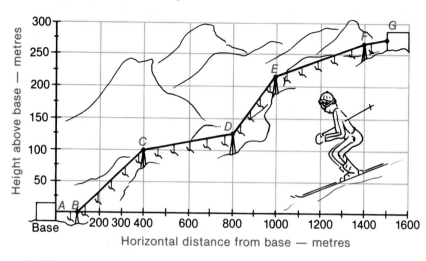

Height above base — metres

Base

Horizontal distance from base — metres

a) List the coordinates of points A, B, C, D, E, F, and G.
b) Find the slopes of the segments: *AB, BC, CD, DE, EF, FG.*
c) Which segment is the steepest?

7. Find the slope of a road that rises 25 m in every kilometre.

C 8. What can be said about the slope of a line parallel to the x-axis?   to the y-axis?

9. Each second after take-off, an aircraft climbs 40 m for every 100 m of horizontal movement. The coordinates of the aircraft after 1 s are shown in the graph.

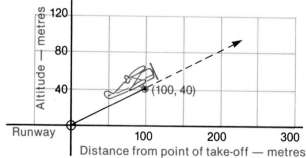

Altitude — metres

Runway

(100, 40)

Distance from point of take-off — metres

a) Draw a graph showing the coordinates of the aircraft after 1 s, 2 s, 3 s, 4 s, and 5 s.
b) What is the slope of the path of the aircraft?
c) How high is the aircraft when it is 10 km from touchdown?
d) How long does it take to reach an altitude of 12 000 m?

10. On its landing approach, an aircraft sinks 3 m for every 100 m of horizontal movement.
    a) Draw a graph showing the aircraft's position up to 1 km from the runway.
    b) What is the slope of the path of the aircraft?
    c) How high is the aircraft when it is 5 km from the runway?

11. On the same grid, draw the line through point $(1, 1)$ with slope $\frac{2}{3}$, and the line through point $(-2, 4)$ with slope $\frac{4}{6}$. What seems to be the relationship between the two lines?

12. On the same grid, draw the line through point $(2, 3)$ with slope $\frac{3}{2}$ and the line through point $(0, 7)$ with slope $\frac{-2}{3}$. What seems to be the relationship between the two lines?

13. a) Locate the vertices and draw the sides of quadrilateral $A(4, -5.5)$, $B(2, 4)$, $C(-6, 9.5)$, $D(-4, 0)$,
    b) Compare the slopes of AB and CD;   of BC and AD.
    c) What kind of quadrilateral is this?

9 - 3   Displaying Relations

Tickets to the Roxy theatre cost $6.00 each. The cost of a number of tickets (less than seven for convenience) can be displayed in several different ways. For example:

In a table

Number of tickets	Total cost
0	$ 0
1	6
2	12
3	18
4	24
5	30
6	36

In a graph

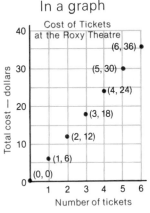

Cost of Tickets at the Roxy Theatre

Points: (0, 0), (1, 6), (2, 12), (3, 18), (4, 24), (5, 30), (6, 36)

Total cost — dollars (vertical axis)
Number of tickets (horizontal axis)

Both the table and the graph relate numbers in pairs. A more concise way of doing this is as a set of ordered pairs:

$$\{(0, 0), (1, 6), (2, 12), (3, 18), (4, 24), (5, 30), (6, 36)\}$$

The first number in each pair is the number of tickets. The second number is the total cost of the tickets. Thus, the ordered pair $(3, 18)$ indicates that 3 tickets cost $18.

A relation is a set of ordered pairs.

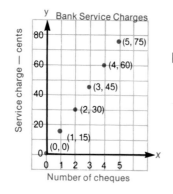

Example 1. The graph shows the relationship between the service charge on a chequing account and the number of cheques written.

a) From the graph, determine
   i) the service charge on 4 cheques;
   ii) the number of cheques handled for a charge of 75¢.

b) What relation is defined by the graph?

Solution. a) i) The ordered pair, $(4, 60)$, indicates that the service charge on 4 cheques is 60¢.
    ii) The ordered pair, $(5, 75)$, indicates that 75¢ is the service charge on 5 cheques.

b) The relation defined by the graph is:
$$\{(0,0), (1, 15), (2, 30), (3, 45), (4, 60), (5, 75)\}.$$

Sometimes there is no simple arithmetical relationship between the numbers in the ordered pairs.

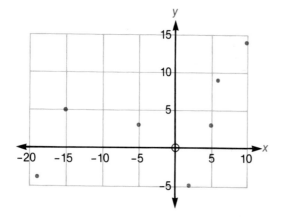

Example 2. What relation is defined by the graph?

Solution. The relation defined by the graph is the following set of ordered pairs:
$$\{(-19, -4), (-15, 5), (-5, 3),$$
$$(10, 14), (6, 9), (5, 3), (2, -5)\}.$$
It does not matter in what order the ordered pairs are listed.

Exercises   9 - 3

1. What relations are defined by the table and the graph?

a)

Number	Cost
1	$0.50
2	0.95
3	1.35
4	1.70
5	2.00

b)

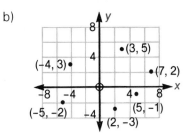

2. State the relation defined by each graph:

a)

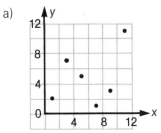

b)

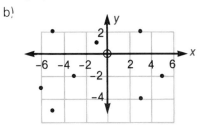

c)

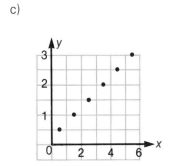

d)

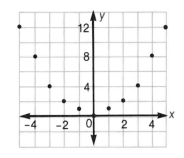

3. State the relation and draw the graph for each of the following:

a)

Rectangle	
Length cm	Width cm
11	1
10	2
9	3
8	4
7	5
6	6

b)

Women	
Height cm	Mass kg
152	47
160	52
167	56
175	63
181	70

c)

Pizza	
Number of choices	Cost $
1	3.90
2	4.40
3	4.90
4	5.40
5	5.90
6	6.40

4. The graph represents the sum of the natural numbers from 1 to
   n (n ≤ 7).

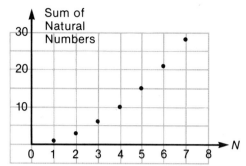

a) From the graph, determine
   i) the sum of the first five numbers;
   ii) how many numbers are required for a sum of 10.
b) What relation is defined by the graph?

5. For the graph shown,
   a) what values of y correspond to these values of x: 2, 6, 10, 13?
   b) what values of x correspond to these values of y: 15, 25, 30?
   c) state the relation between x and y.

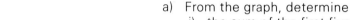

B 6. Some parents decide that their children should receive a weekly
      allowance from age 4 to age 16. The amount is 25¢ the first year
      and increases by 25¢ per year.
      a) Make a table of values for the relation.
      b) Graph the relation.
      c) Find a formula relating the allowance to the child's age.

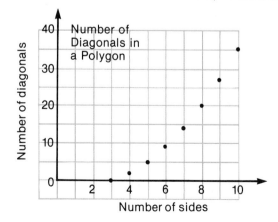

7. The graph shows the relation between the
   number of sides and the number of diagonals
   in a polygon.
   a) How many diagonals do these polygons
      have?
      i) a triangle   ii) a pentagon
      iii) a hexagon   iv) an octagon
   b) How many sides has a polygon that has
      the following number of diagonals?
      i) 2   ii) 9   iii) 27
   c) A decagon has 10 sides. How many
      diagonals does it have?

C   8. A college copy shop advertises as follows:

┌───┐
│   FOR   EACH   ORIGINAL   │
│   │
│ first 5 copies   second 5 copies all additional copies │
│   7¢ each   5¢ each   3¢ each   │
└───┘

   a) Draw up a table of values for up to 15 copies of one original.
   b) Graph the relation between the total cost and the number of copies.

   9. A bank allows 3 free cheques per month but makes a service charge of 18¢ for each cheque after that. Draw a graph of the relation between the service charge and the number of cheques for up to 8 cheques.

  10. Theatre tickets are $5 for adults and $2 for students.
   a) Draw a graph showing the relation between the number of each kind of ticket you could buy for a total expenditure of
       i) exactly $20;   ii) $20 or less.
   b) State the relation for each part of (a).

9 - 4   Relations Having Many Ordered Pairs

Very often, relations have too many ordered pairs for all of them to be listed. When the same relationship exists between the first and second elements of every ordered pair, the relation can be expressed as a formula.

Example 1. The interest, I, on a principal amount of money, P, invested at 11% per annum for 1 year is given by the formula: $I = 0.11P$.
   a) Show, on a graph, the annual interest earned by amounts up to $5000.
   b) Estimate from the graph
       i) the annual interest on $2750;
       ii) the principal required to earn an annual interest of $400.

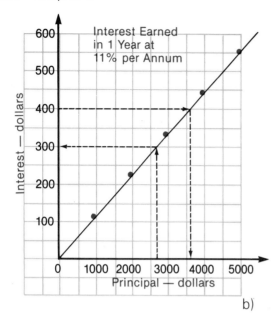

Solution. a) We draw up a table of values by choosing suitable values of *P* and calculating *I*.

P	I
$ 0	$ 0
1000	110
2000	220
3000	330
4000	440
5000	550

The points are plotted on a grid and joined with a line. This displays the relation for any value of *P* up to $5000.

b) i) The graph shows that $2750 earns approximately $300 interest in a year.
   ii) The graph shows that $400 interest is earned in a year by approximately $3650.

We express the relation of the previous example as a set of ordered pairs in the following manner.

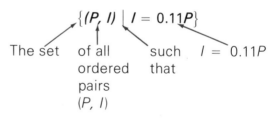

$$\{(P,\ I)\ |\ I = 0.11P\}$$

The set   of all   such $I = 0.11P$
         ordered   that
         pairs
         (*P, I*)

Now let us consider the opening problem of the chapter.

Example 2. The height, *h*, in metres, of a rocket *x* seconds after blast-off is given by the formula:
$h = x^2 + 4x$.

a) Draw a graph of this relation.

b) How high is the rocket 7.5 s after blast-off?

c) After how many seconds does the rocket reach a height of 50 m?

d) Write the relation as a set of ordered pairs.

Solution. a) To draw the graph, we first draw up a table of values.

x	h
1	5
2	12
3	21
4	32
5	45

x	h
6	60
7	77
8	96
9	117
10	140

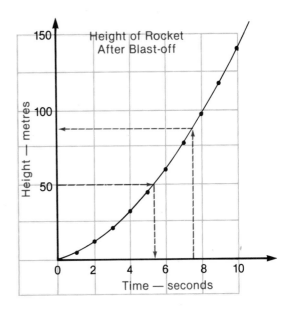

b) From the graph, the rocket's height 7.5 s after blast-off is about 85 m.

c) When $h = 50$, $x \doteq 5.2$. The rocket reaches a height of 50 m after approximately 5.2 s.

d) The relation, expressed as a set of ordered pairs, is: $\{(x, h) \mid h = x^2 + 4x\}$.

Sometimes a relation cannot be expressed as an equation. In which case, the known ordered pairs may be plotted and the graph of the relation sketched by drawing a smooth curve through the points.

Example 3. While Freda was ill, her temperature was taken every 2 h from 8 a.m. to midnight, and at 4 a.m., over a period of 48 h. The readings were plotted on a grid and the points connected by a curve.

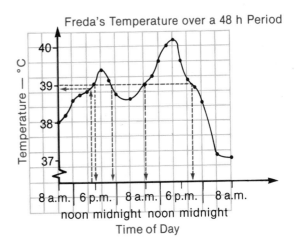

a) Estimate Freda's temperature at 5 p.m. on the first day.

b) When was Freda's temperature 39°C?

c) Estimate the total length of time that Freda's temperature was above 39°C.

Solution. a) Freda's temperature at 5 p.m. on the first day was about 38.9°C.

b) Freda's temperature was 39°C. at approximately:

6 p.m. and 10:30 p.m. on the first day, and 8 a.m. and 9 p.m. on the second day.

c) Freda's temperature was above 39°C for about 4.5 h on the first day and 13 h on the second for a total of about 17.5 h.

Exercises 9 - 4

A   1. In a physical-fitness program, the maximum pulse rate, m (beats per minute), is taken to be related to a person's age, a, by the formula $m = 220 - a$.

a) Draw a graph of this relation for persons between the ages of 18 and 50.

b) From the graph, determine the age of a person whose maximum pulse rate is 180.

c) If the working pulse rate, w, is given by the formula $w = \frac{3}{4}(220 - a)$, graph this relation for $18 \leqslant a \leqslant 50$ and determine from the graph:

i)   the working pulse rate of a person of 40;
ii) the age of a person whose working pulse rate is 150.

2. The height, h, in metres, of a pebble t seconds after being dropped from a height of 335 m is given by: $h = 335 - 4.9t^2$.

a) Graph this relation.

b) How far is the pebble from the ground after 4.5 s?

c) When is the pebble 90 m from the ground?

d) Repeat parts (b) and (c) by substituting in the formula. Which method is easier?

e) Write the relation as a set of ordered pairs.

3. The speed of a vehicle, v, in kilometres per hour, can be estimated from the length, d, in metres, of the skid marks using the formula $v = -7 + 8.2\sqrt{d}$.

a) Graph the relation for $50 \leqslant d \leqslant 300$.

b) How long would be the skid marks of a car travelling at
i) 80 km/h?   ii) 120 km/h?

c) How fast was a car travelling if the length of its skid marks were
i) 92 m?   ii) 210 m?

4. The area, A, of an equilateral triangle of side length x is given by the formula $A = \frac{\sqrt{3}}{4}x^2$. Draw the graph of this relation. If the side length doubles, by what factor does the area increase?

5. The weekly cost of operating a certain model of car is $40 for the fixed costs and $0.05/km for gasoline and oil.
   a) Graph the relation between weekly cost and distance travelled.
   b) What is the total cost for a week's driving of 625 km?
   c) How far can you drive in a week if your car allowance is $75 per week?

6. The temperature on a summer day can be estimated by adding 5 to the number of chirps a cricket makes in 8 s.
   a) Write an equation for the relation between the temperature and the number of cricket chirps in 8 s.
   b) Draw a graph of the relation.
   c) Use the graph to estimate the temperature for the following rates of chirping: i)   17 chirps in 8 s
      ii)   100 chirps in 40 s   iii)   200 chirps in 1 min

7. Of two tame whales, Xen and Yan, Yan is the older by 2 years.
   a) Write an equation that relates their ages.
   b) Graph the relation.
   c) How old will Xen be when Yan is 9 years old?

8. Two numbers are related as follows. Double the first number plus the second number equals 12.
   a) Write the relation as a set of ordered pairs.
   b) Draw a graph of the relation.

9. Two numbers are related as follows. Triple the first number plus the second number equals 18.
   a) Write the relation as a set of ordered pairs.
   b) Draw a graph of the relation.

10. When there are no traffic delays, a taxi company charges 80¢ for the first 0.2 km and an additional 10¢ every 0.2 km after that.
   a) Show the relation of cost and distance on a graph for trips up to 3.0 km.
   b) From the graph, determine
      i)   the cost of a taxi ride of 2.9 km;
      ii)   the distance travelled for a fare of $1.80.

9 - 5   Linear Relations

A **linear relation** is one in which the terms containing the variable are all of the first degree. As the name implies, the graph of a linear relation is a straight line or line segment.

These are linear relations:
$$\begin{cases} \{(x, y) \mid 3x + 5y = 15\} \\ \{(x, y) \mid 5x - 7y = -3\} \\ \{(a, h) \mid h = 17 - \frac{1}{2}a\} \end{cases}$$

These are not linear relations:
$$\begin{cases} \{(x, y) \mid y = x^2 - 5x + 6\} \\ \{(x, y) \mid y = \sqrt{x}\} \\ \{(t, h) \mid th = 3\} \end{cases}$$

The coordinates of only two points are needed to draw the graph of a linear relation. A third point should always be obtained as a check.

Example 1.   Graph the relation $\{(x, y) \mid 2x - 5y = 10\}$.

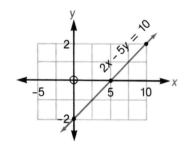

Solution.   When $x = 0$, $y = -\dfrac{10}{5}$,   or -2.

When $y = 0$, $x = \dfrac{10}{2}$,   or   5.

When $x = 10$, $y = \dfrac{-10}{-5}$,   or   2.

We chose a third point with an x-coordinate of 10 because its y-coordinate is an integer. Points with coordinates that are integers are easier to plot.

Example 2.   The number of hours of sleep needed by persons up to age 18 is given by the formula $h = 17 - \dfrac{1}{2}a$, where a is the person's age in years.

a)   Draw the graph of the relation.

b)   Write the relation as a set of ordered pairs.

Solution.   a)   When $a = 0$, $h = 17$.
When $a = 10$, $h = 12$.
When $a = 18$, $h = 8$.
It is not reasonable to substitute 0 for h because everyone needs some sleep.

b) Only reasonable values of a should be used. Therefore the relation must include the restrictions: $0 \leqslant a \leqslant 18$. The relation is:

$$\{(a, h) \mid h = 17 - \tfrac{1}{2}a, \ 0 \leqslant a \leqslant 18\}.$$

When writing and graphing relations, care should be taken that only reasonable values of the variables are included.

Exercises 9 - 5

1. Which of the following graphs represent linear relations?

a)

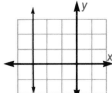

b)

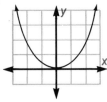

c)

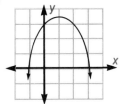

d)

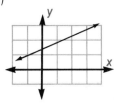

e)

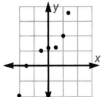

f)

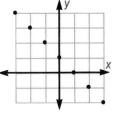

2. Which of these relations are linear?

a) $\{(x, y) \mid 3x - y = 7\}$

b) $\{(x, y) \mid 7x - 5 = 2y\}$

c) $\{(x, y) \mid 8x + 2xy = 11\}$

d) $\{(x, y) \mid 3x^2 + 16y^2 = 48\}$

e) $\{(x, y) \mid y = x^2 - 3\}$

f) $\{(x, y) \mid \dfrac{1}{x} + \dfrac{1}{y} = 1\}$

g)

x	0	1	4	9	16
y	0	±1	±2	±3	±4

h) $\{(-2, 3), (0, 2), (2, 1), (4, 0), (6, -1)\}$

i) $\{(3, -5), (5, 0), (7, 5), (9, 10)\}$

j) $\{(-3, -1), (0, 1), (3, 4), (6, 9)\}$

3. Draw the graph of each relation and determine its slope.

   a) $\{(x, y) \mid 2x + 3y = 18\}$ b) $\{(x, y) \mid 4x - 3y = 12\}$

   c) $\{(x, y) \mid 7x - 4y = -28\}$ d) $\{(x, y) \mid x + 5y = 10\}$

   e) $\{(x, y) \mid y = 2x - 6\}$ f) $\{(x, y) \mid 3x = 9 - 6y\}$

B   4. Draw the graph of each relation and determine its slope.

   a) $\{(p, q) \mid 4p + 5q = 10\}$ b) $\{(m, n) \mid 6m - 8n = 12\}$

   c) $\{(r, s) \mid 3s = 2r - 8\}$ d) $\{(w, z) \mid 5w - 7z = 0\}$

   e) $\{(a, b) \mid b = a - 7\}$ f) $\{(c, d) \mid d = 3\}$

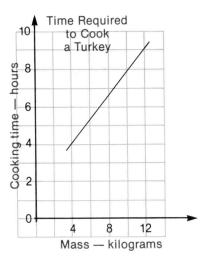

5. What is the slope of the line in *Example 2*? What does the slope represent in terms of a person's age and the number of hours of sleep required?

6. The graph shows the relation between the time required to cook a turkey and the mass of the turkey.

   a) What is the cooking time for
      i) a 5 kg turkey?   ii) an 8 kg turkey?

   b) Find the slope of the line.

   c) What does the slope of the line represent?

7. The graph shows the distance a car has travelled in a given time for a journey lasting 6 h.

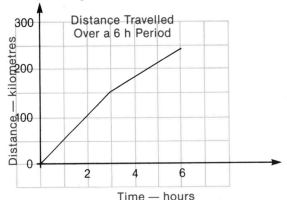

   a) How far had the car travelled in 4 h?

   b) How long did it take to cover the first 125 km?

   c) How much farther did the car travel in the first 3 h than in the last 3 h?

   d) Find the slope of each line segment.

   e) What do the slopes represent?

   f) Does the graph represent a linear relation?

8. A car travels at a constant speed from Edmonton to Calgary. The graph relates the distance of the car from its destination to the time it has been travelling.

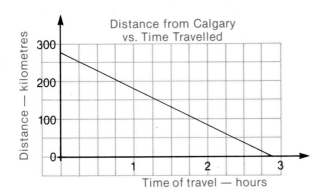

a) How far is Calgary from Edmonton?

b) What was the total travelling time?

c) After 2 h, how far was the car from Calgary?   from Edmonton?

d) Find the slope of the line.

e) What does the slope represent?

9. An old oven is calibrated in Fahrenheit units. The relation between the Fahrenheit temperature scale and the Celsius scale is given by the formula:

$$C = \frac{5}{9}(F - 32).$$

a) Draw the graph of this relation.

b) What should be the Fahrenheit setting when a recipe calls for a setting of:
   i) 90°C?   ii) 120°C?   iii) 200°C?

c) What is the slope of the line?   What does it mean?

10. From the age of 8 until he was 15, the relation between Linus's height and his age was linear. He was 132 cm tall when he was 8 and 156 cm tall when he was 12.

a) Draw a graph of the relation between height and age.

b) Use the graph to estimate
   i) Linus's height when he was 10; .   15;
   ii) Linus's age when his height was 148 cm;   165 cm.

c) Find the slope of the line.

d) State what the slope represents.

9 - 6 Quadratic Relations

A **quadratic relation** is one that has at least one second degree term but no terms of a higher degree. The graph of a quadratic relation is usually a curve.

These are quadratic relations:
$$\begin{cases} \{(x, y) \mid x^2 + y^2 = 25\} \\ \{(x, y) \mid xy = 10\} \\ \{(t, d) \mid d = 4.9t^2\} \end{cases}$$

These are not quadratic relations:
$$\begin{cases} \{(x, y) \mid y = x^3 - 9x\} \\ \{(x, y) \mid y = 2^x\} \\ \{(x, y) \mid x + xy^2 = 4y\} \end{cases}$$

Example 1. The distance, d, in metres, that a stone falls from rest in t seconds is given by P, where $P = \{(t, d) \mid d = 4.9t^2\}$.

a) Draw the graph of this relation.

b) From the graph, estimate how far the stone falls in 2.5 s.

c) About how long does it take to fall 100 m?

Solution. a) We draw up a table of values to obtain the coordinates of points on the graph.

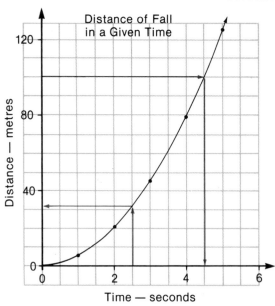

t	d
0	0
1	4.9
2	19.6
3	44.1
4	78.4
5	122.5

b) When $t = 2.5$, $d \doteq 31$. That is, in 2.5 s, the stone falls about 31 m.

c) When $d = 100$, $t \doteq 4.5$. That is, it takes about 4.5 s for the stone to fall 100 m.

The graph of *Example 1* is part of a curve called a **parabola**.

Example 2. Draw the graph of H where $H = \{(x, y) \mid xy = 10\}$.

Solution.   We draw up a table of values to obtain the coordinates of points on the graph.

x	y
1	10
2	5
3	$3\frac{1}{3}$
4	$2\frac{1}{2}$
5	2
10	1

x	y
−1	−10
−2	−5
−3	$-3\frac{1}{3}$
−4	$-2\frac{1}{2}$
−5	−2
−10	−1

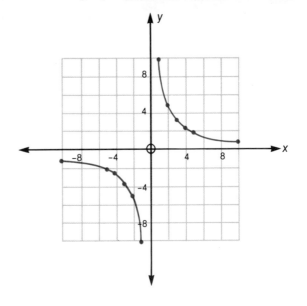

The curve is called a **hyperbola**.

Example 3.   Draw the graph of C where
$C = \{(x, y) \mid x^2 + y^2 = 36\}$.

Solution.   We draw up a table of values to obtain the coordinates of points on the graph.

x	y
0	±6
2	±5.66
−2	±5.66
4	±4.47
−4	±4.47
6	0

x	y
±6	0
±5.66	2
±5.66	−2
±4.47	4
±4.47	−4
0	6

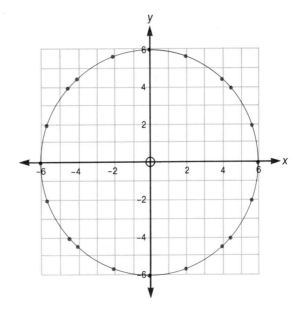

The graph of C is a **circle** with centre (0, 0) and radius 6 units. C is the set of all points 6 units from the origin.

Exercises 9 - 6

A   1. Graph these relations and name each curve.
   a) $\{(x, y) \mid y = x^2\}$
   b) $\{(x, y) \mid x^2 + y^2 = 25\}$
   c) $\{(x, y) \mid xy = 12\}$

2. Graph these relations. If possible, name each curve.
   a) $\{(x, y) \mid y = 4x - x^2\}$ b) $\{(x, y) \mid x^2 + 4y^2 = 16\}$
   c) $\{(x, y) \mid xy = -24\}$

3. Graph these relations and name each curve:
   a) $\{(x, y) \mid y = \sqrt{x}\}$ b) $\{(x, y) \mid 9x^2 + y^2 = 36\}$
   c) $\{(x, y) \mid x^2 - y^2 = 16\}$

4. A car, improperly parked, begins to coast down a hill. The distance it rolls, d, in metres, is given by $d = \frac{1}{2}t^2$, where t is the time in seconds from the start of roll.
   a) Graph the relation for $0 \leqslant t \leqslant 15$, and name the curve.
   b) How far does the car roll in the first 7.5 s?
   c) How long does it take the car to roll 75 m?

B 5. Graph these relations:
   a) $\{(x, y) \mid x^2 + 4y^2 = 36\}$ b) $\{(x, y) \mid 4x^2 + y^2 = 36\}$
   c) $\{(x, y) \mid x^2 - 4y^2 = 36\}$ d) $\{(x, y) \mid 4x^2 - y^2 = 36\}$

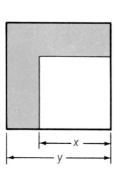

6. When a square is cut from another square, in the manner shown in the diagram, the area of the remaining portion is 144 cm^2.
   a) Find the relation between x and y.
   b) Draw the graph of the relation.

7. At 27°C, the pressure and volume of a 3.9 g sample of oxygen are related by the formula $PV = 300$, where P is the pressure in kilopascals (kPa) and V is the volume in litres.
   a) Graph the relation for $100 \leqslant P \leqslant 1000$.
   b) At what pressure is the volume 1.5 L?
   c) What is the volume when the pressure is 250 kPa?

8. An aircraft travelling at x km/h uses fuel at the rate of y L/h, where $y = 0.01x^2 + 5x + 100$.
   a) Graph the relation for $200 \leqslant x \leqslant 400$, and name the curve.
   b) What is the speed of the aircraft when fuel is being used at the rate of 2000 L/h?
   c) What is the rate of fuel consumption when the speed of the aircraft is 325 km/h?

C 9. Graph these relations:
   a) $\{(x, y) \mid x^2 - y^2 = 0\}$ b) $\{(x, y) \mid (x - 2y)(x + 3y) = 0\}$
   c) $\{(x, y) \mid x^2 + y^2 = 0\}$

*9-7   Other Non-Linear Relations

Many useful relations are neither linear nor quadratic.

Example 1.   The temperature of a cup of hot chocolate is taken every 2 min for 10 min and entered in a table as shown.

Time min	Temperature °C
0	100
2	77
4	61
6	49
8	41
10	35

a)   Graph the relation between temperature and time.

b)   Use the graph to estimate
   i)   the temperature after 5.5 min;
   ii)   when the temperature was 37°C.

c)   What was the drop in temperature in
   i)   the first minute?   ii)   the third minute?

Solution.   b)   i)   After 5.5 min, the temperature was approximately 52°C.
   ii)   The temperature was 37°C after approximately 9.2 min.

c)   i)   In the first minute, the temperature falls from 100°C to approximately 87°C, or about 13°C.
   ii)   In the third minute, the temperature falls from 77°C to approximately 68°C, or about 9°C.

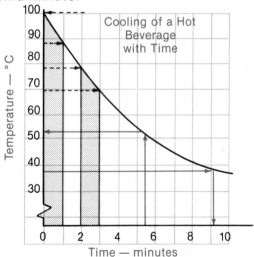

If T is the temperature in degrees Celsius, and t the time in minutes, the equation for this relation is $T = 80e^{-0.17t} + 20$, where e, like π, is an irrational number and is approximately equal to 2.718.

Example 2. Draw the graph of the relation defined by
$\{(x, y) \mid y = x^3 - 9x, \ -4 \leqslant x \leqslant +4\}$.

Solution. We use several values of x between -4 and $+4$ to construct a table of values.

x	0	0.5	1	1.5	2	2.5	3	3.5	4
y	0	-4.4	-8	-10.1	-10	-6.9	0	11.4	28

x	0	-0.5	-1	-1.5	-2	-2.5	-3	-3.5	-4
y	0	4.4	8	10.1	10	6.9	0	-11.4	-28

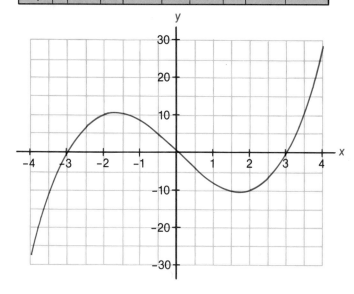

$y = x^3 - 9x$ is a *cubic* equation.

Example 3. In a science experiment, a mass, *m,* is suspended from a beam by a spring, as shown. When it is pulled down and released, the mass moves up and down. From the recorded observation the accompanying graph is drawn showing the heights of the mass above the floor during the first few seconds of motion.

a) What is the height of the mass after 1 s? after 2.8 s?

b) When is the height of the mass 40 cm? 140 cm?

c) How long does it take the mass to move up and down once?

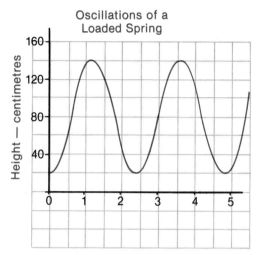

Oscillations of a
Loaded Spring

Height — centimetres

Time — seconds

Solution. a) After 1 s the height of the mass is about 132 cm.
After 2.8 s the height of the mass is about 54 cm.

b) The height of the mass is 40 cm at about 0.3 s, 2.1 s, 2.7 s, 4.5 s, and 5.2 s.
The height of the mass is 140 cm at about 1.2 s and 3.5 s.

c) The height of the mass is 20 cm at 0 s and at 2.4 s. Therefore it takes 2.4 s to move up and down once.

The mass moves up and down in what is called simple harmonic motion. The equation of the relation is $h = 60 \sin 2\pi \left(\dfrac{t - 0.6}{2.4}\right) + 80$.

Exercises 9 - 7

1. Classify each of these graphs as linear, quadratic, or other:

a)

b)

c)

d)

e)

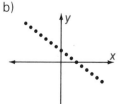
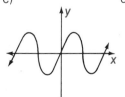
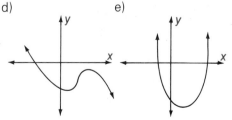

B   2. Draw the graph of each relation for $-3 \leqslant x \leqslant 3$:

   a) $\{(x, y) \mid y = 2^x\}$

   b) $\{(x, y) \mid y = x^3 - 4x\}$

   c) $\{(x, y) \mid y = \dfrac{4}{x^2 + 1}\}$

3. If a ball is dropped from a height of 2 m and allowed to bounce freely, the height, h, in metres, to which it bounces is given by the formula $h = 2(0.8)^n$, where n is the number of bounces.

   a) Draw a graph of the relation for $1 \leqslant n \leqslant 7$.

   b) What height does the ball reach after the sixth bounce?

   c) On what bounce does it reach a height of approximately 0.82 m?

4. The free end of a diving board dips d cm when under a load of x kg. The relation between the dip and the mass is given by the formula $d = 0.000\,01x^2(x + 50)$.

   a) Draw the graph of this relation for $25 \leqslant x \leqslant 125$.

   b) What would be the dip

     i) if *you* stood on the end of the board?

     ii) with someone twice your mass on the end?

5. The table shows the population of Canada since Confederation.

Year	Population (millions)	Year	Population (millions)
1871	3.7	1931	10.4
1881	4.3	1941	11.5
1891	4.8	1951	14.0
1901	5.4	1961	18.2
1911	7.2	1971	21.6
1921	8.8	1979	23.8

   a) Show this information graphically.

   b) In what year was the population 12 000 000?

   c) What was the population in 1920?

   d) Use the graph to estimate the population in 1991.

6. The number, N, of bacteria in a culture n h after midnight is given by $N = 1000 \times 2^n$.

   a) Draw a graph of this relation for $3 \leqslant n \leqslant 9$.

   b) How many bacteria are there at 4:30 a.m.?   at 6:45 a.m.?

   c) When would the number of bacteria in the culture be 500? 10 000?

7. Psychology experiments provide the following data relating to the retentiveness of a person's memory.

Time (days)	1	5	15	30	60
Recall (percent)	84	71	61	56	54

a) Show this information on a graph.

b) After how many days would you expect to recall only 75% of what you had learned for an examination?

c) How much are you likely to *forget* over the March break?

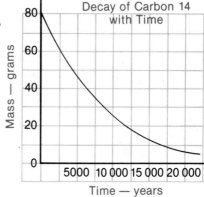

8. The graph approximates the decay curve for carbon 14 over a long period of time.

a) What is the mass of carbon 14 after
   i) 3000 years?   ii) 6000 years?
   iii) 9000 years?   iv) 15 000 years?

b) The half-life of a radioactive substance is the time it takes to lose half its mass. What is the half-life of carbon 14?

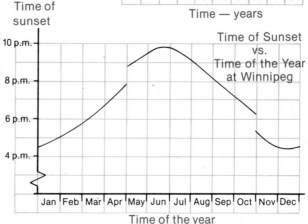

C   9. The graph shows how the time of sunset varies at Winnipeg with the time of year.

a) Use the graph to find the time of sunset on
   i)   March 1;
   ii) August 15;
   iii) December 20.

b) At what time of year does the sun set at
   i)   5 p.m.?
   ii) 6:30 p.m.?
   iii) 8:30 p.m.?

c) Why are there two breaks in the curve?

10. The numbers 3 and 1.5 are unusual because their sum equals their product. That is, $3 + 1.5 = 3 \times 1.5$.

a) State a relation, in terms of x and y, between all numbers whose sum equals their product.

b) Find three more pairs of numbers whose sum equals their product.

c) Draw the graph of the relation.

11. a) State a relation, in terms of x and y, between all numbers whose difference equals their product.

b) Draw a graph of the relation.

Mathematics Around Us

THE RETURN OF HALLEY'S COMET

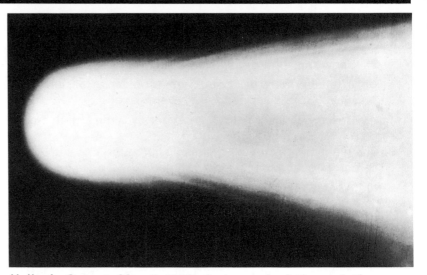

Halley's Comet, May 8, 1910 photographed through the reflecting telescope on Mount Wilson

In 1705, the English astronomer, Edmund Halley, concluded from observations and calculations that the appearances of a comet in 1531, 1607, and 1682 were the return visits of the same comet. He predicted it would return again in 1758, and it did. It was first sighted on Christmas Day of that year.

Astronomers now know that this comet travels in a narrow elliptical orbit around the sun approximately once every 77 years (the time varies slightly owing to the influence of the planets). It is predicted that Halley's comet will be seen again in 1985 and will reach its perihelion (position closest to the sun) February 9, 1986.

Coordinate geometry can be used to show the relationship between the orbit of Halley's comet and those of the planets.

1. Draw a pair of axes and on each axis mark a scale ranging from −40 to +40. The origin (0, 0) represents the position of the sun.

2. With centre (0, 0), draw circles with the following radii to represent the orbits of the planets.

Planet	Earth	Mars	Jupiter	Saturn	Uranus	Neptune
Radius of Orbit	1	1.5	5	9.5	19	30

3. Each of the following points represents the approximate position of Halley's comet in a certain year. Plot the points on your graph and draw a smooth curve through them.

Year	1918	1933	1948	1963	1978
Position	(17, 4.5)	(30, 3.1)	(35, 0)	(30, −3.1)	(17, −4.5)

Year	1984	1985	1986	1987	1988
Position	(9, −4.0)	(6, −3.6)	(−1, 0)	(6, 3.6)	(9, 4.0)

4. At what points in its orbit is the comet moving fastest and when is it moving slowest?

THE SEARCH FOR A POLYNOMIAL
TO GENERATE PRIME NUMBERS

A **prime** number is any whole number, greater than 1, which is divisible by only itself and 1. Since the discovery of prime numbers, mathematicians have attempted to find a polynomial for generating them.

In 1772, the great Swiss mathematician, Leonhard Euler, devised the polynomial

$$n^2 - n + 41$$

for the generation of primes.

Leonhard Euler 1707-1783

If $n = 1, n^2 - n + 41 = 1^2 - 1 + 41$, or 41 — a prime

If $n = 2, n^2 - n + 41 = 2^2 - 2 + 41$, or 43 — a prime

If $n = 3, n^2 - n + 41 = 3^2 - 3 + 41$, or 47 — a prime

If $n = 4, n^2 - n + 41 = 4^2 - 4 + 41$, or 53 — a prime

The polynomial does, in fact, produce primes for $n = 1, 2, 3, 4, \ldots, 40$.

However, if $n = 41$,

$$n^2 - n + 41 = 41^2 - 41 + 41, \text{ or } 1681.$$

This number is divisible by 41 and therefore is not a prime number.

In 1879, E.B. Escott devised the polynomial:
$$n^2 - 79n + 1601$$

1. Test this polynomial for several values of n to see if the results are prime numbers.

2. a) Substitute 80 for n, and show that the result is *not* a prime number.

   b) Find another value of n which gives a result that is not a prime number.

3. Find a value for n for which the following polynomials do not produce a prime number:

   a) $n^2 + n - 1$

   b) $n^2 - n + 17$

   c) $n^2 - n + 41$ (for a value of n other than 41)

At the present time, there is no known polynomial that produces *only* prime numbers. There is also no known polynomial that will produce *all* the prime numbers.

9 - 8   Relations as Mappings

In Section 9-3, we showed the relation between the cost of
theatre tickets and the number of tickets bought by plotting
points on a grid. The graph is reproduced below with arrows
showing how it is read. The use of arrows suggests three
other ways that the relation might be shown.

The graph of Section 9-3

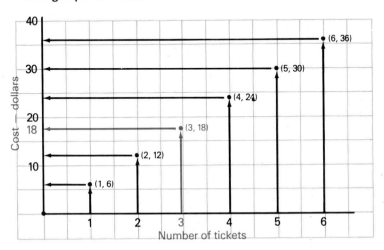

I. Plotted points omitted

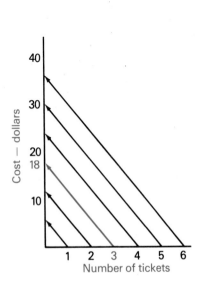

II. Points matched on
   parallel lines

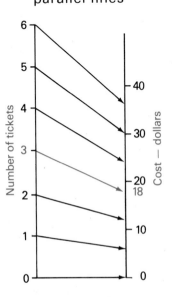

III. Points matched
   in sets

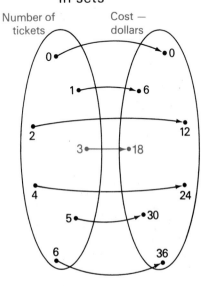

Graphs that relate numbers in one set to numbers in another set by means of arrows are called **mapping diagrams**. The relation is referred to as a **mapping**.

Example 1. On a mapping diagram, graph the relation defined by {(−6, −2), (−2, 1), (−1, −2), (1, 5), (4, 5)}.

Solution.

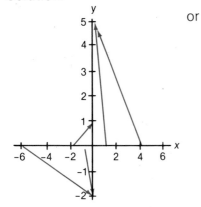 or 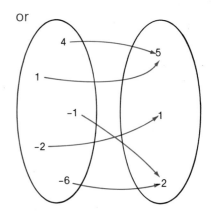 or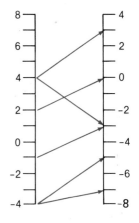

Example 2. Write the relation in the mapping diagram as a set of ordered pairs.

Solution. The tails of the arrows indicate the first elements of the ordered pairs, the heads the second. Therefore the relation is {(4, 3), (4, −3), (2, 0), (−1, −3), (−4, −5), (−4, −7)}.

It is impractical to draw mapping diagrams for relations containing many ordered pairs; there would be too many arrows. However, when a relation can be expressed as an equation it can also be described as a mapping.

In Section 9-4, we studied the relation between the principal amount of money P and a year's interest, I, when the annual rate of interest was 11%. The equation relating P and I is $I = 0.11P$. Instead of writing the equation, we may write

$$P \rightarrow 0.11P \qquad (P \geqslant 0).$$

This way of writing a relation is called a **mapping notation**. We say that "P is mapped onto $0.11P$". The value of the principal is mapped onto the value of the interest. If we start with a value of P, we calculate the value of $0.11P$.

If $P = 3000$, then $0.11P = 330$. We write $3000 \rightarrow 330$.
If $P = 5000$, then $0.11P = 550$. We write $5000 \rightarrow 550$.

Example 3. Write these relations using mapping notation.

a) $R = \{(x, y) \mid y = x^2 - 5\}$

b) $S = \{(x, y) \mid 2x - y = -1\}$

Solution. a) In R, the value of x is squared and then decreased by 5. We write:

$$x \to x^2 - 5$$

b) In S, we rearrange the equation to get $y = 2x + 1$. The value of x is doubled and then increased by 1. We write:

$$x \to 2x + 1$$

Exercises 9 - 8

A 1. List the relations represented by these mapping diagrams:

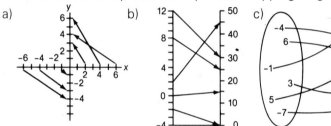

a)   b)   c)

2. Draw three different mapping diagrams for each of these relations:

a) $\{(1, 3), (2, 5), (3, 7), (4, 9), (5, 11)\}$

b) $\{(-3, 2), (-1, 4), (1, 2), (-3, 7), (5, -1), (6, 5)\}$

c) $\{(-2, 4), (-1, 1), (1, 1), (2, 4), (3, 9)\}$

3. Draw a mapping diagram for each of these relations for integral values of x:

a) $\{(x, y) \mid y = 2x, \quad -3 \leqslant x \leqslant 3\}$

b) $\{(x, y) \mid y = 2x - 1, \quad -3 \leqslant x \leqslant 3\}$

c) $\{(x, y) \mid 3x + 2y = 12, \quad -2 \leqslant x \leqslant 4\}$

d) $\{(x, y) \mid y = 2x^2 - 1, \quad -2 \leqslant x \leqslant 2\}$

e) $\{(x, y) \mid y = 10 - \frac{2}{3}x^2, \quad -3 \leqslant x \leqslant 3\}$

4. Write the relations in Exercise 3 using mapping notation.

5. Give five ordered pairs belonging to each of these relations:

a) $x \to 3x - 2$ b) $x \to x^2 - 4x$ c) $y \to 3y^2 + y - 2$

d) $m \to \frac{1}{2}m + 5$ e) $y \to 3y^2 - 7$ f) $z \to \frac{1}{3}z^2 - 3$

6. Draw a mapping diagram using the five ordered pairs in each of the relations in Exercise 5.

7. Write each of these relations using mapping notation and draw a mapping diagram for at least six points:
   a) $\{(x, y) \mid 2x - 3y = 18\}$
   b) $\{(x, y) \mid x^2 - 5x + y = 2\}$
   c) $\{(1, 2), (2, 4), (3, 6), (4, 8), (5, 10)\}$
   d)   e)

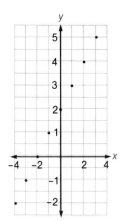

Review Exercises

1. Plot the points: $K(-4, -1), L(-1, -4), M(5, 2), N(2, 5)$. Draw the line segments: KL, LM, MN, NK. What familiar polygon have you drawn?

2. Where are the points with
   a) first coordinate -2?   b)   second coordinate $+3$?
   c) x-coordinate positive, y-coordinate negative?

3. Find the slope of the line passing through the points:
   a) $(6, -1)$ and $(2, 7)$;   b) $(-5, -1)$ and $(7, -2)$;
   c) $(2, 7)$ and $(-4, 4)$;   d) $(6, 3)$ and $(-5, -2)$.

4. On a coordinate system, draw the line through the point
   a) $(1, -1)$ with slope 3;   b) $(1, -1)$ with slope -3;
   c) $(-2, -4)$ with slope $\frac{1}{3}$;   d) $(2, 1)$ with slope $-\frac{3}{4}$.

5. State the relation defined by each graph:

a)

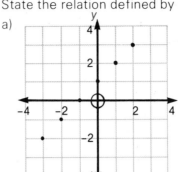

b)

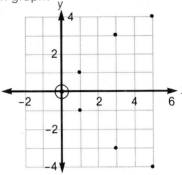

6. The first of two numbers equals three times the second.
   a) Express the relation as an equation.
   b) Write the relation as a set of ordered pairs.
   c) Draw a graph of the relation.

7. Draw the graph of each relation and determine its slope:
   a) $\{(x, y) \mid 3x + y = 9\}$ b) $\{(x, y) \mid 5x - 3y = 15\}$
   c) $\{(x, y) \mid y = 5x\}$ d) $\{(r, s) \mid s = 7 - 2r\}$

8. The number of students, s, and teachers, t, in a school are determined by the relation: $t = \frac{1}{23}s + 5$.
   a) Graph the relation.
   b) How many teachers are required for a school with 460 students?
   c) How many students are there in a school having 10 teachers?

9. Graph these relations and name each curve:
   a) $\{(x, y) \mid y = 2x^2\}$ b) $\{(x, y) \mid y^2 = 4x\}$
   c) $\{(x, y) \mid x^2 + y^2 = 4\}$ d) $\{(x, y) \mid xy = 16\}$

10. When a square is cut from another square, in the manner shown in the diagram, the area of the remaining portion is 36 cm^2.

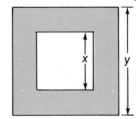

   a) State the relation between x and y.
   b) Draw the graph of the relation for $0 \leqslant x \leqslant 10$.

11. The diagram shows two squares, each with side length 6 cm, overlapping to form a rectangle with dimensions x and y.

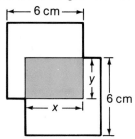

a) Write an expression for the perimeter of the combined figures.

b) If the perimeter of the combined figures is twice the perimeter of the rectangle, find the relation between x and y.

c) Draw the graph of the relation using reasonable values of x and y.

12. The diagram shows an annulus of inner radius x and outer radius y. Its area is 9π cm².

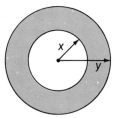

a) Find the relation between x and y.

b) Draw the graph of the relation using reasonable values of x and y.

13. Draw the graph of each relation for $-3 \leqslant x \leqslant 3$:

a) $\{(x, y) \mid y = x^2 - 1\}$ b) $\{(x, y) \mid y = 2^x - 2\}$

14. Draw three different mapping diagrams for the relation: $\{(2, 3), (2, -3), (-2, 3), (-2, -3)\}$.

15. Draw a mapping diagram for each relation for integral values of x:

a) $\{(x, y) \mid y = 3x - 2, \quad -4 \leqslant x \leqslant 4\}$

b) $\{(x, y) \mid y = x^2 - 1, \quad -3 \leqslant x \leqslant 3\}$

16. Give five ordered pairs belonging to each relation:

a) $x \rightarrow 2x + 5$ b) $x \rightarrow 3x^2 - 10$

CALCULATOR POWER

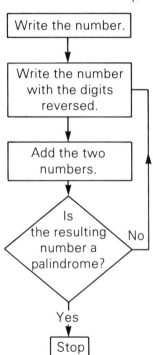

Write the number.

↓

Write the number
with the digits
reversed.

↓

Add the two
numbers.

↓

Is
the resulting
number a
palindrome? — No

↓ Yes

Stop

Numerical Symmetry

The English language has words and phrases in which the letters appear in the same sequence when read in either direction. Such words and phrases are called **palindromes**.

Examples of palindromes are:

Noon　　　　Level　　　　Deed　　　　Kayak

Madam, I'm Adam　　　A man, a plan, a canal, Panama

Numbers can be palindromic as well. That is, they are symmetric about their middle digit or digits. Some examples are:

66　　　121　　　5005　　　38 783　　　246 642

Palindromic numbers can be generated from any number by the following method:

Example 1

Key in	Read
39	*39*
⊞ 93	*93*
⊞	*132*
231	*231*
⊞	*363*

363 is the palindrome generated from 39.

Example 2

Key in	Read
289	*289*
⊞ 982	*982*
⊞	*1271*
1721	*1721*
⊞	*2992*

2992 is the palindrome generated from 289.

The calculator is particularly well-suited for finding palindromic numbers because while some palindromic numbers are found quickly, finding others may take many reversals and additions. Try to find a palindrome for each of these starting numbers:

1. 146	2. 834	3. 165	4. 724
5. 384	6. 508	7. 127	8. 916
9. 628	10. 283	11. 481	12. 4286

If your calculator displays eight digits, the palindrome for 365 will stretch it to the limit. Find the palindrome starting with 365.

Cumulative Review (Chapters 7 - 9)

1. In a survey of twenty families, the number of children in each family was found to be as follows:

   2 1 6 1 1 4 2 3 5 2
   0 2 2 3 0 3 4 5 2 1

   a) Make a frequency table.

   b) Display the data on a frequency graph.

2. Find the measures of central tendency for the data in Exercise 1.

3. Find the mean, median, and mode for each set of data:

   a) 12, 14, 10, 11, 14, 16, 13, 17, 11, 14

   b) 2.6, 3.1, 2.9, 3.2, 2.7, 3.1, 3.1, 2.5, 3.0, 2.9

4. What is the probability that the number on a ball selected at random from 12 balls numbered from 1 to 12 is

   a) odd? b) prime? c) a multiple of 5?

   d) a two digit-number? e) divisible by 3?

5. A thumbtack is tossed 500 times and lands "point up" 275 times. About how many times should it land "point up" if it is tossed 5200 times?

6. A bag contains 2 black balls and 3 red balls. Find the probability of drawing 2 red balls in succession if

   a) the first ball is replaced before the second is drawn;

   b) the first ball is not replaced.

7. Simplify:

   a) $(3a^2 + 6a - 8) + (-5a^2 - a + 4)$

   b) $(2x + 3) + (6x - 1) - (x - 6)$

   c) $(3 - 4c + c^2) - (2c - 3c^2) + (4 - 3c + 6c^2)$

   d) $(18x - 27) + (35x + 23) + (25x + 14)$

8. Find the product:

   a) $c(4c + 1)$ b) $s^2(2 - 5s)$ c) $-4a(2a - 1)$

   d) $3a^2(2 - 5a - a^2)$ e) $(12a - 5)3a$ f) $5x^2(2x^2 - x - 6)$

9. Simplify:

   a) $\dfrac{x^3 + 4x}{x}$ b) $\dfrac{3m^4 - 6m^3}{3m^2}$ c) $\dfrac{8x^5 - 4x^4 + 2x^3}{2x^2}$

   d) $\dfrac{-12y^3 + 9y^2}{-3y^2}$ e) $\dfrac{18a^6 - 9a^5 - 6a^3}{-3a^3}$ f) $\dfrac{10b^3 + 15b^4 - 25b^5}{5b^3}$

10. Factor:
    a) $3a^2 + 9a$
    b) $3y^2 + 12y - 6$
    c) $16c^2 - 32c + 24c^3$
    d) $8a^2b - 32ab^2 + 16a^2b^2$
    e) $3x^4 - 6x^5$
    f) $3m^3n^3 - 12mn + 9m^2n^2 - 6m^4n$

11. Find the product:
    a) $(m + 3)(m + 4)$
    b) $(a + 9)(a - 7)$
    c) $(5x - 2)(2x - 1)$
    d) $3(y - 3)(3 - y)$
    e) $5(5a + 1)(4a - 3)$

12. Find the product:
    a) $(x + 2)(x + 2)$
    b) $(c - 4)(c - 4)$
    c) $(4x - 3)(4x + 3)$
    d) $5(3y + 2)(3y - 2)$
    e) $2m(3 - m)(3 - m)$

13. Factor:
    a) $x^2 + 8x + 7$
    b) $a^2 + 13a + 40$
    c) $m^2 - 6m + 5$
    d) $14 - 9y + y^2$
    e) $2a^2 + 12a + 10$
    f) $a^3 - 2a^2 - 15a$

14. Factor:
    a) $x^2 - 8x + 16$
    b) $x^2 + 12x + 36$
    c) $25 - 10a + a^2$
    d) $t^2 - 100$
    e) $49 - \frac{1}{4}x^2$
    f) $3x^2 - 75$

15. A rectangular field is x m by $(3x + 5)$ m.
    a) Write expressions for the area and perimeter of the field.
    b) Find the area and perimeter when $x = 10$.

16. Find the slope of the line passing through
    a) $(5, -3)$ and $(-4, 6)$;
    b) $(-4, -3)$ and $(6, -8)$.

17. On a grid, draw a line through
    a) $(3, -2)$ with slope 2;
    b) $(-1, 4)$ with slope $- 2$.

18. State the relation defined by the graph shown.

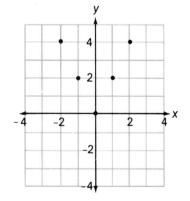

19. Draw the graph and determine the slope:
    a) $\{(x, y) \mid 4x + y = 8\}$
    b) $\{(x, y) \mid y = -3x + 1\}$

20. Graph each relation and name the curve:
    a) $\{(x, y) \mid y = -2x^2\}$
    b) $\{(x, y) \mid y^2 = 8x\}$
    c) $\{(x, y) \mid x^2 + y^2 = 49\}$
    d) $\{(x, y) \mid xy = 24\}$

21. Draw a mapping diagram for this relation for integral values of
    x: $\{(x, y) \mid y = 2x^2 - 1, \quad -4 \leqslant x \leqslant 4\}$

10 Geometry

John and Jane have a measuring tape and a compass. How can they measure the width of a river they cannot cross? (See *Example 3*, Section 10-5.)

10 - 1 What Is Geometry?

Geometry is everywhere in the world around us, as the illustrations on this page show. Study each illustration and the question that accompanies it. The questions and your answers will involve some of the ideas and language of geometry.

Architecture

What shapes do you see?

Nature

What is the shape of the horns?

Sports

Why do the runners start at different places?

Navigation

What measurements does a navigator make?

Art

How is the feeling of "depth" achieved?

Building and Decorating

What is the purpose of the weighted string?

Your answers to the above questions required knowledge of the following basic geometric figures.

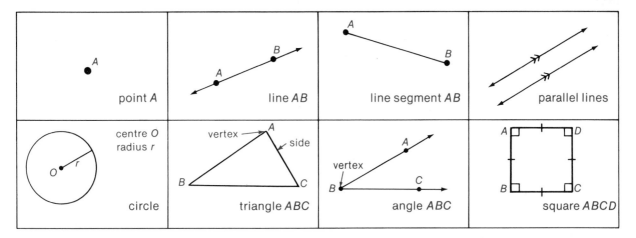

point A	line AB	line segment AB	parallel lines
centre O radius r / circle	vertex / side / triangle ABC	vertex / angle ABC	square ABCD

The study of these figures, and others made from them, is called geometry. The ancient Egyptians and Babylonians used geometric ideas to determine areas of fields and volumes of buildings such as temples and pyramids. The word "geometry" comes from two Greek words meaning *earth measure*. It was about 300 B.C. that the Greeks began to study geometry systematically. Their contributions have influenced the study of the subject to the present time.

Example 1.   What line segments can be named in the given line?

Solution.   Line segments *AB*, *BC, and AC* can be named in the line.

Problems in geometry usually involve combinations of the basic figures.

Example 2.   In how many different ways can a line and a circle intersect?

Solution.   There are three different ways, as the diagrams show.

Two points of intersection   One point of intersection   No points of intersection

Exercises 10 - 1

A 1. Collect three pictures that show how geometry arises in the world around us.

2. Name all the line segments in this line.

3. Name all the angles in the figures:

a) 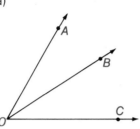   b)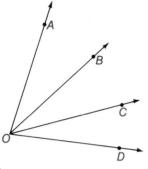

4. Name all the triangles in this figure.

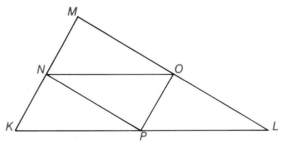

B 5. In how many different ways can two circles intersect if
   a) the circles have equal radii?
   b) the circles have different radii?

6. In how many different ways can two lines intersect?

7. In how many different ways can the following intersect?
   a) a line segment and a line
   b) a line and a triangle

8. What is the greatest number of points in which each of the following can intersect?
   a) a triangle and a circle
   b) two triangles
   c) a square and a circle

Mathematics Around Us

MISLEADING DIAGRAMS

When geometric figures are combined, the resulting diagrams can sometimes deceive the eye. This can happen in different ways.

Optical Illusions

An optical illusion can lead to a false conclusion. Are the red circles the same size?

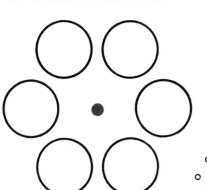

Reversing Diagrams

Some diagrams can be seen in different ways. What do *you* see?

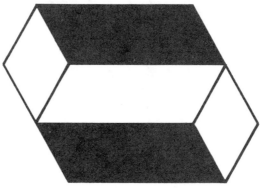

Impossible Objects

A two-dimensional diagram can be drawn of a three-dimensional object that cannot exist. Do you think you could make this object?

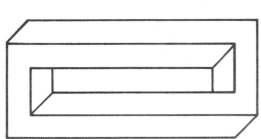

Subjective Contours

Sometimes the outline of a figure is clearly visible when it is not really there. Do you see a white triangle? Is it really there?

These examples show how careful we must be when drawing conclusions from a diagram.

Questions

1. Are the diagonal lines parallel?

2. Can you make a physical model of this drawing?

3. Is there really a white square?

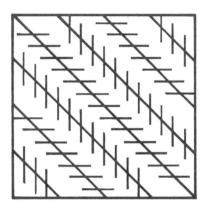

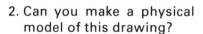

4. Do you see a white triangle? Are the line segments equal?

5. Can you make a model from this diagram?

6. Do you see two heads or a birdbath?

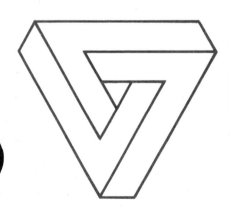

7. Are the horizontal lines bent?

8. Does this box open two ways?

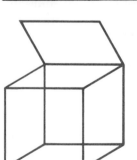

10 - 2 Angles and Intersecting Lines

Angles are classified according to their measures.

Measure	Kind of Angle	Examples
Less than 90°	acute	
90°	right	
between 90° and 180°	obtuse	
180°	straight	

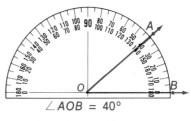

∠AOB = 40°

One degree, 1°, is $\frac{1}{360}$ of a complete rotation.

Two intersecting lines form two pairs of angles called **vertically opposite angles**:

   ∠AOD and ∠BOC   ∠AOC and ∠BOD

By drawing several pairs of intersecting lines and measuring the vertically opposite angles with a protractor, you can verify the following statement.

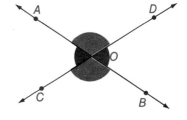

> When two lines intersect, the vertically opposite angles are equal.

Example 1. Find the value of x:

a)   b)

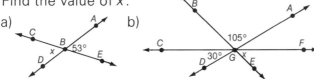

Solution. a) Since ∠CBD and ∠EBA are vertically opposite angles, x = 53°.

b) Since ∠DGE and ∠AGB are vertically opposite angles, ∠DGE = 105°.
Since ∠CGF is a straight angle,
30° + 105° + x = 180°.
     x = 180° − 135°
     x = 45°

In *Example 1(b)*, can you suggest other ways to find *x*?

Example 2. Plot the points $A(5, 7)$, $B(-3, -1)$, $C(-5, 5)$, and $D(7, 1)$ on a grid.
  a) Draw line segments AB and CD.
  b) Measure the vertically opposite angles formed by AB and CD.

Solution. a)

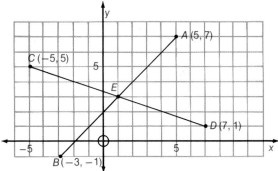

  b) If AB and CD intersect at E, then
$$\angle AED = 63°, \qquad \angle AEC = 117°,$$
$$\angle BEC = 63°, \qquad \angle BED = 117°.$$

A special case of intersecting lines occurs when they intersect at right angles. We say that the lines are perpendicular, and write: $AB \perp CD$.

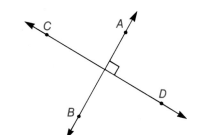

Exercises 10 - 2

A 1. In each figure, name two pairs of vertically opposite angles:

a)

b)

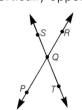

2. Find the values of *x* and *y*:

a)   b)   c)

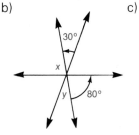

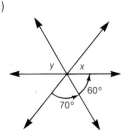

3. Plot each set of four points on a grid. Draw line segments *PQ* and *RS*. Measure the vertically opposite angles formed by *PQ* and *RS*.

a) *P*(8, 2), *Q*(0, −2), *R*(2, 4), *S*(6, −4)

b) *P*(2, 9), *Q*(−4, −2), *R*(−3, 7), *S*(3, −3)

4. Classify each angle:

a)   b)   c) 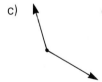   d)

5. Draw each set of three points on a grid. Draw and classify ∠*ABC*:

a) *A*(4, 0), *B*(−3, −2), *C*(0, 5)

b) *A*(7, −1), *B*(2, 2), *C*(−5, 0)

c) *A*(−4, 8), *B*(−1, 2), *C*(7, 6)

B 6. If two lines intersect to form four equal angles, what can you say about the lines?

7. How many intersection possibilities are there for a line and two parallel lines?

8. Two angles have the same vertex and a side in common. In how many ways can they be drawn?

C 9. Find the value of *x*:

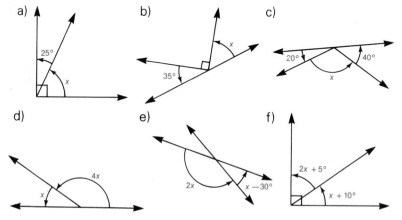

10. In how many different ways can the following intersect?

a) two parallel lines and a circle

b) two parallel lines and a line segment

10 - 3 Angles and Parallel Lines

Two lines in the same plane that never meet are called **parallel lines**. A line which intersects two or more lines is called a **transversal**.

When a transversal intersects two parallel lines, certain pairs of angles are given special names.

There are two pairs of **alternate angles.**

There are four pairs of **corresponding angles.**

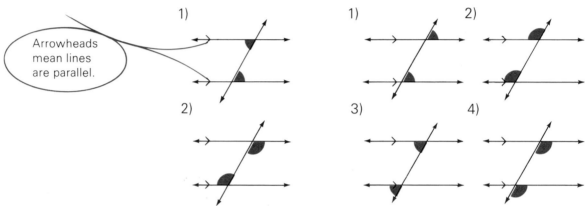

Arrowheads mean lines are parallel.

Draw several pairs of parallel lines, each pair intersected by a transversal. With a protractor, measure pairs of alternate angles to verify the following statement:

> When a transversal intersects two parallel lines, the alternate angles are equal.

In a similar fashion, measure pairs of corresponding angles and verify this statement:

> When a transversal intersects two parallel lines, the corresponding angles are equal.

Example 1. Find the values of x and y:

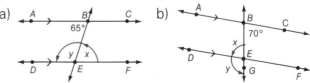

Solution. a) Since $\angle ABE$ and $\angle FEB$ are alternate angles,
$$x = 65°.$$
Since $\angle DEF$ is a straight angle,
$$y + 65° = 180°.$$
$$y = 115°$$

b) Since $\angle CBE$ and $\angle DEB$ are alternate angles,
$$x = 70°.$$
Since $\angle GEB$ is a straight angle,
$$70° + y = 180°.$$
$$y = 110°$$

Example 2. Find the value of x:

a)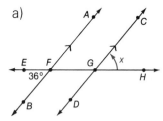

b)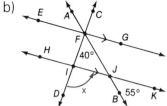

Solution. a) Since $\angle EFB$ and $\angle FGD$ are corresponding angles,
$$\angle FGD = 36°.$$
Since $\angle FGD$ and $\angle CGH$ are vertically opposite angles,
$$x = 36°.$$

b) Since $\angle KJB$ and $\angle GFJ$ are corresponding angles,
$$\angle GFJ = 55°.$$
$$\angle GFI = 55° + 40°$$
$$= 95°$$
Since $\angle GFI$ and $\angle JID$ are corresponding angles,
$$x = 95°.$$

In *Example 2*, can you suggest other ways of finding x?

Exercises   10 - 3

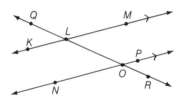

A 1. In the figure opposite, name
   a) two pairs of alternate angles;
   b) four pairs of corresponding angles.

2. Find the value of x:

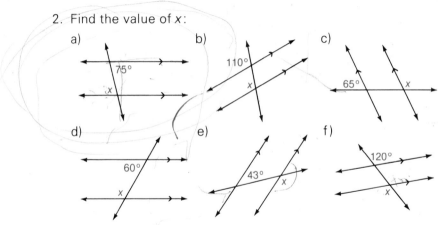

a) 75° x

b) 110° x

c) 65° x

d) 60° x

e) 43° x

f) 120° x

B 3. Find the angle measure indicated by each letter:

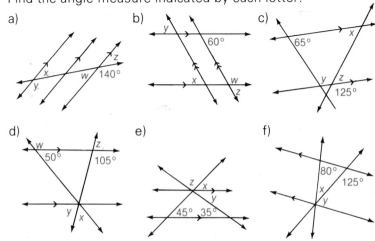

a) x y z w 140°

b) y 60° x w z

c) 65° x y z 125°

d) w 50° z 105° y x

e) z x y 45° 35°

f) 80° 125° x y

4. Can two intersecting lines both be parallel to a third line? Draw a diagram to support your answer.

5. Can two intersecting lines both be perpendicular to a third line in the same plane? Draw a diagram to support your answer.

6. In how many different ways can three lines intersect?

C 7. From the information given in the diagram, find the sum of the angles in $\triangle ABC$.

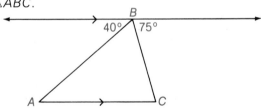

B 40° 75° A C

10 - 4   Angles and Triangles

A triangle has three sides and three angles. Draw some triangles like those shown below. With a protractor, measure
each angle. You will find that the sum of the angles is always
the same.

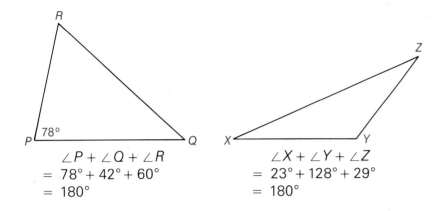

$$\angle P + \angle Q + \angle R$$
$$= 78° + 42° + 60°$$
$$= 180°$$

$$\angle X + \angle Y + \angle Z$$
$$= 23° + 128° + 29°$$
$$= 180°$$

> The sum of the angles in any triangle is 180°.

Example 1.   On a grid, draw the triangle with vertices $A(8, 4)$,
$B(1, 2)$, and $C(5, -3)$. Measure the three angles
and find their sum.

Solution.

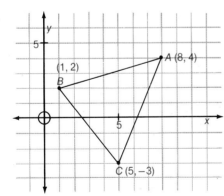

$$\angle A = 51°$$
$$\angle B = 67°$$
$$\angle C = \underline{\ 62°}$$
$$\text{Sum} = 180°$$

Example 2.   Find the value of x:

a)

b)

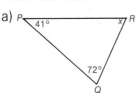

Solution.   a) The sum of the angles in $\triangle PQR$ is 180°.

That is, $x + 41° + 72° = 180°$
$$x = 180° - 113°$$
$$x = 67°$$

b) Since $\angle DBE$ and $\angle ABC$ are vertically oppo-
site angles,
$$\angle ABC = 44°.$$
Since the sum of the angles in $\triangle ABC$ is 180°,
$$\angle ABC + \angle A + \angle ACB = 180°.$$
$$44° + 53° + \angle ACB = 180°$$
$$\angle ACB = 180° - 97°,$$
$$\text{or } 83°$$
Since $\angle BCF$ is a straight angle,
$$83° + x = 180°$$
$$x = 97°$$

Triangles may be classified by the measures of their angles.

Description	Kind of Triangle	Examples
one angle is 90°	right	
one angle is obtuse	obtuse	
all angles are acute	acute	

Exercises 10 - 4

A 1. On a grid, draw the triangle with vertices $A(0, -5)$, $B(-8, 0)$, and $C(6, 3)$. Measure its angles and find their sum.

2. Classify each triangle:

a)   b)   c)   d)

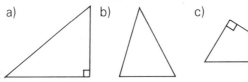

3. On a grid, draw and classify each triangle having vertices with coordinates as follows:

a) (4, 4), (7, 0), (0, −2)   b) (5, 0), (−4, 6), (−3, 1)

c) (−1, 6), (7, 4), (2, 1)   d) (−4, 0), (1, 4), (2, −2)

4. Listed below are the measures of two angles of a triangle. In each case, find the third angle and state whether the triangle is acute, right, or obtuse.

a) 35°, 65°   b) 70°, 75°   c) 40°, 25°   d) 60°, 30°

5. Find the value of *x*:

a)   b)   c)

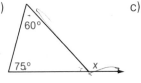

B 6. Cut any triangle *ABC* from a piece of paper. Tear off corners *A* and *B* and fit them at *C*, as shown. Explain how this shows that ∠*A* + ∠*B* + ∠*C* = 180°.

7. Find the value of *x*:

a)   b)   c)

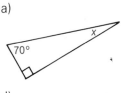
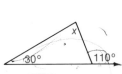
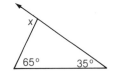

d)   e)   f)

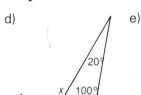
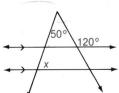

g)   h)   i)

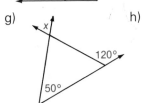
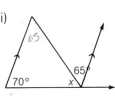
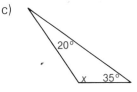

8. Giving reasons for any "no" answers, say whether a triangle can be drawn having

a) 2 acute angles;   b) 3 acute angles;

c) 2 right angles;   d) 2 obtuse angles;

e) a straight angle

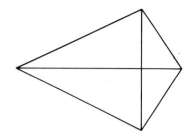

9. Explain why every triangle must have at least two acute angles.

10. a) How many triangles are in the figure opposite?
    b) How many of these triangles appear to be
       i) acute? ii) right? iii) obtuse?

11. In this figure,

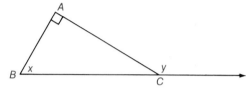

a) find *y* if: i) *x* = 30°; ii) *x* = 50°;
b) find *x* if: i) *y* = 110°; ii) *y* = 160°;
c) find the equation relating *x* and *y*.

C 12. Find the angle measure indicated by each letter:

a)   b)   c)

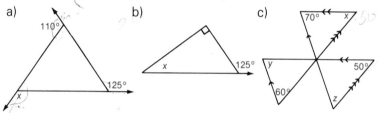

13. Find the value of *x*:

a)   b)   c)   d)

14. Find the sum of the shaded angles:

a)   b)   c)

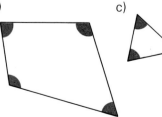

10 - 5   Isosceles and Equilateral Triangles

In Section 10-4, triangles were classified according to the measures of their angles. Triangles are also classified according to the lengths of their sides.

Description	Kind of Triangle	Examples
3 sides equal	equilateral	
at least 2 sides equal	isosceles	
no sides equal	scalene	

In $\triangle PQR$, $PQ = PR$. Measuring $\angle Q$ and $\angle R$ with a protractor, we find that they are both 66°. Draw several isosceles triangles and measure the angles opposite the equal sides. You will find that they are equal. Such measurements verify the following statement:

> In an isosceles triangle, the angles opposite the equal sides are equal.

Example 1.   Find the value of x:

a)

b)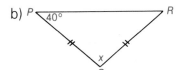

Solution.   a)   Since $AB = AC$, $\angle C = x$.
$$x + x + 40° = 180°$$
$$2x = 140°$$
$$x = 70°$$

   b)   Since $QP = QR$, $\angle R = 40°$.
   The sum of the angles in $\triangle PQR$ is 180°.
$$x + 40° + 40° = 180°$$
$$x = 180° - 80°$$
$$x = 100°$$

Example 2. Find the size of each angle in an equilateral triangle.

Solution. In equilateral △*ABC*,

since $CA = CB$, $\angle A = \angle B$, and
since $BA = BC$, $\angle A = \angle C$,
all three angles are equal and can be represented by x.
Since the sum of the angles is 180°,
$$x + x + x = 180°.$$
$$3x = 180°$$
$$x = 60°$$
Each angle of an equilateral triangle is 60°.

Consider now the problem on the first page of the chapter.

Example 3. John and Jane have a measuring tape and a compass. How can they measure the width of a river they cannot cross?

Solution. Let the width of the river be one of the equal sides of a right isoceles triangle with the 90° angle where John and Jane are standing. Then the other equal side is along the bank. The angles opposite the equal sides are 45°. John and Jane stand opposite a prominent tree on the far bank and take a reading with the compass. It is 30° east of north. Jane walks along the bank on a compass bearing of 60° west of north until the sighting of the tree is 75° east of north. With the tape they measure the distance between them. This is the width of the river.

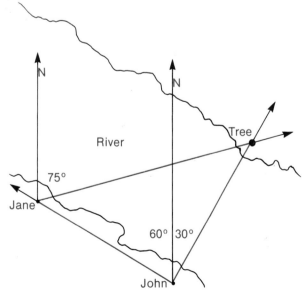

Exercises   10 - 5

A 1. Find the value of *x*:

a)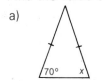

b)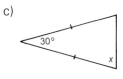

25°   *x*

c)
30°
x

70°   *x*

d)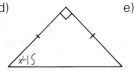

x+15

e)
x
50
65°

f)
60
x

g)

150
x
75°

h)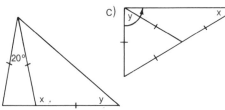

x
120

i)
20°
x 100

2. a) Are all equilateral triangles isosceles?

   b) Are all isosceles triangles equilateral?

3. In *Example 3*, if the sighting angle of the tree had been the following angles from the north direction:

   i) 10° east,   ii) 45° east,   iii) 30° west,

   a) in what direction would Jane walk to make a right isosceles triangle?

   b) what would be the sighting angle of the tree?

B · 4. Find the values of *x* and *y*:

a)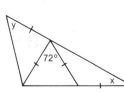

y
72°   *x*

b)

20°
x.   *y*

c)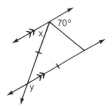

y   *x*

d)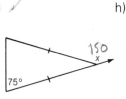

y
72°
x

e)
x
30°
y

f)
70°
x
y

5. One angle of an isosceles triangle is given. Find the remaining angles.

   a) 30°   b) 40°   c) 80°   d) 90°

6. Draw an example of
   a) an isosceles right triangle;
   b) an isosceles obtuse triangle;
   c) a scalene right triangle;
   d) a scalene obtuse triangle;
   e) an isosceles acute triangle.

7. In isosceles $\triangle ABC$,

   a) find y if: i) $x = 70°$;   ii) $x = 25°$;
   b) find x if: i) $y = 80°$;   ii) $y = 110°$;
   c) find the equation relating x and y.

8. In $\triangle XYZ$,

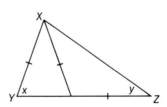

   a) find y if: i) $x = 60°$;   ii) $x = 40°$;
   b) find x if: i) $y = 40°$;   ii) $y = 25°$;
   c) find the equation relating x and y.

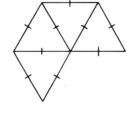

C 9. A number of equilateral triangles are joined together with whole sides touching. The diagram shows a shape formed with four equilateral triangles. How many different shapes can be formed if the number of equilateral triangles used is:

   a) 3?   b) 4?   c) 5?

10. You have a right, isosceles, plastic triangle and a tape measure. Explain how you would use them to find the height of a tree.

10 - 6 Geometric Constructions

Geometric constructions are generally performed with straightedge and compasses only. For each construction shown below, study the arcs and their centres and the lines used.

Bisector of an angle

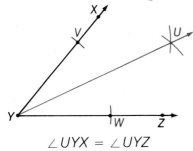

∠UYX = ∠UYZ

Perpendicular to a line at a point, *P*, on the line

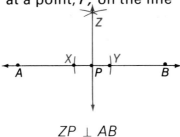

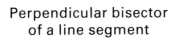

ZP ⊥ AB

Perpendicular to a line through a point, *P*, not on the line

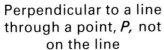

PQ ⊥ LM

Perpendicular bisector of a line segment

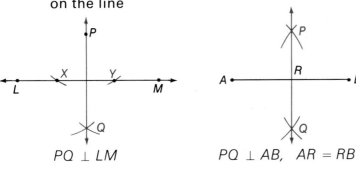

PQ ⊥ AB,   AR = RB

Line through a point, *P*, parallel to a given line

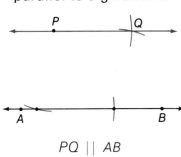

PQ || AB

Angle of 60°

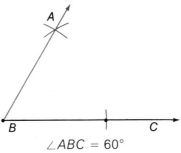

∠ABC = 60°

Does the setting of the compasses need to be changed in any of these constructions?

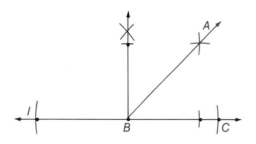

Example 1. Using straightedge and compasses, construct a 45° angle.

Solution. Draw a line, *l*, and construct a perpendicular to *l* at any point *B* on *l*. Bisect the 90° angle formed. ∠*ABC* = 45°.

Example 2. a) Construct △*ABC* such that *AB* = 6.0 cm, *BC* = 5.0 cm, and *AC* = 4.0 cm.

b) Find the midpoint, *M,* of *AB* and draw *CM*.

c) Measure the length of *CM*.

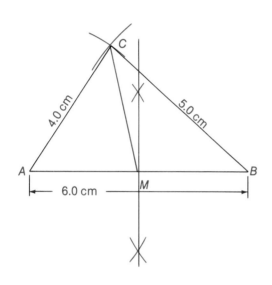

Solution. a) Draw line segment *AB* 6.0 cm long. With centre *B* and radius 5.0 cm, draw an arc. With centre *A* and radius 4.0 cm, draw an arc to intersect the first arc at *C*. Draw *CB* and *CA*.

b) Construct the perpendicular bisector of *AB*. Draw *CM,* where *M* is the point of intersection of segment *AB* and its bisector.

c) By measurement, *CM* ≐ 3.4 cm.

Exercises 10 - 6

A 1. a) Draw an acute angle and bisect it.

b) Draw an obtuse angle and bisect it.

2. Draw a line, *l*. Choose any point *A* on *l*. Construct the perpendicular to *l* at *A*.

3. Draw a line, *l*. Choose any point *B*, not on *l*. Construct the perpendicular from *B* to *l*.

4. Draw a line segment, *AB*. Construct the perpendicular bisector of *AB*.

5. Draw a line, *l*. Choose any point *P*, not on *l*. Construct a line through *P* parallel to *l*.

6. Construct an angle of

   a) 30°;

   b) $22\frac{1}{2}°$;

   c) 135°;

   d) 120°;

   e) 75°;

   f) $67\frac{1}{2}°$.

7. a) Construct $\triangle ABC$ with $BC = 8.0$ cm, $\angle B = 45°$, and $\angle C = 30°$.

   b) Find the midpoint, *M,* of *BC.* Draw *AM.*

   c) Measure: i) *AM*;   ii) $\angle BAC$;   iii) $\angle AMC$.

8. Construct $\triangle PQR$ in which $\angle Q = 90°$, $PQ = QR$, and $PQ = 5.0$ cm. Measure $\angle P$ and $\angle R$.

9. Construct $\triangle RST$ in which $\angle S = 60°$, $ST = 6.0$ cm, and $RS = 10.0$ cm. Measure $\angle R$, $\angle T$, and *RT.*

B 10. a) Construct $\triangle XYZ$ such that $XY = 4$ cm, $YZ = 5$ cm, and $XZ = 6$ cm.

   b) Construct the line through *Z* parallel to *XY.*

   c) Construct the line through *X* perpendicular to *YZ.*

   d) Construct the bisector of $\angle Y$.

11. a) Draw any angle and divide it into four equal parts.

    b) Draw any line segment and divide it into four equal parts.

12. Triangle *ABC* has vertices at $A(7, 9)$, $B(1, 1)$, $C(11, 3)$.

    a) Draw $\triangle ABC$ on a grid.

    b) Using ruler and compasses, locate *P*, the midpoint of *AB*, and *Q*, the midpoint of *AC*.

    c) Compare the lengths of *PQ* and *BC.*

C 13. Mark two points, *A* and *B,* on your paper. Construct a circle having *AB* as a diameter.

14. Mark two points, *A* and *B,* on your paper. Construct a square where *A* and *B* are

    a) the endpoints of one side;

    b) the endpoints of a diagonal;

    c) the midpoints of two opposite sides;

    d) the midpoints of two adjacent sides.

15. a) Try to construct a triangle with side lengths of
    i) 3 cm, 4 cm, 7 cm; ii) 5 cm, 5 cm, 12 cm.
    b) What conclusions do you draw?

16. Using only a straightedge, construct three angles with a sum of 180°.

17. Make a diagram similar to the one shown. Using only a straightedge, construct a third angle equal to the sum of ∠B and ∠C.

18. Make a diagram similar to the one shown.
    a) Construct *OD,* the bisector of ∠COB.
    b) Construct *OE,* the bisector of ∠AOC.
    c) Measure ∠DOE.
    d) Can you explain why the measure of ∠DOE will always be the same?

19. Which of the six constructions of this section can be done using only
    a) both sides of a ruler?
    b) a straightedge and a plastic triangle?
    c) a straightedge and compasses, with the compasses at a fixed setting?
    d) paper folding?

10 - 7 Lines and Triangles

Some special ways that a line can intersect a triangle are shown below.

The line may bisect one of the angles.

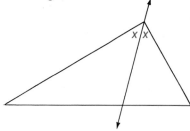

The line may be the perpendicular bisector of a side.

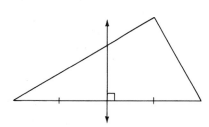

The line may pass through a vertex and be perpendicular to the opposite side.

The line may pass through a vertex and the midpoint of the opposite side.

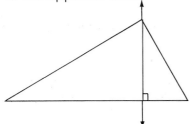

That part of the line joining the vertex to the opposite side is called an **altitude.**

That part of the line joining the vertex to the opposite side is called a **median.**

Example 1. Draw any scalene triangle and construct the perpendicular bisector of each side. What property do the three perpendicular bisectors have?

Solution.

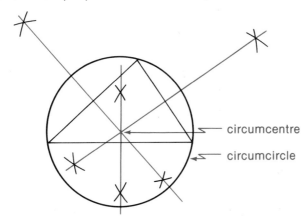

circumcentre

circumcircle

The perpendicular bisectors of the three sides of a triangle intersect at a common point called the **circumcentre.**

   Using the circumcentre and the distance to any vertex as a radius, a circle can be drawn passing through all three vertices of the triangle. This circle is called the **circumcircle.**

Example 2. Show that an altitude of a triangle may be

   a)   a side of the triangle;

   b)   outside the triangle.

Solution. a) An altitude of a
right triangle is a
side of the triangle.

b) An altitude of an ob-
tuse triangle is out-
side the triangle.

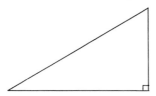
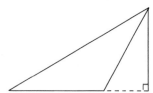

In each case in *Example 2,* where are the other two altitudes?

Exercises 10 - 7

A 1. Draw any acute triangle *ABC.*

　a) Construct the median from *B* to *AC.*

　b) Construct the altitude from *A* to *BC.*

　c) Construct the bisector of ∠ *C.*

　d) Construct the perpendicular bisector of *AB.*

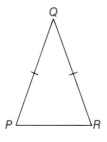

2. a) Construct an isosceles triangle *PQR,* such that *PQ* = *RQ.*

　b) Construct *QS* ⊥ *PR* at *S.*

　c) Does *QS* bisect ∠ *Q?*

　d) Is *QS* a median?

B 3. a) Draw any scalene triangle and construct the three angle bisectors.

　b) What do you notice about the bisectors?

　c) The point of intersection of the angle bisectors is called the **incentre**. Using the incentre as the centre, draw a circle that just touches all three sides. This circle is called the **inscribed circle** of the triangle.

4. a) Draw any acute triangle and construct the three altitudes. What do you notice? The point of intersection of the altitudes is called the **orthocentre.**

　b) Draw an obtuse triangle. Construct the three altitudes and find the orthocentre. What do you notice?

　c) What is the condition for the orthocentre to be outside the triangle?　　on the triangle?

5. a) Draw any scalene triangle and construct the three medians. What do you notice? The point of intersection of the medians is called the **centroid.**

   b) Paste the triangle on a piece of thin cardboard and cut it out. You should be able to balance the triangle at the centroid.

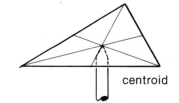

centroid

6. a) Construct an isosceles triangle *ABC,* such that *AC* = *BC.*

   b) Extend *AC* to *D.*

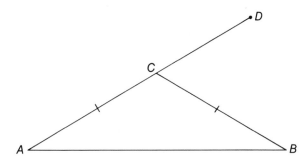

   c) Construct *CE,* the bisector of ∠*BCD.*

   d) What can be said about *CE* and *AB?* Explain.

C 7. a) Draw △*ABC* with *AB* = 11.5 cm, *BC* = 13.0 cm, and *AC* = 9.0 cm.

   b) Construct the perpendicular bisector of each side and call the circumcentre *O.*

   c) Construct the three altitudes and label the orthocentre *H.*

   d) Construct the three medians and label the centroid *G.*

   e) If you have worked carefully, you should be able to draw a line through *O, H,* and *G.* It is called the **Euler line.**

   f) Determine whether the incentre lies on the Euler line.

8. An altitude, a median, and an angle bisector extended to meet the opposite side are drawn from the same vertex of a scalene triangle. Show which of the three divides the triangle into two equal areas.

9. Construct △*XYZ* with base *XY* = 4 cm and altitude *ZH* = 3 cm. How many different triangles can be drawn?

10. Construct △*XYZ* with base *XY* = 4 cm and median *ZN* = 3 cm. How many different triangles can be drawn?

11. Under what condition would the triangle in Exercise 9 and the triangle in Exercise 10 be the same triangle?

DRAWING CIRCLES THROUGH SETS OF POINTS

Since ancient times, mathematicians have investigated the circumstances under which circles can be drawn through various numbers of points. For example:

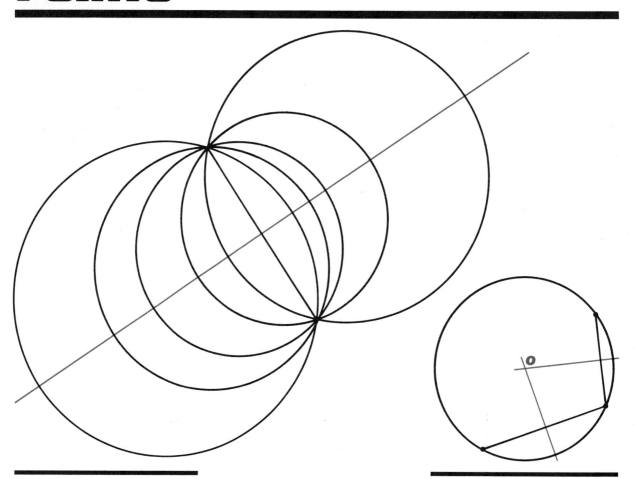

Two Points. Any number of circles can always be drawn through two points. The centres of the circles lie on the perpendicular bisector of the line segment joining the points.

Three Points. One circle can always be drawn through three non-collinear points. Its centre is the intersection of the perpendicular bisectors of the line segments joining the points.

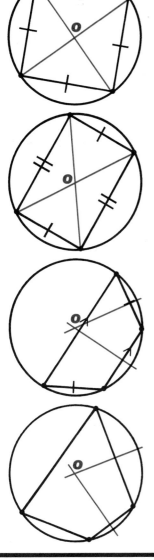

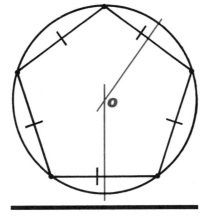

An interesting series of constructions leads to a circle that can be drawn through *nine* seemingly unrelated points:

1. On a large piece of paper, draw $\triangle ABC$ with $AB =$ 22.9 cm, $AC =$ 24.4 cm, and $BC =$ 18.4 cm.

2. Construct the perpendicular bisector of each side and label the circumcentre O.

3. Construct the altitude from each vertex and label the orthocentre H.

4. Locate the midpoints of line segments HA, HB, and HC.

5. Draw line segment OH and locate its midpoint, F.

6. Mark these nine points clearly:
   a) the midpoints of the three sides of $\triangle ABC$;
   b) the three points where the altitudes meet the opposite sides;
   c) the midpoints of HA, HB, and HC.

7. If your constructions are accurately done, using F as the centre and the distance to any of the nine points as a radius, you should be able to draw a circle that passes through all nine points. It is called the **nine-point circle** of $\triangle ABC$.

Repeat the nine-point circle construction using a large, scalene, obtuse triangle.

Four Points. A circle can be constructed through four points if, and only if:
 i) they are the vertices of a square or a rectangle; or
ii) they are the vertices of an isosceles trapezoid, or a quadrilateral whose opposite angles are supplementary.

Five and More Points. A circle can be constructed to pass through five points if they are the vertices of a regular pentagon. In fact, a circle can be constructed to pass through n points if they happen to be the vertices of a regular polygon.

10 - 8 Congruent Triangles

If two triangles have the same size and shape, they are said to be congruent. Congruent triangles can be made to coincide because their corresponding angles and sides are equal.

In triangles ABC and PQR,

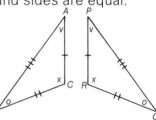

$$\angle A = \angle P \qquad AB = PQ$$
$$\angle B = \angle Q \qquad BC = QR$$
$$\angle C = \angle R \qquad AC = PR$$

Therefore, $\triangle ABC \cong \triangle PQR$

≅ is read:
"is congruent to"

Example 1. $\triangle DEF \cong \triangle GHK$

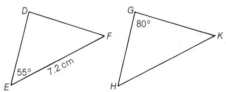

a) Find the length of HK.

b) Find the measure of the remaining angles in each triangle.

Solution. a) Since $\triangle DEF \cong \triangle GHK$,
$$EF = HK, \quad HK = 7.2 \text{ cm}.$$

b) $\angle E = \angle H, \quad \angle H = 55°$
$\angle D = \angle G, \quad \angle D = 80°$
$\angle F = \angle K,$ and $\angle F = 180° - 55° - 80°$
$$= 45°$$
$\angle K = 45°$

If the three sides and three angles of two triangles are equal, the triangles are congruent. However, it is not necessary to know this much information to show that two triangles are congruent.

In triangles ABC and DEF,
$$AC = DF, \qquad \angle A = \angle D, \qquad AB = DE.$$
If A is placed on D and AC made to coincide with DF and AB with $DE,$ then BC will coincide with $EF.$
$$\triangle ABC \cong \triangle DEF$$

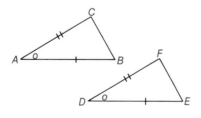

If two sides and the contained angle of one triangle are equal to the corresponding two sides and contained angle of another triangle, the triangles are congruent.

In a similar manner, it can be shown that there are other conditions under which triangles are congruent.

In triangles *ABC* and *DEF*,

$$\angle A = \angle D, \qquad AB = DE, \qquad \angle B = \angle E,$$

$$\triangle ABC \cong \triangle DEF$$

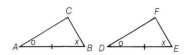

> If two angles and the contained side of one triangle are equal to the corresponding angles and the contained side of another triangle, the triangles are congruent.

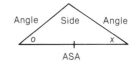

In triangles *ABC* and *DEF*,

$$AB = DE, \qquad AC = DF, \qquad BC = EF.$$

$$\triangle ABC \cong \triangle DEF$$

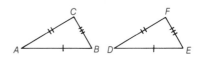

> If three sides of one triangle are equal to three sides of another triangle, the triangles are congruent.

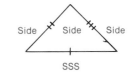

Example 2. Show that any point on the bisector of an angle is equidistant from the sides of the angle.

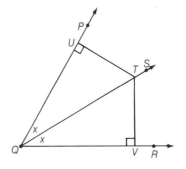

Solution.   *T* is any point on the bisector, *QS*, of ∠*PQR*.

$$TU \perp PQ, \quad \text{and} \quad TV \perp QR.$$

In △*TUQ* and △*TVQ*,

since ∠*TUQ* = ∠*TVQ* and ∠*TQU* = ∠*TQV*,

then, ∠*QTU* = ∠*QTV*.

Side *TQ* is common to both triangles.

Therefore, △*TUQ* ≅ △*TVQ*.................ASA

and,   *TU* = *TV*

That is, any point on the bisector of an angle is equidistant from the sides of the angle.

Exercises 10 - 8

A   1. Name the angles and sides that are equal:

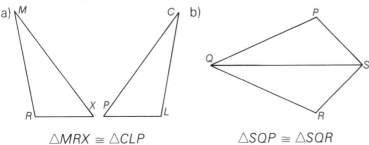

$$\triangle MRX \cong \triangle CLP \qquad\qquad \triangle SQP \cong \triangle SQR$$

2. If $PQRS \cong XWZY$, list the pairs of equal sides and equal angles.

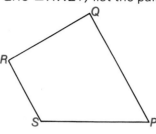
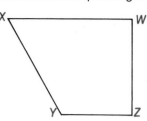

3. State which pairs of triangles are congruent. For those that are, state whether the congruence is known by SAS, ASA, or SSS.

a)

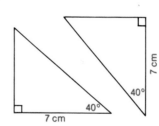

b)

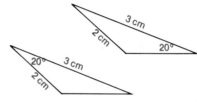

c)

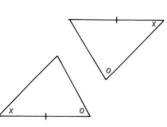

d)

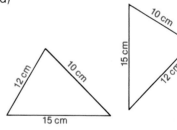

e)

f)

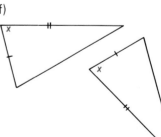

g)

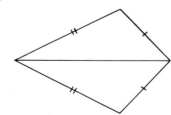

h)

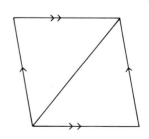

4. Find pairs of congruent triangles and state the condition for congruence.

a) i)   ii)   iii)   b) i)   ii)   iii)

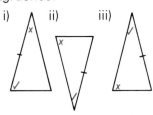   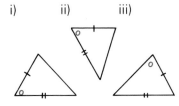

c) i)   ii)   iii)   d) i)   ii)   iii)

  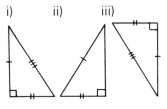   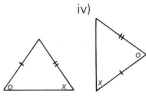 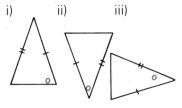

e) i)   ii)   iii)   iv)

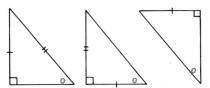 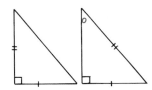

f) i)   ii)   iii)   iv)   v)

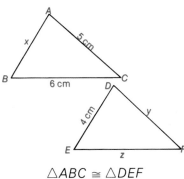 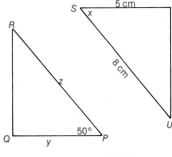

5. Find the values of x, y, and z:

a)   b)

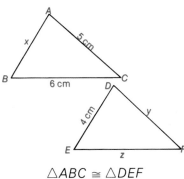

$\triangle ABC \cong \triangle DEF$ $\triangle PQR \cong \triangle STU$

c)

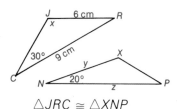

$\triangle JRC \cong \triangle XNP$

d)

$\triangle FSN \cong \triangle TRC$

e)

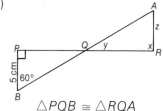

$\triangle BNA \cong \triangle AMB$

f)

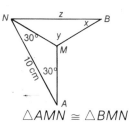

$\triangle DXY \cong \triangle EXY$

g)

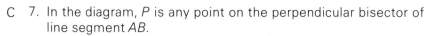

$\triangle PQB \cong \triangle RQA$

h)

$\triangle AMN \cong \triangle BMN$

B 6. Use congruent triangles to explain why these constructions are valid:

   a) Bisector of an angle

   b) Perpendicular to a line at a point on the line

   c) Perpendicular to a line from a point not on the line

   d) Perpendicular bisector of a line segment

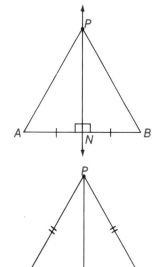

C 7. In the diagram, P is any point on the perpendicular bisector of line segment AB.

   a) Explain why $\triangle PNA \cong \triangle PNB$.

   b) Explain why $PA = PB$.

   c) State a conclusion about any point on the perpendicular bisector of a line segment.

8. In the diagram, N is the midpoint of line segment AB and P is any other point such that $PA = PB$.

   a) Explain why $\triangle PNA \cong \triangle PNB$.

   b) Explain why $\angle PNA = \angle PNB = 90°$.

   c) State a conclusion about any point that is equidistant from the endpoints of a line segment.

10 - 9 Polygons

Closed figures like those shown are called **polygons.** Their sides are line segments that do not cross.

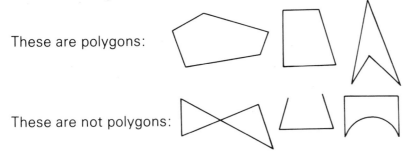

These are polygons:

These are not polygons:

Polygons are named according to the number of sides they have. Some examples are shown in the table.

Name	Number of sides	Examples
Triangle	3	
Quadrilateral	4	
Pentagon	5	
Hexagon	6	
Octagon	8	
Decagon	10	

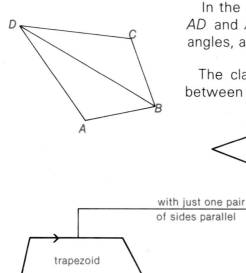

In the quadrilateral *ABCD*, *AB* and *CD* are *opposite* sides, *AD* and *AB* are *adjacent* sides, ∠*A* and ∠*C* are *opposite* angles, and *DB* is a *diagonal*.

The classification of quadrilaterals involves relationships between sides, as shown in the table that follows.

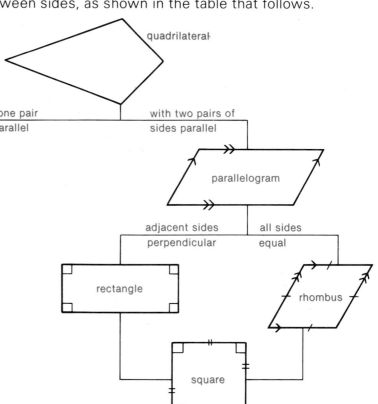

Any figure in the chart has all the properties of any figure above it, but not below it.

Example 1. a) Is a square a parallelogram? Why?
b) Is a parallelogram a square? Why?

Solution. a) A square is a parallelogram because both pairs of opposite sides are parallel.

b) A parallelogram is not necessarily a square, because adjacent sides need not be perpendicular or equal.

Example 2. Show that the opposite sides and opposite angles of a parallelogram are equal.

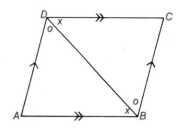

Solution. In parallelogram *ABCD,* draw diagonal *DB.*
Since $DC \parallel AB$,
$$\angle CDB = \angle ABD \ldots \text{alternate angles}$$
Since $AD \parallel BC$,
$$\angle ADB = \angle CBD \ldots \text{alternate angles}$$
$$DB = BD$$
Therefore, $\triangle ADB \cong \triangle CBD \ldots$ ASA
and, $\angle A = \angle C$, $AB = CD$, $AD = CB$.
By drawing diagonal *AC,* it can be shown that
$\angle D = \angle B$. That is, the opposite sides and opposite angles of a parallelogram are equal.

Example 3. Find the values of *x* and *y*:

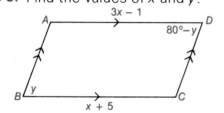

Solution. Since *ABCD* is a parallelogram,

$3x - 1 = x + 5$	$y = 80° - y$
$2x = 6$	$2y = 80°$
$x = 3$	$y = 40°$

Exercises 10 - 9

A 1. Give reasons for your answers to the following questions:

  a) Is a square a rectangle?
  Is a rectangle a square?

  b) Is a square a rhombus?
  Is a rhombus a square?

  c) Is a rectangle a parallelogram?
  Is a parallelogram a rectangle?

  d) Is a rhombus a parallelogram?
  Is a parallelogram a rhombus?

  e) Is a rectangle a trapezoid?
  Is a trapezoid a rectangle?

  f) Is a trapezoid a quadrilateral?
  Is a quadrilateral a trapezoid?

2. Find the values of *w*, *x*, *y*, and *z*:

a)

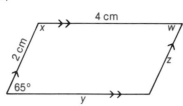

b)

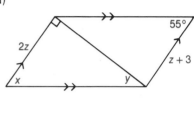

c)

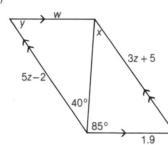

d)

3. A **kite** is defined as a quadrilateral that has two distinct pairs of adjacent sides equal in length. Give reasons to support these statements:

B

a) A kite has one pair of equal angles.

b) Diagonals of a kite intersect at right angles.

4. Find the values of *x* and *y*:

a)   b)   c)

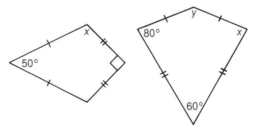
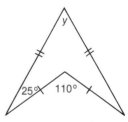

5. How many i) squares, ii) rectangles are in these figures?

a)   b)   c)

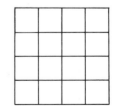

6. Draw any quadrilateral. Find the midpoint of each side. Join these midpoints to form another quadrilateral. What properties does this second quadrilateral appear to have?

C 7. Any polygon can be divided into triangles by joining vertices. The diagram shows a pentagon divided into three triangles.

a) Copy and complete this table:

Polygon	Number of Sides	Number of Triangles	Sum of Angles
Triangle	3	1	180°
Quadrilateral	4		
Pentagon	5	3	
Hexagon			
Octagon			
Decagon			

b) Write a simple formula for the sum of the angles in a polygon with n sides.

8. A **regular polygon** is one that has all sides the same length and all angles equal.

a) Make an additional column to your table of Exercise 7 giving the measure of each angle of a regular polygon.

b) Write a simple formula for the measure of each angle in a regular polygon with n sides.

9. Find the value of x:

a)   b)   c)

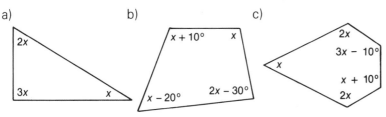

10. Draw a quadrilateral having one diagonal inside the figure and one diagonal outside. Is it possible for a quadrilateral to have both diagonals outside?

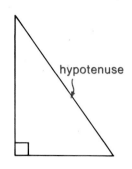

10 - 10 The Pythagorean Theorem

In the 6th century B.C., a Greek named Pythagoras proved an important property of right triangles. He showed that the areas of the squares drawn on each side of a right triangle are related in a certain way.

The diagram below shows two right triangles with squares drawn on each side. The areas of the squares on the hypotenuse are found by dividing them into smaller parts.

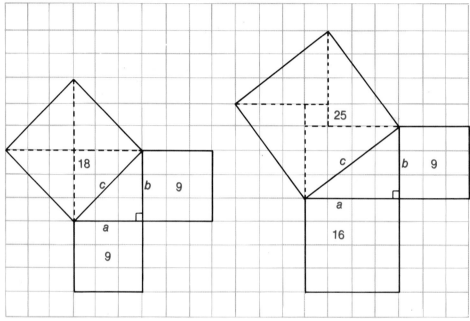

Areas of squares on sides of right angle		
$a^2 = 3 \times 3,$	or	9
$b^2 = 3 \times 3,$	or	9

Area of square on hypotenuse		
$c^2 = 4(\frac{1}{2})(3)(3),$	or	18

$a^2 = 4 \times 4,$	or	16
$b^2 = 3 \times 3,$	or	9
$c^2 = 4(\frac{1}{2})(4)(3) + 1,$	or	25

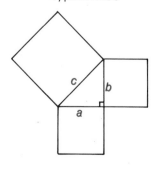

In both cases, the area of the square on the hypotenuse is equal to the sum of the areas of the squares on the other two sides.

Since the area of a square is the square of a side, the Pythagorean theorem is usually stated in terms of the sides of a right triangle.

> For any right triangle with sides a, b, and c, where c is the hypotenuse, $c^2 = a^2 + b^2$.

Example 1. Calculate the value of x:

a)

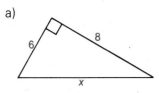

b)

Solution. a) $x^2 = 6^2 + 8^2$

$= 36 + 64$

$= 100$

$x = \sqrt{100}$, or 10

b) $17^2 = x^2 + 15^2$

$289 = x^2 + 225$

$289 - 225 = x^2$

$64 = x^2$

$x = \sqrt{64}$, or. 8

The Pythagorean theorem can be used to calculate the distance between two points on a grid.

Example 2. Plot the points $A(-2, 3)$ and $B(3, 6)$ on a grid and calculate the length of the line segment AB.

Solution. Draw a right triangle with horizontal and vertical sides and with AB as the hypotenuse. Let the length of AB be x units.

$x^2 = 5^2 + 3^2$

$= 25 + 9$, or 34

$x = \sqrt{34}$, or approximately 5.83

The length of AB is approximately 5.83 units.

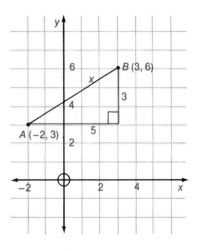

Exercises   10 - 10

A 1. Use the Pythagorean theorem to find x:

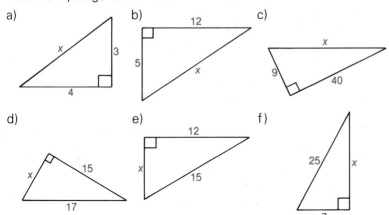

2. Use the Pythagorean theorem to find *x* correct to two decimal places:

a)

b)
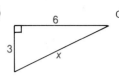

c)

d)

e)
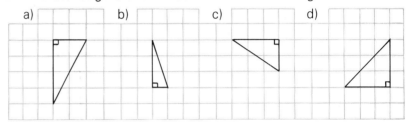

f)
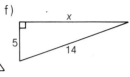

3. Find the lengths of all three sides of each triangle:

a)   b)   c)   d)

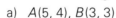

4. Plot each pair of points on a grid. Draw the line segment *AB* and calculate its length.

   a) *A*(5, 4), *B*(3, 3)   b) *A*(3, 0), *B*(5, −3)
   c) *A*(−2, 3), *B*(−4, −1)   d) *A*(−2, −3), *B*(1, 1)
   e) *A*(4, 5), *B*(−3, 4)   f) *A*(−3, −6), *B*(2, −1)

B 5. How long must a guy wire be to anchor a tower if it is attached 50 m up the tower and 12 m from the base?

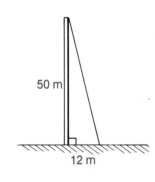

6. A ladder, 8.2 m long, is placed with its foot 1.8 m from a wall. How high up the wall will the ladder reach?

7. Find the length of the rafters for a building 12 000 mm wide and having the peak of the roof 3000 mm above the ceiling.

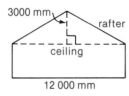

8. Can an umbrella 1.3 m long be packed flat in a box 1.1 m by 0.3 m? Give reasons for your answer.

9. When on a hike, Jeanne cuts diagonally across a large rectangular field, 1.6 km by 3.0 km, instead of keeping to the sides. What distance does she save?

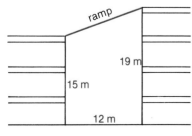

10. A ramp is to be built from the top level of one parking garage to another. Calculate the length of the ramp.

11. Find the lengths indicated by the letters:

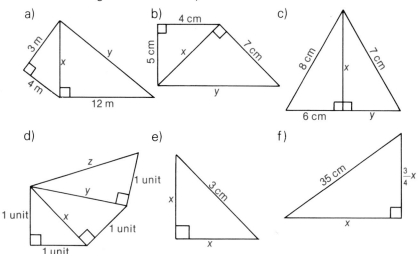

a)

b) 4 cm

c)

d)

e)

f)

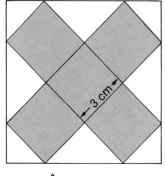

C 12. Each colored square has sides 3 cm long.

   a) Find the lengths of the sides of the outer square.

   b) What percent of the outer square is covered by the five colored squares?

13. A TV set has a screen 66 cm across a diagonal. If the screen is 1.2 times as wide as it is high, what is its width and height?

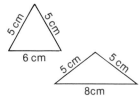

14. a) Show that the two isosceles triangles have the same area.

   b) Find another pair of isosceles triangles that have the same area.

15. A set of three positive integers that represent the lengths of the sides of a right triangle is known as a Pythagorean triple. Some examples are: 3, 4, 5, and 5, 12, 13. Others may be found by substituting positive integers for s and $t(s > t)$ in:

$$s^2 - t^2 \qquad 2st \qquad s^2 + t^2$$

   a) Copy and complete this table by finding five more Pythagorean triples:

s	t	$s^2 - t^2$	$2st$	$s^2 + t^2$
2	1	3	4	5
3	1	8	6	10
3	2	5	12	13
4	1	15	8	17

   b) Show that: $(s^2 - t^2)^2 + (2st)^2 = (s^2 + t^2)^2$.

CALCULATOR POWER

Using the Pythagorean Theorem

Expressions such as $a^2 + b^2$ and $c^2 - a^2$ arise when the Pythagorean theorem is applied. A simple hand calculator can be used to evaluate $3^2 + 4^2$, but the numbers cannot be keyed in as they appear. If they are, we get:

3 ⊠ 3 ⊞ 4 ⊠ 4 ▭   *5 2*   instead of 25.

This happens because the hand calculator adds the first 4 to the product 9 before multiplying by the second 4. However, $a^2 + b^2$ can be rewritten as:

$$\left(\frac{a \times a}{b} + b\right)b.$$

The sequence: **a** ⊠ **a** ⊡ **b** ⊞ **b** ⊠ **b** ▭

gives the correct result. For $3^2 + 4^2$, this gives:

3 ⊠ 3 ⊡ 4 ⊞ 4 ⊠ 4 ▭ *2 5*

What would be the sequence for $c^2 - a^2$?

Example 1. Simplify: $\sqrt{5^2 + 12^2}$

Sequence: 5 ⊠ 5 ⊡ 12 ⊞ 12 ⊠ 12 ▭ √̄   Display: *12.999999*

or 13

If there is no √̄ key on the calculator, the square root can be found by systematic trial, Newton's method, or from a table.

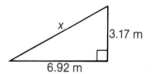

Example 2. For the right triangle shown, find x.

$$x^2 = (6.92)^2 + (3.17)^2$$
$$x = \sqrt{(6.92)^2 + (3.17)^2}$$

Sequence: 6.92 ⊠ 6.92 ⊡ 3.17 ⊞ 3.17 ⊠ 3.17 ▭ √̄

Display: *7.611523*

$$x \doteq 7.61 \text{ m}$$

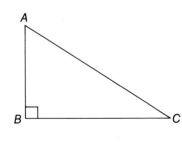

Exercises

Find the length of the unknown side of $\triangle ABC$, given:

1. $AB = 2.37$ cm, $BC = 4.19$ cm;
2. $AB = 7.66$ m, $BC = 7.66$ m;
3. $BC = 44.9$ cm, $AC = 59.3$ cm;
4. $AB = 1.92$ m, $AC = 3.06$ m.

10-11   The Circle

From your earlier work, you are already familiar with the formulas for the circumference, C, and area, A, of a circle of radius r.

$$C = 2\pi r$$
$$A = \pi r^2$$

radius OA chord CD diameter EF

The circle, and lines associated with the circle, have many properties. Some of them are considered in this section. The most important property is stated below.

All points on a circle are the same distance from the centre.

Example 1.   In the circle shown, if BC is a diameter, find x.

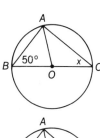

Solution.   Since O is the centre of the circle,
$$OA = OB = OC.$$
That is, $\triangle OAB$ and $\triangle OAC$ are both isosceles.
Therefore, $\angle OAB = 50°$ and $\angle OAC = x$.
Since the sum of the angles in $\triangle ABC$ is 180°,
$$50° + 50° + x + x = 180°.$$
$$2x = 180° - 100°$$
$$= 80°$$
$$x = 40°$$

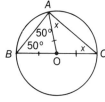

A chord of a circle has an interesting property. If you draw any chord of a circle and construct the perpendicular bisector, it passes through the centre of the circle.

The perpendicular bisector of a chord passes through the centre of the circle.

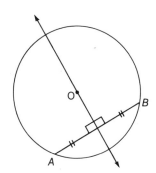

The use of this property often involves the Pythagorean theorem.

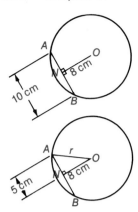

Example 2. For the circle shown, use the information given to calculate, rounded to one decimal place:

a) the radius;   b) the diameter;

c) the circumference;   d) the area.

Solution. a) Since $ON \perp AB$, ON is the perpendicular bisector of AB. Join OA and let the radius be r. Using the Pythagorean theorem:

$$r^2 = 5^2 + 8^2$$
$$= 25 + 64$$
$$= 89$$
$$r = \sqrt{89}, \text{ or approximately } 9.43$$

The radius of the circle is approximately 9.4 cm.

b) Since $d = 2r$, $d \doteq 2 \times 9.43$, or 18.86. The diameter of the circle is approximately 18.9 cm.

c) $C = 2\pi r$ d) $A = \pi r^2$
  $\doteq 2 \times 3.14 \times 9.43$ $\doteq 3.14 \times (\sqrt{89})^2$
  $\doteq 59.22$ $\doteq 3.14 \times 89$,
The circumference   or 279.46
is approximately   The area is approx-
59.2 cm.   imately 279.5 cm².

Exercises 10 - 11

In these exercises, use 3.14 as the value of π.

A 1. Give reasons for your answers to these questions:

a) Is a diameter a chord?   b) Is a radius a chord?

c) What is the longest chord that can be drawn in a circle?

2. Find the value of x:

a)   b)   c)

d)   e)   f)

3. Find the values of x rounded to one decimal place:

a)

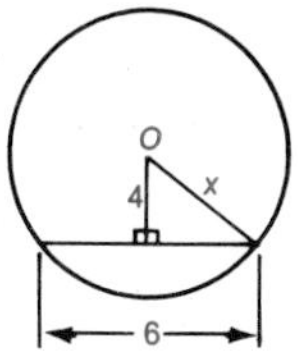

b)

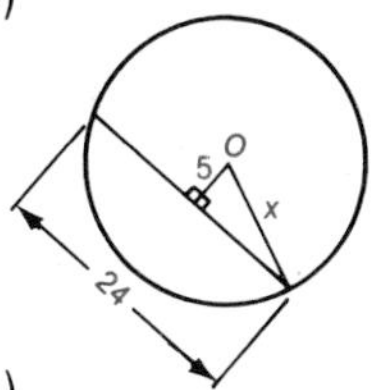

c)

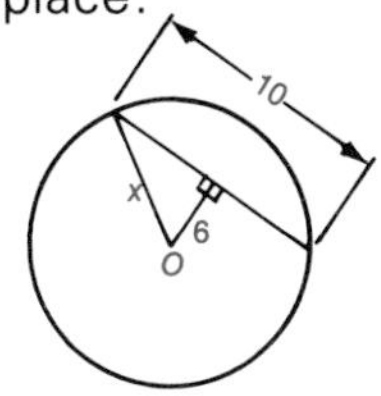

d)

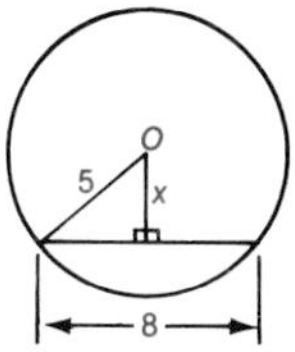

e)

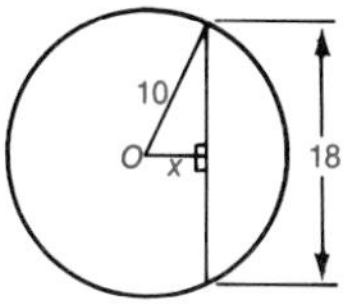

f)

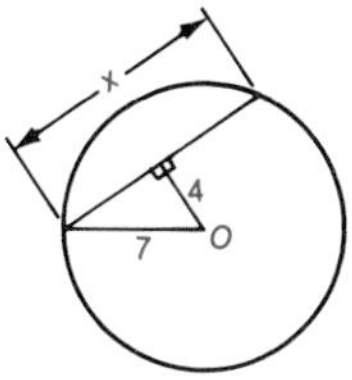

4. For parts (a), (b), and (c) of Exercise 3, find:
   a) the length of the diameter;
   b) the circumference;
   c) the area.

B 5. a) Explain how to construct a circle that passes through the three given points A, B, and C.
   b) Can a circle always be drawn through A, B, and C no matter what their positions? Explain.

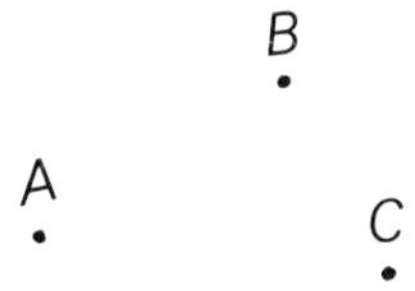

6. Plot each set of points on a grid and, for each set, determine the coordinates of the centre of the circle that can be drawn through the points:
   a) $A(3, 2)$, $B(3, -2)$, $C(-3, -2)$
   b) $A(0, 6)$, $B(0, 0)$, $C(-8, 0)$
   c) $A(1, 3)$, $B(7, 3)$, $C(7, 7)$

7. Find the value of x rounded to one decimal place:

a)

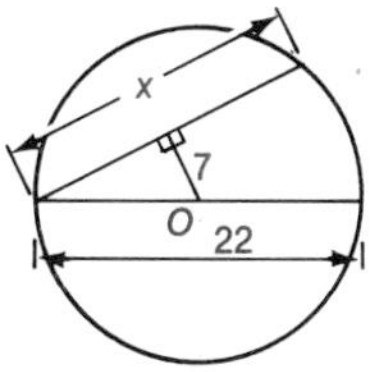

b)

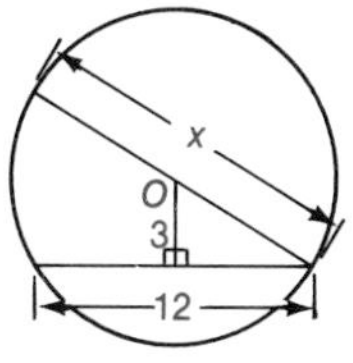

c)

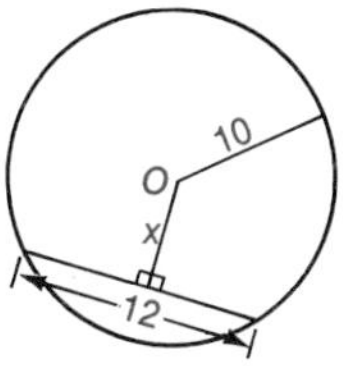

d)

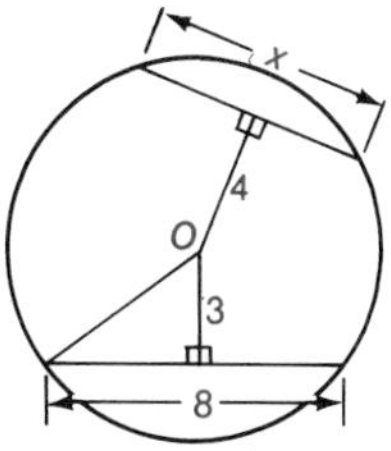

e)

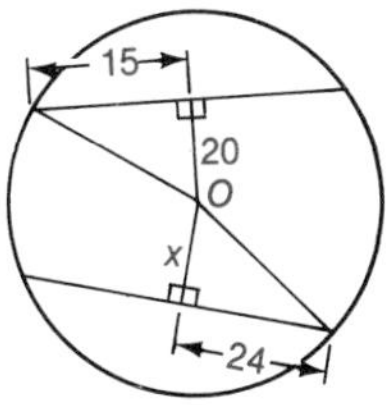

f)

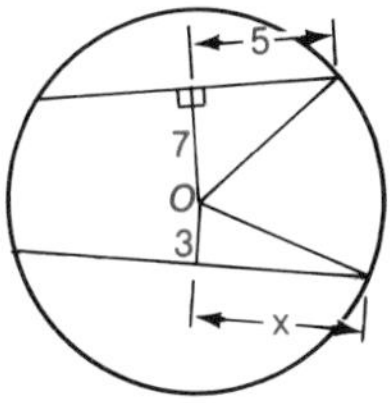

8. *O* is the centre of a circle of radius 5 cm, and *OABC* is a rectangle with vertex *B* on the circle. Find the length of the diagonal *AC*.

9. Find the value of *x* rounded to one decimal place:

a)

b)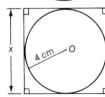

c)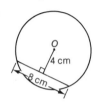

d)

10. For each circle, calculate, rounding your answers to one decimal place: i) the diameter; ii) the circumference; iii) the area.

a)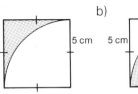

b)

c)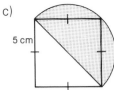

C 11. For the diagram shown,
   a) find *y* if: i) *x* = 30°: ii) *x* = 18°;
   b) find *x* if: i) *y* = 75°; ii) *y* = 57°;
   c) find the equation relating *x* and *y*.

12. Find the areas of the shaded regions:

a)

b)

c)

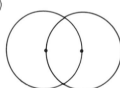

13. If all circles have a radius of 10 cm, find the perimeter of each figure:

a)

b)

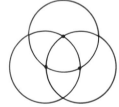

Mathematics Around Us

GEOMETRY ON THE SURFACE OF THE EARTH

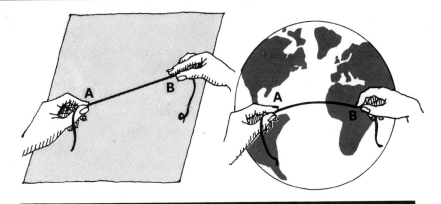

The geometry you have been studying is called plane geometry because the figures were all drawn on a plane. Geometry on the surface of Earth is called **spherical geometry** because the figures are drawn or visualized on a sphere.

The shortest distance between two points can be represented by stretching a thread between them. On a plane, this is a straight line. On a globe representing Earth's surface, the shortest distance between two points is part of a **great circle**—the circle that is formed when a plane passes through the two points and Earth's centre.

The shortest distance between two cities is an arc of a great circle, but it does not look that way on most maps. On the map (right) the straight line joining Winnipeg and Bombay crosses the Atlantic Provinces and North Africa. This route, however, is actually much longer than the great-circle route which passes near the North Pole. The shortest route from Winnipeg to Hong Kong passes along the north coast of Alaska.

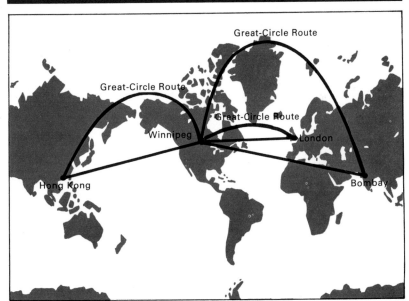

Great-Circle Routes from Winnipeg   Mercator Projection

Other maps can be drawn on which certain great circles appear as straight lines. For example, the shortest distance from Winnipeg to any point on Earth is shown as a straight line on this map. The map, however, is not useful for cities other than Winnipeg.

Triangles on a plane have three sides, each side being a straight line. The three sides of a **spherical triangle** are arcs of great circles.

On a globe representing Earth, A is the North Pole, and B and C are points on the Equator. The sum of the angles of spherical triangle ABC is:

$$90° + 90° + 45° = 225°$$

If A and C are fixed and B moves along the Equator, how does the sum of the angles in spherical $\triangle ABC$ change?

Great-Circle Routes from Winnipeg   Azimuthal Equidistant Projection

Questions

1. On a globe, using tape and thread, show the great circle routes from Winnipeg to Hong Kong, London, and Bombay. Compare your routes with those shown on the maps.

2. Would commercial aircraft actually fly these routes? Why?

3. Which city is farther from Winnipeg than any other city?

4. On a globe, with tape and thread, show the spherical triangles with vertices at the following cities. Measure their angles with a protractor. What is the sum for each?

   a) St. John's (Newfoundland), Vancouver (B.C.), Miami (Florida)

   b) Winnipeg (Manitoba), Cairo (Egypt), Rio de Janeiro (Brazil)

   c) Honolulu (Hawaii), Caracas (Venezuela), Nairobi (Kenya)

5. These statements are true for plane geometry. Are they true for spherical geometry? Explain.

   a) When two lines intersect, the vertically opposite angles are equal.

   b) Parallel lines never meet.

   c) Each angle of an equilateral triangle measures 60°.

   d) The angles opposite the equal sides of an isosceles triangle are equal.

Review Exercises

1. Find the value of each variable:

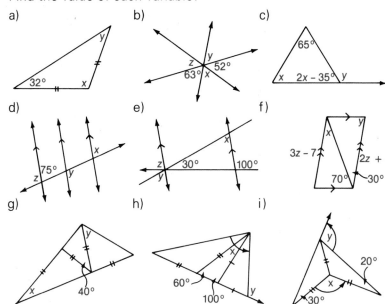

a)

b)

c)

d)

e)

f)

g)

h)

i)

2. Plot each set of points on a grid. Join consecutive points and name the resulting figure. List its geometric properties.

a) $A(-2, -1)$, $B(4, 0)$, $C(5, -6)$

b) $P(-3, 1)$, $Q(-1, -2)$, $R(5, -1)$, $S(3, 2)$

c) $W(4, -1)$, $X(1, 4)$, $Y(-4, 1)$, $Z(-1, -4)$

3. Construct an angle of 135°.

4. Draw a line segment 8 cm long. Mark any point P about 3 cm from the segment. Construct

a) the perpendicular through P to the line segment;

b) a line through P parallel to the segment.

5. Draw a triangle with sides 5 cm, 6 cm, and 7 cm long. Construct the perpendicular bisectors of the sides and use their point of intersection to draw the circumcircle.

6. In each set, find a pair of congruent triangles and state the condition for congruence:

a) i)   ii)   iii)   b) i)   ii)   iii)

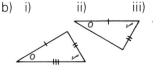

7. Use the Pythagorean theorem to find x and y to one decimal place:

a)

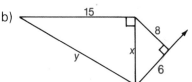

b)

8. The size of a television screen is given by the length of its diagonal. Determine the size of a screen that is 34 cm by 40 cm.

9. Will a sheet of plywood 1200 mm wide fit into the rear of a station wagon if the door opening is 950 mm wide and 700 mm high?

10. The longest side of an isosceles right triangle is 50 cm. How long are each of the other two sides?

11. Find the value of each variable:

a)

b)

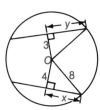

c)

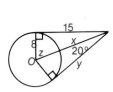

12. Find the radius, circumference, and area of the circle shown.

13. Find the value of x rounded to one decimal place:

a)

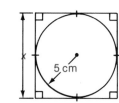

b)

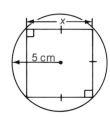

c)

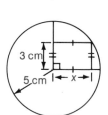

d)

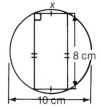

14. The points in each set lie on a circle. Determine the coordinates of the centre of the circle.

a) (1, 3), (4, 0), (6, 3)   b) (−1, −5), (2, −2), (−1, 1)

11 Transformations

Two oil storage tanks, several kilometres apart, are on the same side of a main pipeline. Where, along the pipeline, should a pumping station be located to serve both tanks so that the total length of pipe is a minimum? (See *Example 1*, Section 11-8.)

11-1 What Are Transformations?

Whenever the shape, size, appearance, or position of an object is changed, it has undergone a **transformation**. Under a transformation, some of the characteristics of an object may be changed while others remain the same. Those characteristics that are unchanged are said to be **invariant**.

The illustrations below show some common transformations. For each one, identify some characteristics that are invariant as well as some that are not.

Reflections

How else could you shave or apply make-up but by a reflection of your face in a mirror?

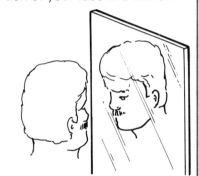

Enlargements/Reductions

The image on the drive-in screen is projected from a negative only 35 mm long.

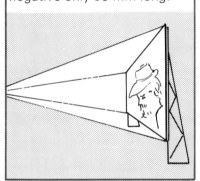

Translations

Each move on a chess board is the translation of a chess piece from one position to another.

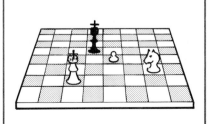

Rotations

A rotating disc enables you to hear your favorite pop star.

In Chapter 9, we studied relations which mapped one set of numbers onto another set of numbers. A transformation is a relation that maps points onto points.

> A transformation is a relation that maps every point, *P*, onto an image point, *P′*.

Transformation geometry is the study of transformations. Most transformations change the size and shape of a figure. We shall confine ourselves to three kinds of transformations all of which preserve both size and shape. This makes them special in the set of all transformations. The three kinds are:

Rotation   Reflection   Translation

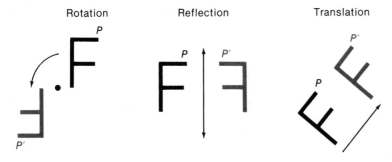

Example 1. What transformation maps each figure onto the image shown?

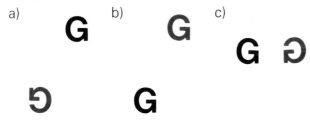

a)   **G**   b)   **G**   c)

G Ә

Ꮐ   **G**

Solution. a) A rotation b) A translation c) A reflection

Example 2. The four parts into which the figure is divided are congruent. Name the transformation that maps

    a) I onto IV;   b) II onto IV;

    c) I onto III;   d) III onto IV.

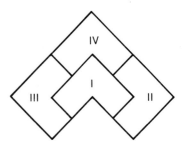

Solution. a) A translation   b) A reflection

           c) A rotation   d) A reflection

Exercises 11 - 1

A 1. Name the transformations required to map each figure onto the images shown.

a)

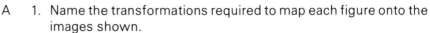

b)

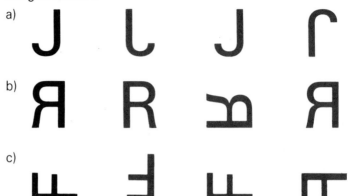

c)

2. For figure A below, name the transformation that maps

a) IV onto I; b) III onto IV; c) IV onto II;

d) I onto II; e) III onto I; f) II onto III.

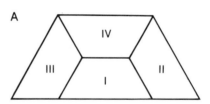
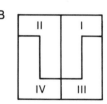

3. For figure B above, name the transformation that maps
a) III onto I; b) I onto II; c) IV onto III.

B 4. Copy the diagrams below. Divide each one into two parts of the same size and shape. Name the transformation needed to map one part onto the other.

a)   b)   c)

d)   e)   f)

11 - 2 Rotations

When a shape is turned about a fixed point it is said to have undergone a **rotation**. The fixed point is called the **turn centre**. The illustration shows $\frac{1}{4}$-turn, $\frac{1}{2}$-turn, $\frac{3}{4}$-turn, and full-turn images of A. In this chapter, all turns will be counterclockwise.

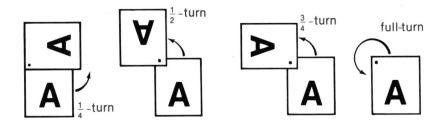

The examples that follow present the images of familiar letters and shapes under various rotations.

Example 1.   Sketch the images of the letters E, H, J, R, and N under rotations of $\frac{1}{4}$-turn, $\frac{1}{2}$-turn, $\frac{3}{4}$-turn, and full turn.

Solution.

Images				
Letter	$\frac{1}{4}$-turn	$\frac{1}{2}$-turn	$\frac{3}{4}$-turn	Full Turn
E	ɯ	Ǝ	ɯ	E
H	I	H	I	H
J	⌐	ſ	⌐	J
R	ᴚ	ᴚ	ᴚ	R
N	И	N	И	N

Example 2. For each shape, determine the amount of rotation, if any, to give the image shown.

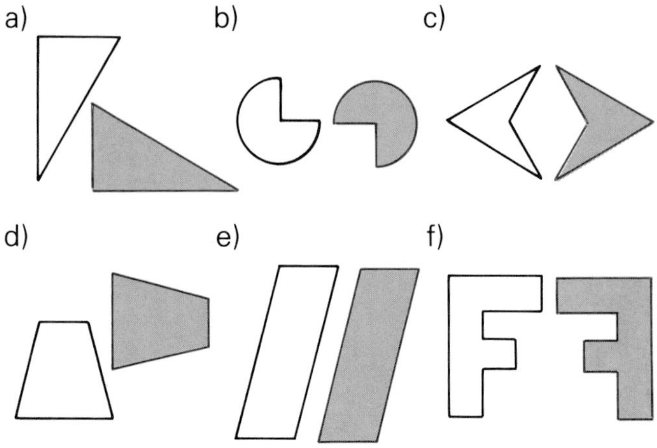

Solution. a) $\frac{1}{4}$-turn   b) $\frac{1}{2}$-turn   c) $\frac{1}{2}$-turn

d) $\frac{3}{4}$-turn   e) 0-turn, $\frac{1}{2}$-turn, or full turn

f) Under the $\frac{1}{4}$-, $\frac{1}{2}$-, $\frac{3}{4}$-, and full turns, the **F** looks like this — **ш**, **ⅎ**, **╥**, **F**, but not like this — **ꟻ**. There is no rotation that maps **F** onto **ꟻ**.

The above answers can be verified by using tracing paper.

Images resulting from $\frac{1}{2}$-turns are easily located on a grid.

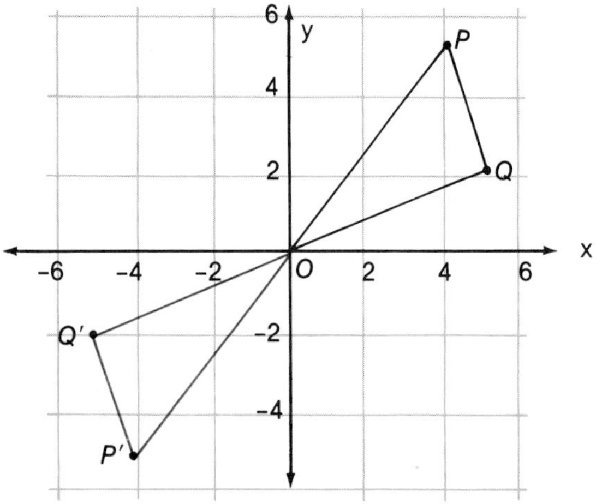

Example 3. A triangle has vertices $P(4, 5)$, $O(0, 0)$, $Q(5, 2)$. Draw its image and give the coordinates of its vertices under a $\frac{1}{2}$-turn about O.

Solution. The image of $\triangle POQ$ under a $\frac{1}{2}$-turn about O is the triangle with vertices $P'(-4, -5)$, $O(0, 0)$, $Q'(-5, -2)$.

Example 4. Sketch the image of each figure under a $\frac{1}{4}$-turn, $\frac{1}{2}$-turn, $\frac{3}{4}$-turn, and full turn about turn centre O.

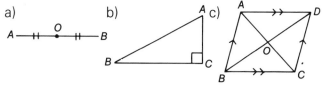

a)   b)   c)

Solution.

	Figure	$\frac{1}{4}$-turn	$\frac{1}{2}$-turn	$\frac{3}{4}$-turn	Full Turn
a)					
b)					
c)					

From *Example 4* we see that:

- Under a $\frac{1}{2}$-turn about its centre, a line segment maps onto itself.

- Under a $\frac{1}{2}$-turn about its centre, a parallelogram maps onto itself.

In other words, line segments and parallelograms are unchanged when rotated through a $\frac{1}{2}$-turn about their centres.

We can use half turns to verify some important geometrical properties.

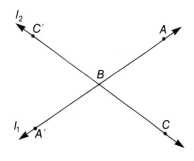

Example 5. If two lines intersect, show that the vertically opposite angles are equal.

Solution. In the diagram, l_1 and l_2 intersect at *B*. Also, $BA = BA'$, and $BC = BC'$.
Under a half turn about *B*: $BA \longrightarrow BA'$ and $BC \longrightarrow BC'$.
Therefore, $\angle ABC$ maps onto $\angle A'BC'$.
That is, $\angle ABC = \angle A'BC'$
The vertically opposite angles are equal.

In *Example 5*, name another pair of equal vertically opposite angles.

Exercises   11 - 2

A   1. Copy and complete this table:

Figure	Images			
	$\frac{1}{4}$-turn	$\frac{1}{2}$-turn	$\frac{3}{4}$-turn	Full Turn
K	**⋉**			
7			**⌐**	
Y		**⋋**		
5	**⎁**			

2. For each shape, determine the amount of rotation to give the image shown:

a)   b)   c)

d)   e)

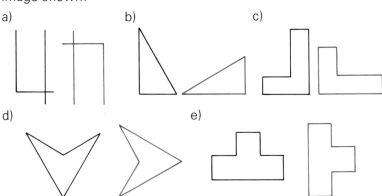

3. A triangle has vertices $A(1, 4)$, $B(5, 5)$ and $O(0, 0)$. Draw its image and give the coordinates of its vertices under a $\frac{1}{2}$-turn about O.

4. A quadrilateral has vertices $A(1, 3)$, $B(4, 7)$, $C(6, 4)$, and $D(3, 1)$. Draw its image and give the coordinates of its vertices under a $\frac{1}{2}$-turn about the origin $(0, 0)$.

B 5. A triangle has vertices $P(-3, 5)$, $Q(1, -7)$, and $R(2, 1)$. Draw its image and give the coordinates of its vertices under a $\frac{1}{2}$-turn about the point $(0, 0)$.

6. A quadrilateral has vertices $K(-3, 4)$, $L(1, 6)$, $M(6, -1)$, and $N(-4, -3)$. Draw its image and give the coordinates of its vertices under a $\frac{1}{2}$-turn about the point $(0, 0)$.

7. a) What can be said about the line segment joining point P on a figure and its $\frac{1}{2}$-turn image, P'?

   b) How do the coordinates of a point and its $\frac{1}{2}$-turn image about $(0, 0)$ compare?

8. a) In your workbook, draw a triangle similar to that in (i).

   b) Draw its $\frac{1}{2}$-turn image about the turn centre indicated.

   c) Name the figure formed by the triangle and its image.

   d) Repeat parts (a), (b), and (c) for each of the other triangles in turn.

i)   ii)   iii)

iv) 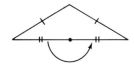   v)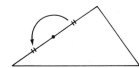

9. Sketch the image of each figure under a $\frac{1}{4}$-turn, $\frac{1}{2}$-turn, $\frac{3}{4}$-turn, and a full turn about turn centre O. Which figures are unchanged when rotated through a $\frac{1}{2}$-turn?

a) b) c)

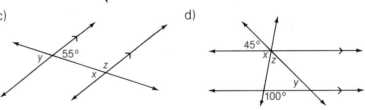

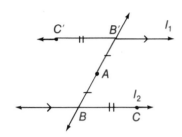

10. In the diagram, l_1 and l_2 are parallel lines, A is the midpoint of the line segment BB', and $BC = B'C'$.

 a) Show that alternate angles $\angle ABC$ and $\angle AB'C'$ are equal.

 b) Name another pair of equal alternate angles.

11. Find the values of x, y, and z:

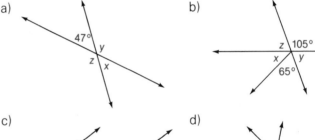

a)

b)

c)

d)

C 12. a) Graph the equation: $y = x + 3$.

 b) Draw the image of the graph under a $\frac{1}{2}$-turn about $(0, 0)$.

13. a) Graph the equation: $y = 2x$.

 b) Draw the image of the graph under a $\frac{1}{2}$-turn about $(0, 0)$.

 c) Explain the result.

14. What properties remain invariant under a rotation?

11 - 3   Reflections

When you look into a mirror you see an image of yourself. The image appears to be as far behind the mirror as you are in front of it. The transformation that relates points and their images in this way is called a **reflection**.

A reflection in line *l* is a transformation that maps every point *P* onto an image point *P'* such that:

- *P* and *P'* are equidistant from line *l*;
- *PP'* is perpendicular to line *l*.

The reflection line is the perpendicular bisector of the line segment *PP'*.

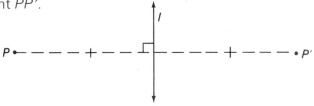

The examples that follow present the images of familiar letters and shapes under various reflections.

Example 1. Sketch the images of the letters F, R, H, and S under reflections in a vertical, a horizontal, and an oblique line.

Solution.

| | | Reflection Images in | |
Letter	A Vertical Line	A Horizontal Line	An Oblique Line
F	F⫶Ⅎ	F / Ⅎ	F / Ⅎ
R	R⫶Я	R / Я	R / Я
H	H⫶H	H / H	H / H
S	S⫶Ƨ	S / Ƨ	S / Ƨ

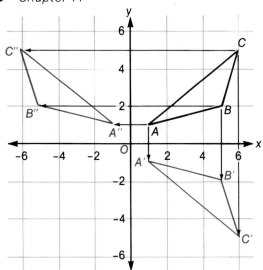

Example 2. Triangle *ABC* has vertices *A*(1, 1), *B*(5, 2) and *C*(6, 5). Sketch the image of △*ABC* under a reflection in:

a) the *x*-axis;   b) the *y*-axis.

Solution. a) Under a reflection in the *x*-axis:

$$A(1, 1) \longrightarrow A'(1, -1)$$
$$B(5, 2) \longrightarrow B'(5, -2)$$
$$C(6, 5) \longrightarrow C'(6, -5)$$

b) Under a reflection in the *y*-axis:

$$A(1, 1) \longrightarrow A''(-1, 1)$$
$$B(5, 2) \longrightarrow B''(-5, 2)$$
$$C(6, 5) \longrightarrow C''(-6, 5)$$

Example 3. Determine the line of reflection for each shape and its reflected image:

a)

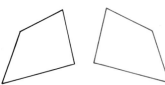

b)

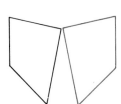

c)

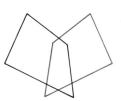

Solution. The reflection line is the perpendicular bisector of the line segment joining any point, *P*, and its image, *P'*.

a)

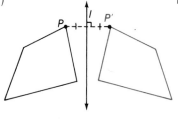

b)

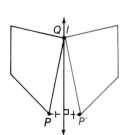

c)

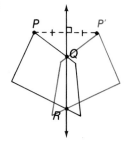

The solution to *Example 3b* suggests another property of reflections. The reflection line, *l*, passes through *Q*, the intersection of PQ and its image $P'Q$. In general, any point common to a line and its reflected image lies on the line of reflection. In *Example 3c*, both Q and R lie on the reflection line.

Exercises 11 - 3

A 1. Sketch the images of A, 3, K, and 5 under reflections in a vertical and a horizontal line.

2. Triangle PQR has vertices $P(2, 6)$, $Q(6, 4)$, and $R(3, 2)$. Draw the image of $\triangle PQR$ under a reflection in:

   a) the x-axis;   b) the y-axis.

3. Quadrilateral $ABCD$ has vertices $A(2, 5)$, $B(6, 5)$, $C(9, 1)$, and $D(2, 1)$. Find the coordinates of the image of quadrilateral $ABCD$ under a reflection in:

   a) the x-axis;   b) the y-axis.

4. State the time shown on the mirror image of each clockface.

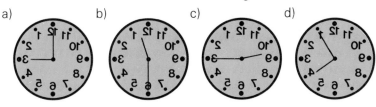

a)   b)   c)   d)

5. Draw the clockface shown opposite as it would appear in a mirror.

6. Triangle ABC has vertices $A(4, 7)$, $B(7, 2)$, $C(3, 3)$. Its image under a reflection is $A'(-2, 7)$, $B'(-5, 2)$, $C'(-1, 3)$. Graph both triangles and determine the line of reflection.

7. If $\triangle ABC$ in Exercise 6 has another image at $A''(4, -11)$, $B''(7, -6)$, $C''(3, -7)$, determine the line of reflection.

8. Draw a line l and a line segment PQ like the example below. Sketch the image of PQ under a reflection in l.

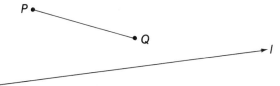

9. Copy the given line segment AB and its image under a reflection, $A'B'$. Determine the line of reflection.

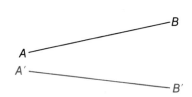

B 10. Draw triangles, as shown, with one side extended.

   a) Draw the image of each triangle using the extended side as the line of reflection.

   b) Name the figure formed by the triangle and its image.

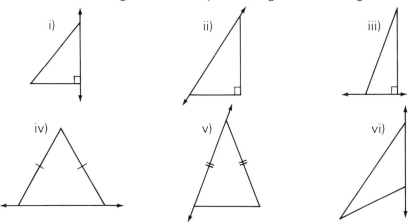

11. a) Draw angles like those shown. In each case *OP′* is the reflected image of *OP*.

    b) Determine the lines of reflection.

    c) Verify that in each case the reflection line is the bisector of the angle.

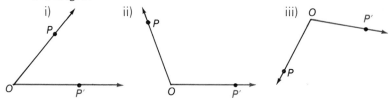

12. Triangle *PQR* has vertices $P(-2, 4)$, $Q(4, 2)$, $R(1, -2)$. Draw the image of $\triangle PQR$ and give the coordinates of the vertices under a reflection in

    a) the *x*-axis;   b) the *y*-axis.

13. Quadrilateral *ABCD* has vertices $A(-2, 6)$, $B(4, 3)$, $C(3, -3)$, $D(-5, -2)$. Give the coordinates of the vertices of its image under a reflection in

    a) the *x*-axis;   b) the *y*-axis.

C 14. a) Graph the equation: $3x + 2y = 12$.

       b) Draw the image of the graph under a reflection in

          i) the *x*-axis;   ii) the *y*-axis.

15. What properties remain invariant under a reflection?

*11 - 4 Constructions With a Plastic Mirror

In Section 11-3, the following properties of reflections and reflection lines were observed:

- If a line segment AB is reflected onto itself so that $A \rightarrow B$, the line of reflection is the perpendicular bisector of AB.

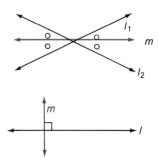

- The bisector, m, of the angle between two intersecting lines, l_1 and l_2, is the line of reflection under which l_1 is mapped onto l_2.

- If a line, l, is mapped onto itself under a reflection in another line, m, then $l \perp m$.

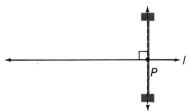

Knowledge of these properties enables us to use a plastic mirror to make various constructions.

1. **The perpendicular bisector of a line segment**

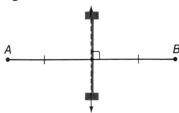

2. **A perpendicular to a line at a point on the line**

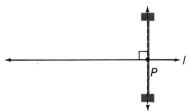

Place the mirror across AB so that the image of A maps onto B. The line drawn along the edge of the mirror is the required line.

Place the mirror across l at P and adjust until l maps onto itself. The line drawn along the edge of the mirror is the required line.

3. **A perpendicular to a line through a point not on the line**

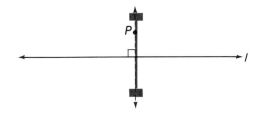

4. **A parallel to a line through a point not on the line**

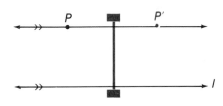

Place the mirror across l and on P and adjust until l maps onto itself. The line drawn along the edge of mirror is the required line.

Place the mirror across l so that l maps onto itself. Locate P', the image of P. The line through PP' is a parallel to l.

5. **The bisector of an angle.**

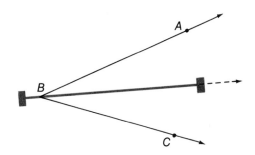

6. **An equilateral triangle with line segment *AB* as a side.**

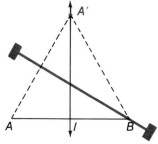

Place the mirror between *A* and *C* at *B*. Adjust until *BA* maps onto *BC*. The line drawn along the edge of the mirror bisects ∠*ABC*.

Construct *l*, the perpendicular bisector of *AB*. Place the mirror on *B* so that *A′*, the image of *A*, is on *l*. Mark *A′*. △*AA′B* is the required triangle.

One or more of these basic constructions with the plastic mirror will be needed in working the exercises.

Exercises 11-4

Use only a plastic mirror to do the following constructions.

A 1. Do each of the constructions 1-6.

2. Construct a line parallel to two given parallel lines midway between them.

3. Divide a line segment into four equal parts.

4. Construct an angle with a measure of: a) 45°; b) 30°.

5. Draw a line segment *AB* and mark a point *P* not on *AB*. Construct a parallelogram with *AB* as a side and *P* as a vertex.

B 6. Draw a line segment *XY* and mark a point *Q* not on *XY*. Construct an isosceles triangle with *XY* as the base and one of the two equal sides passing through *Q*.

7. Construct an angle of $22\frac{1}{2}^{\circ}$.

8. Construct an isosceles right triangle.

C 9. Use the lid of a can or jar to draw a circle. Locate the centre of the circle.

10. On a large piece of paper, draw an acute triangle and construct:

   a) the circumcentre;   b) the centroid;

   c) the incentre;   d) the orthocentre.

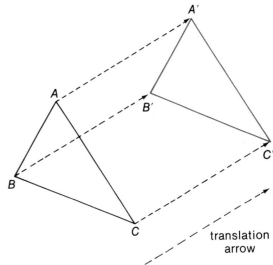

11 - 5   Translations

In Sections 11-2 and 11-3, we studied rotations and reflections. Translations are a third type of transformation.

The **translation** in the diagram is defined by the mapping: $A \rightarrow A'$, $B \rightarrow B'$, $C \rightarrow C'$. Line segments AA', BB', CC' are all equal in length and parallel. Their length and direction are represented by the **translation arrow.**

Under a translation, any figure and its image are identical in all respects except location.

Example 1. Draw the image of line segment AB under the translation that maps A onto A'.

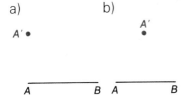
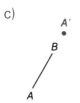

Solution. To find the translation image of AB, we draw the arrow from A to A'. We then draw the arrow of the same length and direction from B. The endpoint of the second arrow is B', and $A'B'$ is the required image.

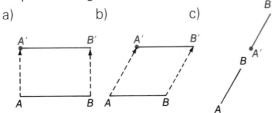

From *Example 1*, it is clear that under a translation a line segment is mapped onto a parallel line segment. In general, a translation maps any line onto a parallel line or onto itself. For any pair of parallel lines, l_1 and l_2, there is always a translation that maps l_1 onto l_2.

We can make use of this property of translations to verify an important angle relationship involving parallel lines.

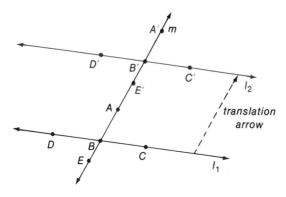

Example 2. Line m intersects parallel lines l_1 and l_2. Show that corresponding angles are equal.

Solution. Label the diagram as shown. Under a translation parallel to line m that maps line l_1 onto line l_2,

$$A \rightarrow A', \quad B \rightarrow B', \quad C \rightarrow C'.$$

Therefore, $\angle ABC$ maps onto $\angle A'B'C'$. That is, corresponding angles are equal.

In *Example 2,* name three other pairs of equal corresponding angles.

In the next example, a translation and the properties of parallel lines are used to show that the sum of the angles of any triangle is 180°.

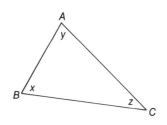

Example 3. In $\triangle ABC$, show that $x + y + z = 180°$.

Solution. There is a translation that maps $\triangle ABC$ onto $\triangle A'CC'$.

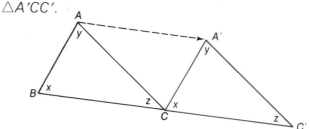

Under the translation, $BA \parallel CA'$.
Since $\angle ACA'$ and $\angle BAC$ are alternate angles,
  then, $\angle ACA' = y$.
Since BCC' is a straight line,
  $x + \angle ACA' + z = 180°$.
Therefore, $x + y + z = 180°$.
That is, the sum of the angles of any triangle is 180°.

In the next example, we use the results we have verified.

Example 4. Find the values of x, y, and z.

a)   b)   c)

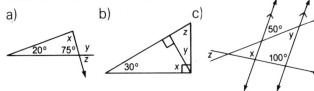

Solution.   a) x = 180° − 75° − 20°, or 85°
              y = 180° − 75°, or 105°
              z = 75°

            b) x = 180° − 90° − 30°, or 60°
              y = 90° − 60°, or 30°
              z = 180° − 90° − 30°, or 60°

            c) x = 100°   y = 50°
              z. = 180° − 100° − 50°, or 30°

Translations can also be shown on a grid.

Example 5. Draw the image of quadrilateral *ABCD* according to the given translation arrow. Then compare the quadrilateral and its image with respect to lengths of sides and measures of angles.

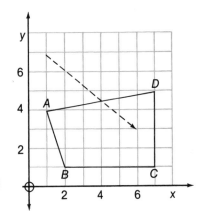

Solution. The translation arrow indicates that each point slides 5 units to the right and 4 units down. Thus, the x-coordinate of each point increases by 5 and the y-coordinate decreases by 4.

A(1, 4) → A'(6, 0)   B(2, 1) → B'(7, −3)
C(7, 1) → C'(12, −3)   D(7, 5) → D'(12, 1)

By plotting A', B', C', D', we obtain the image of quadrilateral *ABCD*. Since *AD* and its image, *A'D'*, are each the hypotenuse of a right triangle with sides of 1 unit and 6 units, then *AD* = *A'D'*. In a similar manner, the other pairs of corresponding sides can be shown to be equal. Also, since the corresponding sides of quadrilateral *ABCD* and its image are parallel, the corresponding angles are equal.

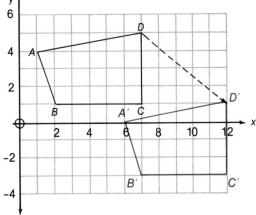

Exercises 11 - 5

A   1. A translation maps the point (1, −2) onto (3, 0).

   a) Plot the two points on a grid and draw the translation arrow.

   b) Find the images of the following points under this translation:

   i) (0, 2)   ii) (5, 6)   iii) (−3, −1)   iv) (−4, 2)

2. The graph shows points A, B, C, D, E and a translation arrow.

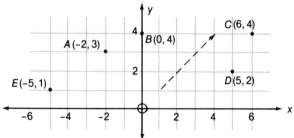

   a) On graph paper, plot the points shown and their images under the translation indicated by the arrow.

   b) Draw line segments AA′ and DD′, and compare them with the translation arrow.

   c) Measure and compare the lengths of line segments DE and D′E′.

   d) Measure and compare the sizes of ∠AED and ∠A′E′D′.

3. Find the values of x, y, and z.

   a) 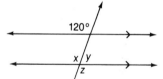   b)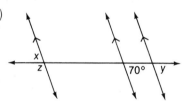

4. A translation maps the point (−2, −3) onto (4, 2).

   a) Plot the points and draw the translation arrow.

   b) Find the images of A(−1, 1), B(1, 4), C(2, −3) under this translation.

   c) Describe the effect of this translation on △ABC.

B   5. A parallelogram has vertices at A(−2, 2), B(2, 1), C(4, −4), D(0, −3). A translation maps points 4 units to the right and 1 unit down. Draw:

   a) the parallelogram;   b) the translation arrow;

   c) the image of the parallelogram under this translation.

6. Find the values of x, y, and z.

a)

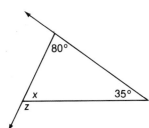

b)

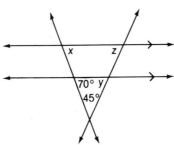

7. A translation maps the point (2, 5) onto (5, −2).

   a) Draw the translation arrow.

   b) If P′(−3, −1), Q′(−1, 3), R′(−5, 0) are the images of P, Q, R under this translation, find the coordinates of P, Q, and R.

C   8. a) Graph the equation: $2x + y = 6$.

   b) Draw the image of the graph under the translation that maps the point (3, 0) onto (0, 0).

9. a) Graph the equation: $5x - 2y = 10$.

   b) Draw the image of the graph under the translation that maps the point (−2, 1) onto (0, 6).

   c) Explain the result.

10. Translation T_1 maps the point (4, 1) onto (2, 3). Translation T_2 maps (−2, −3) onto (3, 0).

   a) Plot △ABC, as shown, and its image, △A′B′C′, under T_1.

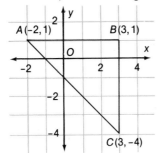

   b) Plot the image of △A′B′C′ under T_2. Call it △A″B″C″.

   c) Draw a translation arrow for the single translation that maps △ABC onto △A″B″C″.

   d) Investigate whether T_1 followed by T_2 gives the same result as T_2 followed by T_1.

11. What properties remain invariant under a translation?

11 - 6 Congruence

Figures that are identical in shape and size are said to be **congruent.** Exact identity never really occurs in the world around us, but many things are so nearly identical that they are considered to be congruent. A very close approximation to congruence is a necessity when duplicating keys, for example, otherwise the duplicate key will not open the lock.

> If there is a translation, rotation, or reflection under which one geometric figure is the image of another, the two figures are congruent.

Since any two congruent figures have the same size and shape, they must also have the same area.

The examples that follow develop some important relationships using the properties of these transformations.

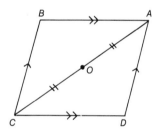

Example 1. Show that the diagonal of a parallelogram divides it into two congruent triangles.

Solution. A $\frac{1}{2}$- turn rotation maps $\triangle ABC$ onto $\triangle CDA$ when the turn centre is the midpoint, O, of the diagonal AC. Therefore, $\triangle ABC$ is congruent to $\triangle CDA$.

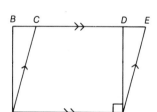

Example 2. Show that the area of a parallelogram is equal to the area of a rectangle with the same base and height.

Solution. Let A represent area.

$$A_{parallelogram\ ACEF} = A_{trapezoid\ ACDF} + A_{\triangle FDE}$$

$$A_{rectangle\ ABDF} = A_{trapezoid\ ACDF} + A_{\triangle ABC}$$

The translation, $A \rightarrow F,\ B \rightarrow D,\ C \rightarrow E$, makes $\triangle ABC$ congruent to $\triangle FDE$, and the area of $\triangle ABC$ equals the area of $\triangle FDE$.

Therefore, $A_{parallelogram\ ACEF} = A_{rectangle\ ABDF}$

The area of a parallelogram is equal to the area of a rectangle with the same base and height.

Example 3.   Show that the angles opposite the equal sides of
an isosceles triangle are equal.

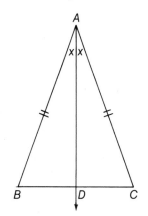

Solution.   Isosceles $\triangle ABC$ has $AB = AC$.
Let AD be the bisector of $\angle BAC$.
Since AD bisects $\angle BAC$, AD is the reflection line
for the reflection that maps AC onto AB. The
reflection also maps $\triangle ACD$ onto $\triangle ABD$.
Therefore, $\triangle ACD$ is congruent to $\triangle ABD$,
and, $\angle ACD = \angle ABD$.

The angles opposite the equal sides of an isosceles triangle are equal.

Exercises   11 - 6

A   1.   State the transformation necessary to map each shaded polygon onto its congruent image.

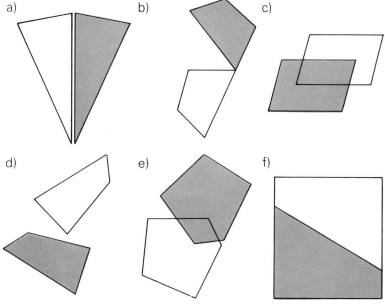

a)

b)

c)

d)

e)

f)

2.   In the quadrilateral $ABCD$, AC is the perpendicular bisector of
BD. Use a reflection to show that $\triangle ABC$ is congruent to $\triangle ADC$.

3.   The diagonals of parallelogram $ABCD$ intersect at O.

a)   Using a $\frac{1}{2}$-turn, show that $\triangle AOD$ is congruent to $\triangle COB$.

b)   Explain why $OA = OC$ and $OB = OD$.

4.   Using the translation that maps A onto B, B onto B', and C onto
C', show that: $\angle CBB' = \angle A + \angle ACB$.

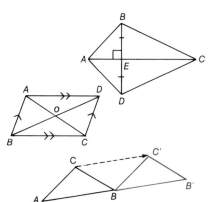

B 5. Find the values of x, y, and z:

a)

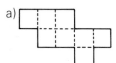

b)

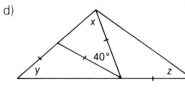

c)

d)

6. Copy the diagrams. Divide each one into two congruent parts. Name the transformation that maps one part onto the other.

a)

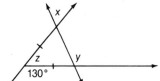

b)

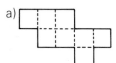

c)

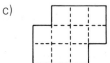

d)

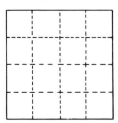

e)

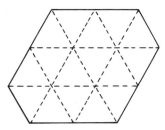

f)

7. In how many different ways can each figure be divided into two congruent parts? (Assume that the figures are divided along the broken lines.)

a)

b)

C 8. Each house in the design on the cover of this book can be identified by the color of its side wall.

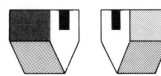

a) Identify the transformation that relates
   i) any yellow house to the orange house;
   ii) any green house to the orange house;
   iii) any red house to the orange house.

b) How are the blue houses related to the orange house?

c) Identify the transformation that relates any two houses of the same color.

11 - 7 Symmetry

Many objects in art and nature seem to be divided, by a line we can visualize, into two matching parts. Such objects are said to have **line symmetry**. Perfect line symmetry rarely occurs in nature. For example, the right wing of the butterfly does not exactly match the left. In mathematics, however, we define line symmetry exactly.

> Any figure that can be mapped onto itself by a reflection is said to have **line symmetry**. The reflection line is called a **line of symmetry**.

Each of the four shapes below has more than one line of symmetry. One line is shown for each; how many others can you see?

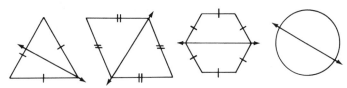

Example 1. Identify all the lines of symmetry in a regular hexagon.

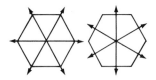

Solution. There are three lines of symmetry joining opposite vertices and three joining the midpoints of the opposite sides. A regular hexagon has a total of six lines of symmetry.

The illustration shows how, with a semi-transparent mirror, a figure having line symmetry can be drawn from any shape.

Just as reflections can be used to create shapes with symmetry, so also can rotations.

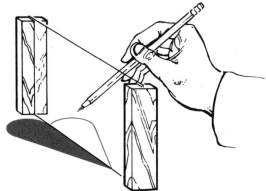

> Any figure that maps onto itself when rotated less than a full turn is said to have **rotational symmetry**.

The following shapes have rotational symmetry because they map onto themselves when rotated the amount indicated about the turn centres shown.

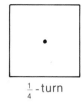 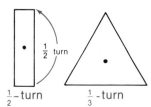 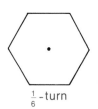

$\frac{1}{2}$-turn $\frac{1}{3}$-turn $\frac{1}{4}$-turn $\frac{1}{6}$-turn

The shapes have rotational symmetry of order 2, 3, 4, and 6 respectively.

Many objects around us appear to have rotational symmetry. The trillium, for example, appears to have rotational symmetry of order 3.

Point symmetry is a special kind of rotational symmetry.

> A figure has **point symmetry** if a $\frac{1}{2}$-turn maps the figure onto itself.

Each of the figures below has point symmetry about the turn centre indicated.

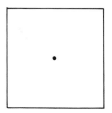  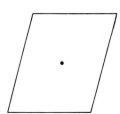

Example 2. Under what angles of rotation will a regular pentagon map onto itself?

Solution. Since the pentagon has rotational symmetry of order 5, five equal rotations will map it onto itself and return it to its original position. The angle of each rotation is therefore:

$$360° \div 5 = 72°.$$

And rotations of 72°, 2 × 72° or 144°, 3 × 72° or 216°, 4 × 72° or 288°, 5 × 72° or 360° will map the pentagon onto itself.

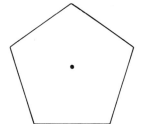

Exercises 11 - 7

A 1. Copy these figures and draw all their lines of symmetry:

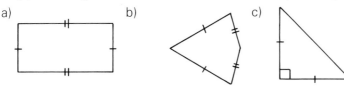

a) b) c)

2. Copy these regular polygons and draw all their lines of symmetry.

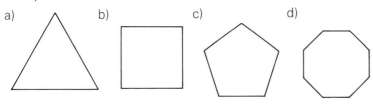

a) b) c) d)

3. A regular decagon has ten equal sides and ten equal angles. How many lines of symmetry does it have?

4. State the order of rotational symmetry of each logo:

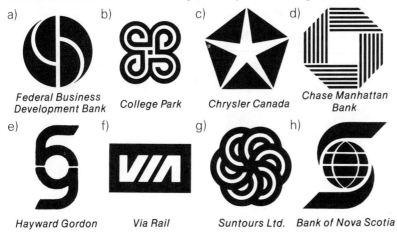

a) b) c) d)

Federal Business
Development Bank   College Park   Chrysler Canada   Chase Manhattan
Bank

e) f) g) h)

Hayward Gordon   Via Rail   Suntours Ltd.   Bank of Nova Scotia

5. a) State the order of rotational symmetry for each of these regular polygons:

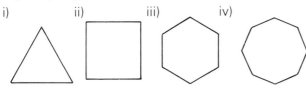

i) ii) iii) iv)

b) A regular dodecagon has twelve equal sides and twelve equal angles. What is its order of rotational symmetry?

6. For each of the regular polygons of Exercise 5, what is the smallest angle of rotation that maps the polygon onto itself?

B 7. On a grid, locate and identify the quadrilateral with vertices at $A(0, 3)$, $B(5, -2)$, $C(4, 5)$, $D(-1, 10)$. Show the lines of symmetry.

8. Draw a figure that has
   a) only two lines of symmetry;
   b) point symmetry but no line of symmetry;
   c) rotational symmetry, order 6, and a line of symmetry;
   d) rotational symmetry, order 6, but no line of symmetry.

C 9. Many crossword puzzles have rotational symmetry.
   a) Identify the order of rotational symmetry of the puzzle shown below (left).

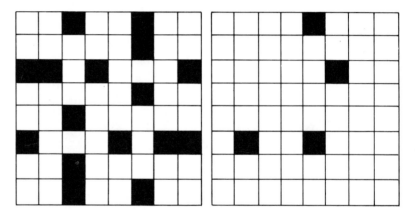

   b) Copy the puzzle above (right) and shade in the least number of squares in order for the puzzle to have rotational symmetry of order 4.
   c) Examine crossword puzzles in newspapers and magazines. Try to find a puzzle that does not have rotational symmetry. Try to find a puzzle that has line symmetry.

11 - 8 Applications of Transformations

A knowledge of transformations can often make an apparently difficult problem easy to solve.

Example 1. A pumping station is to be built somewhere along pipeline *l* to serve tanks at points *A* and *B*. At what point, *P*, on *l* should the pumping station be located so that the necessary length of pipe, *AP* + *BP*, is a minimum?

A •   •*B*

_____ *l*

Solution. Locate *B'*, the reflection image of *B* in *l*.
The shortest distance between *A* and *B'* is the line segment joining them. Let *AB'* intersect *l* at *P*. Draw *PB*. *PB* is the reflection image of *PB'* in *l*. Since *AP* + *PB'* is the shortest distance from *A* to *B'*, then *AP* + *PB* is the shortest distance from *A* to *l* to *B*.

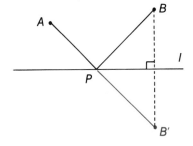

When using transformations to solve problems, we make use of the fact that reflections, rotations, and translations preserve lengths, angles, and areas. *Example 1* involved the invariance of length under a reflection. *Example 2* involves the invariance of lengths and angles under a rotation.

Example 2. For right $\triangle ABC$, show that $a^2 + b^2 = c^2$.

Solution. The $\frac{1}{4}$ -, $\frac{1}{2}$ -, and $\frac{3}{4}$ - turn images of right $\triangle ABC$ about a turn centre are such that vertex $A \rightarrow$ vertex $B \rightarrow$ vertex $B' \rightarrow$ vertex B'', producing the figure shown.

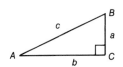

Because lengths and angles are preserved, the result is a large square and a smaller square. The area of the large square equals the area of the small square plus four times the area of the triangle.

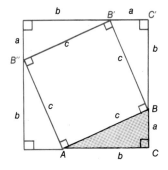

$$(a + b)^2 = c^2 + 4 \times \frac{1}{2}ab$$
$$a^2 + 2ab + b^2 = c^2 + 2ab$$
$$a^2 + b^2 = c^2$$

Exercises 11 - 8

P•

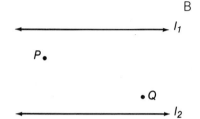

•Q

l

A 1. Using a diagram like the one opposite, show how to locate point R on line *l* such that *PR* + *QR* is a minimum.

2. a) Plot the points *A*(3, 5) and *B*(7, 2) on a grid.

   b) Show the location of point *M* on the *x*-axis such that *AM* + *MB* is a minimum.

   c) Show the location of point *N* on the *y*-axis such that *AN* + *NB* is a minimum.

B 3. Using points *A* and *B* of Exercise 2, locate point *Y* on the *y*-axis and point *X* on the *x*-axis such that *AY* + *YX* + *XB* is a minimum.

l₁

P•

•Q

l₂

4. Draw a diagram like the one shown.

   a) Show the shortest path from *P* to *l₁* to *l₂* to *Q*.

   b) Does it matter which you go to first, *l₁* or *l₂*?

5. The *x*- and *y*-axes on a grid represent two cushions of a billiard table and points *B*(3, 6) and *W*(12, 6) represent the positions of the brown ball and a white ball respectively.

   a) Find the coordinates of a point on the *x*-axis at which the white ball should be aimed so as to rebound and hit the brown ball.

   b) Find the coordinates of the points on the *x*-axis and *y*-axis at which the white ball should be aimed so as to rebound from the *x*-axis to the *y*-axis and hit the brown ball.

l₁

•*A*

l₂

6. A fly lands at the point (4, 6) on a grid. It walks to the *y*-axis, then to the *x*-axis, and finally stops at the point (8, 3). What is the shortest distance the fly could have walked?

7. The promoters of a dirt-bike rodeo are laying out the course. They want the bikers to ride from point *A* to a point on boundary *l₁*. From there they are to ride to a point on boundary *l₂* and return to *A*. Copy the diagram and find the positions of points *B* and *C* on *l₁* and *l₂* respectively so that the length of the course is a minimum.

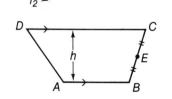

8. *E* is the midpoint of *BC* in trapezoid *ABCD*. Using *E* as the centre of a $\frac{1}{2}$-turn, show that the area of trapezoid *ABCD* is $\frac{1}{2}h\,(AB + CD)$.

Mathematics Around Us

Other Kinds of Transformations

All the transformations in this chapter preserve both the size and the shape of figures. There are many other transformations which change the size of a figure, its shape, or both.

Enlargements and Reductions

The simplest transformations that change the size of a figure are enlargements and reductions, often referred to as dilatations. All dimensions in a dilatation are multiplied by the same number, called the **scale factor**.

Enlargements or reductions of drawings can easily be made with grid lines. The illustration (right) shows an enlargement in which all dimensions of the original drawing are doubled. This is done by spacing the lines on the second grid twice as far apart as they are on the first.

Distortions

Distortions can be produced in many different ways by changing the grids used. A few examples are shown at right and on the next page.

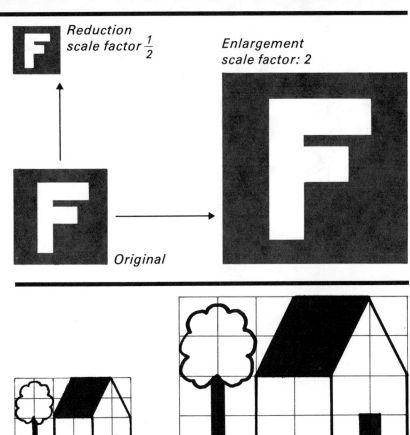

Reduction scale factor $\frac{1}{2}$

Enlargement scale factor: 2

Original

Grid lines need not be the same distance apart.

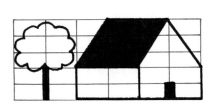

Grid lines do not have to be
perpendicular or parallel.

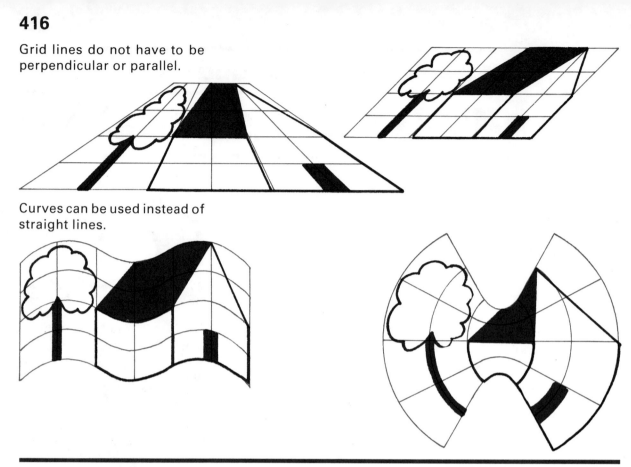

Curves can be used instead of
straight lines.

To make a grid, there must be
two overlapping sets of lines
or curves. Some possibilities
are shown below.

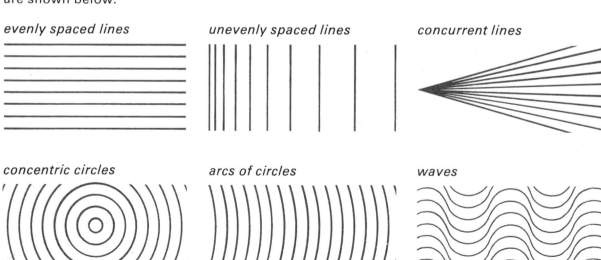

evenly spaced lines

unevenly spaced lines

concurrent lines

concentric circles

arcs of circles

waves

Distortions tend to emphasize certain features in an object, or give it a new perspective. Graphic designers, artists, and advertising people often use these ideas in their work. Look for examples in newspapers, magazines, and on television presentations.

Questions

1. Use grid lines to make an enlargement of (a) and a reduction of (b).

a)

b)

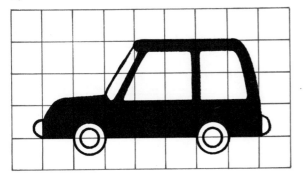

2. Make an enlargement or a reduction of a drawing of your own.

3. Use combinations of sets of lines as grids to create distortions of these figures:

a)

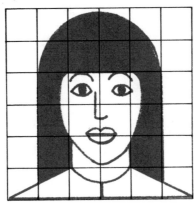

b)

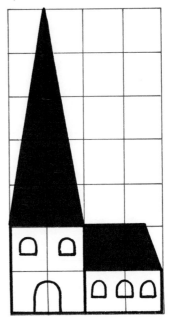

4. Use two sets of lines or curves to make a grid, and create a distortion of a drawing of your own.

Review Exercises

1. What kind of image of △PQR, rotation, reflection, or translation, is each of the triangles shown?

a)

b)

c)

d)

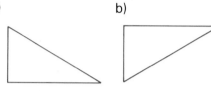

e)

f)

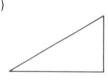

g)

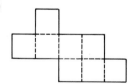

h)

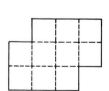

2. Copy these diagrams. Divide each, along the broken lines, into two congruent parts. Name the transformation needed to map one part onto the other.

a)

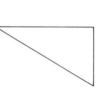

b)

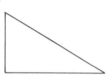

c)

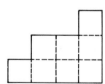

d)

3. Find the values of x, y, and z:

a)

b)

c)

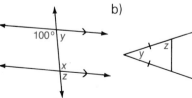

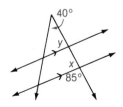

d)   e)   f)

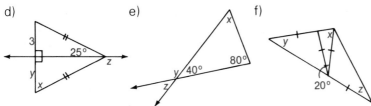

4. A translation maps $(-3, -1)$ onto $(2, -3)$.
   a) Find the images of $A(-2, 4)$, $B(1, -5)$, $C(4, 1)$.
   b) If $P'(2, -3)$, $Q'(0, 4)$, and $R'(-5, 1)$ are image points under this translation, find the object points.

5. Find the images of $A(2, 5)$, $B(-3, 1)$, and $C(1, -4)$ under a reflection in: a) the x-axis;   b) the y-axis.

6. Find the images of $X(-2, -3)$, $Y(4, -3)$, and $Z(4, 5)$ under a $\frac{1}{2}$-turn about $(0, 0)$.

7. The reflection images of $A(1, 4)$, $B(3, -2)$, and $C(5, 1)$ are $A'(-3, 4)$, $B'(-5, -2)$, and $C'(-7, 1)$. Determine the reflection line.

8. State the transformation necessary to map each shaded polygon onto its image.
   a)   b)   c)

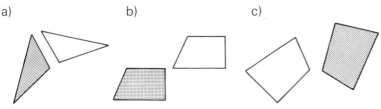

9. Draw a diagram like the one shown. Show the shortest path from P to l to m to Q.

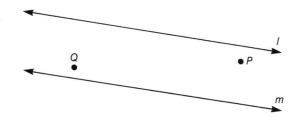

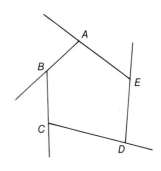

10. Draw a diagram like the one shown. Translate the exterior angles at vertices B, C, D, and E to vertex A. What is the sum of the exterior angles of the polygon?

Table of Square Roots

n	$\sqrt{n}$	n	$\sqrt{n}$	n	$\sqrt{n}$	n	$\sqrt{n}$
1.0	1.000	5.5	2.345	10	3.162	55	7.416
1.1	1.049	5.6	2.366	11	3.317	56	7.483
1.2	1.095	5.7	2.387	12	3.464	57	7.550
1.3	1.140	5.8	2.408	13	3.606	58	7.616
1.4	1.183	5.9	2.429	14	3.742	59	7.681
1.5	1.225	6.0	2.449	15	3.873	60	7.746
1.6	1.265	6.1	2.470	16	4.000	61	7.810
1.7	1.304	6.2	2.490	17	4.123	62	7.874
1.8	1.342	6.3	2.510	18	4.243	63	7.937
1.9	1.378	6.4	2.530	19	4.359	64	8.000
2.0	1.414	6.5	2.550	20	4.472	65	8.062
2.1	1.449	6.6	2.569	21	4.583	66	8.124
2.2	1.483	6.7	2.588	22	4.690	67	8.185
2.3	1.517	6.8	2.608	23	4.796	68	8.246
2.4	1.549	6.9	2.627	24	4.899	69	8.307
2.5	1.581	7.0	2.646	25	5.000	70	8.367
2.6	1.612	7.1	2.665	26	5.099	71	8.426
2.7	1.643	7.2	2.683	27	5.196	72	8.485
2.8	1.673	7.3	2.702	28	5.292	73	8.544
2.9	1.703	7.4	2.720	29	5.385	74	8.602
3.0	1.732	7.5	2.739	30	5.477	75	8.660
3.1	1.761	7.6	2.757	31	5.568	76	8.718
3.2	1.789	7.7	2.775	32	5.657	77	8.775
3.3	1.817	7.8	2.793	33	5.745	78	8.832
3.4	1.844	7.9	2.811	34	5.831	79	8.888
3.5	1.871	8.0	2.828	35	5.916	80	8.944
3.6	1.897	8.1	2.846	36	6.000	81	9.000
3.7	1.924	8.2	2.864	37	6.083	82	9.055
3.8	1.949	8.3	2.881	38	6.164	83	9.110
3.9	1.975	8.4	2.898	39	6.245	84	9.165
4.0	2.000	8.5	2.915	40	6.325	85	9.220
4.1	2.025	8.6	2.933	41	6.403	86	9.274
4.2	2.049	8.7	2.950	42	6.481	87	9.327
4.3	2.074	8.8	2.966	43	6.557	88	9.381
4.4	2.098	8.9	2.983	44	6.633	89	9.434
4.5	2.121	9.0	3.000	45	6.708	90	9.487
4.6	2.145	9.1	3.017	46	6.782	91	9.539
4.7	2.168	9.2	3.033	47	6.856	92	9.592
4.8	2.191	9.3	3.050	48	6.928	93	9.644
4.9	2.214	9.4	3.066	49	7.000	94	9.695
5.0	2.236	9.5	3.082	50	7.071	95	9.747
5.1	2.258	9.6	3.098	51	7.141	96	9.798
5.2	2.280	9.7	3.114	52	7.211	97	9.849
5.3	2.302	9.8	3.130	53	7.280	98	9.899
5.4	2.324	9.9	3.146	54	7.348	99	9.950
		10.0	3.162			100	10.000

Glossary

Acute angle: an angle which measures less than 90°.

Acute triangle: a triangle with three acute angles.

Alternate angles: $\angle ABC$ and $\angle BCD$ are alternate angles.

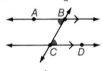

Altitude: a line segment from a vertex perpendicular to the opposite side of a triangle; the length of such a segment (also called the **height**).

The point of intersection of the lines containing the three altitudes of a triangle is called the **orthocentre**.

Angle bisector: a line which divides an angle into two equal angles.

The point of intersection of the bisectors of the three angles of a triangle is called the **in-centre**. This point is the centre of the **incircle**, which touches the three sides of the triangle.

Bar graph: a graph in which data is represented by bars.

Broken-line graph: a graph in which plotted points are joined by line segments. Only the end points of the segments represent actual data.

Circle: a curve consisting of all points in a plane at a given distance, the **radius**, from a fixed point, the **centre**. A line segment joining any two points on a circle is a **chord**. The **diameter** is the chord passing through the centre.

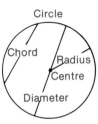

Circle graph: a graph in which a complete set of data is represented by the area of a circle. Various parts of the data are represented by the areas of sectors of the circle.

Circumference: the distance around a circle.

Coefficient: the numerical part of a term, such as 3 in $3x$.

Common factor: a number or expression which is a factor of two or more numbers or expressions. Example: $2x$ is a common factor of $6x^2$ and $10x$.

Congruent: figures having the same size and shape.

Continuous-line graph: a graph that displays the value of one variable, such as speed, corresponding to the value of another variable, such as stopping distance, for all values over a given interval. All points on a continuous line graph represent actual data.

Coordinates: an ordered pair of numbers that represents a point on a plane.

Corresponding angles: $\angle ABC$ and $\angle DEF$ are corresponding angles.

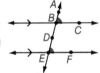

Cube: any expression of the form x^3. The numbers 1, 8, 27, 64, ..., n^3, where n is a natural number, are called **perfect cubes**.

A polyhedron with squares for faces.

Cube root: a cube root of a number is a number which, when cubed, produces the given number. Examples: $\sqrt[3]{8} = 2$ and $\sqrt[3]{-125} = -5$.

Diagonal: a line segment joining any two non-adjacent vertices of a polygon.

Distributive law: a law relating multiplication to addition and subtraction. Examples: $a(b + c) = ab + ac$, and $a(b - c) = ab - ac$.

Equilateral triangle: a triangle with three equal sides.

Exponent: see Power.

Expression: any combination of numbers, variables, and symbols.
Examples: $3x + 5$, $\sqrt{x^2 + y^2}$.

Factor: any number or expression that divides a given number or expression. Example: a is a factor of $3a^2b$.

Frequency: the number of times an event occurs.

Histogram: a graph that records the number of times a variable has a value in a particular interval.

Image: the figure obtained by a transformation of a given figure.

Inequality: a mathematical sentence that uses the sign $<$ or $>$ to relate two expressions.

Integers: the set of numbers:
$\{\ldots, -3, -2, -1, 0, 1, 2, 3, \ldots\}$

Inverse operations: addition and subtraction are inverse operations. Multiplication and division are inverse operations.

Irrational number: see **rational number**.

Isosceles triangle: a triangle with two equal sides.

Like terms: terms that have exactly the same variable. Example: $3x$ and $7x$.

Linear relation: a relation in which the terms containing the variable are all of the first degree.
Example: $\{(x, y) \mid 2x + 3y = 12\}$

Line symmetry: any figure that can be mapped onto itself by a reflection is said to have line symmetry. The reflection line is called a **line of symmetry**.

Mapping diagram: a graph that relates numbers in one set to numbers in another set by means of arrows.

Mapping notation: a method of writing a relation. Example: $x \rightarrow 2x - 1$.

Mean: the sum of the numbers in a set divided by the number of numbers.

Measure of central tendency: any one of the mean, median, or mode.

Median: the middle number when a set of numbers is arranged in order.
A line segment joining a vertex of a triangle to the mid point of the opposite side.
The point of intersection of the three medians is called the **centroid**.

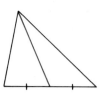

Mode: the most frequently occurring number, or numbers, in a set of numbers.

Natural numbers: the set of numbers:
$\{1, 2, 3, 4, \ldots\}$

Obtuse angle: an angle which measures greater than 90° but less than 180°.

Obtuse triangle: a triangle with one obtuse angle.

Ordered pair: two numbers written in a specific order. Example: $(2, -5)$

Parallel lines: two lines in the same plane that do not intersect.

Parallelogram: a quadrilateral with two pairs of sides parallel.

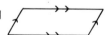

Perpendicular lines: two lines intersecting at an angle of 90°.

Perpendicular bisector: a line which passes through the midpoint of a line segment and is perpendicular to the segment.

The point of intersection of the perpendicular bisectors of the three sides of a triangle is called the **circumcentre**. This point is the centre of the **circumcircle**, which passes through the three vertices of the triangle.

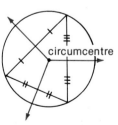

circumcentre

Pictograph: a graph that uses a symbol to represent a certain amount.

Polygon: a closed plane figure formed by line segments that do not cross each other. The segments are called **sides**. The points where two sides meet are the **vertices**. They are named by the number of sides they have.

A **regular polygon** has all sides equal and all angles equal.

Polyhedron: a solid bounded by polygons. The polygons are the **faces**. The faces meet at **edges**. The points where three or more edges meet are **vertices**.

Hexagonal prism

Tetrahedron

Polynomial: the sum or difference of two or more terms in which the variable occurs to positive integral powers only. If there is only one variable, the highest power determines the **degree**. Example: the polynomial $5x^3 - 7x^2 + 4x - 1$ has degree 3.

Population: the set of all things being considered.

Power: an expression of the form a^n. a is called the **base** and n is called the **exponent**.

Probability: the likelihood of the outcome of an experiment. If an experiment has n equally likely outcomes of which r are favorable to event A, then the probability of event A is $P(A) = \dfrac{r}{n}$.

Proportion: a statement that two ratios are equivalent. Example: $a:b = c:d$

or $\dfrac{a}{b} = \dfrac{c}{d}$.

Pythagorean theorem: see **right triangle**.

Quadratic relation: a relation that has at least one second degree term but no terms of higher degree.

Quadrilateral: a polygon with four sides.

Radical sign: the symbol $\sqrt{}$.

Random numbers: a set of single digit numbers that is generated in such a way that each number has an equal chance of occurring each time.

Random sample: a sample that is chosen in such a way that each item in the population has an equal chance of being selected.

Rate: a comparison of two quantities having different units. Example: speed of a car in kilometres per hour.

Ratio: a comparison of quantities measured in the same units.

Rational number: any number that can be written in the form $\dfrac{m}{n}$, where m and n are integers, and $n \neq 0$. Rational numbers can also be expressed either as terminating or repeating decimals. Examples:

$$\frac{3}{5} = 0.6$$

$$-\frac{4}{3} = -1.3333\ldots.$$

Any number that cannot be written in such a form is called an **irrational number**. The decimal form of an irrational number neither terminates nor repeats. Examples:

$$\pi = 3.141\,592\,653\,589\ldots,$$

$$\sqrt{2} = 1.414\,213\,562\,373\ldots$$

Real number: any member of the set of rational or irrational numbers.

Reciprocals: two numbers with a product of 1. Example: $\dfrac{3}{2}$ and $\dfrac{2}{3}$.

Rectangle: a parallelogram with adjacent sides perpendicular.

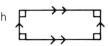

Reflection: a reflection in a line l is a transformation that maps every point P onto an image point P' such that:
• P and P' are equidistant from line l.
• PP' is perpendicular to line l.
The line l is called the **reflection line**.

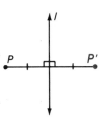

Relation: a set of ordered pairs.

Relative frequency: if outcome, *A*, occurs *r* times in *n* repetitions of an experiment, the relative frequency of *A* is $\frac{r}{n}$.

Rhombus a parallelogram with all sides equal.

Right angle: an angle which measures 90°.

Right triangle: a triangle with one right angle. The side opposite the right angle is called the **hypotenuse**. The **Pythagorean theorem** relates the lengths of the sides of a right triangle: $AC^2 = AB^2 + BC^2$.

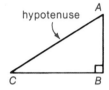

hypotenuse

Rotation: when a shape is turned about a fixed point it is said to have undergone a rotation. A rotation through 180° is called a **half-turn**.

Rotational symmetry: any figure that maps onto itself when rotated less than a full turn is said to have **rotational symmetry**. A figure has **point symmetry** if a half-turn maps the figure onto itself.

Sample: a representative portion of a population.

Scalene triangle: a triangle with all three sides of different length.

Scientific notation: a number is written in scientific notation when it is expressed as the product of:
- a number equal to or greater than 1 but less than 10, and
- a power of 10.

Example: 3.72×10^8.

Slope: the ratio: $\frac{\text{rise}}{\text{run}}$.
In a coordinate system,
$$\text{slope} = \frac{\text{difference in } y\text{-coordinates}}{\text{difference in } x\text{-coordinates}}.$$

rise

run

Sphere: a surface consisting of all points in space at a given distance, the **radius**, from a fixed point, the **centre**.

Square: any expression of the form x^2. The numbers 1, 4, 9, 16, ..., n^2, where *n* is a natural number, are called **perfect squares**.
A rectangle with four equal sides.

Square root: a **square root** of a number is a number which, when squared, produces the given number. The **radical sign,** $\sqrt{}$, denotes the positive square root. Example: $\sqrt{36} = 6$.

Statistics: the branch of mathematics which deals with the collection, organization, and interpretation of data.

Straight angle: an angle which measures 180°.

Supplementary angles: two angles with a sum of 180°. Example: angles of 72° and 108° are supplementary.

Tetrahedron: see **polyhedron**.

Transformation: whenever the shape, size, appearance, or position of an object is changed, it has undergone a transformation.

Translation: a translation is defined by the mapping $A \rightarrow A'$, $B \rightarrow B'$, $C \rightarrow C'$ in the diagram. The segments AA', BB' and CC' are all equal in length and parallel. The **translation arrow** represents the length and direction of these segments.

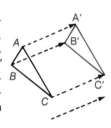

Transversal: a line which intersects two or more lines.

Trapezoid: a quadrilateral with just one pair of parallel sides.

Vertex: see **polygon** and **polyhedron**.

Vertically opposite angles: $\angle AOB$ and $\angle COD$ are vertically opposite angles.

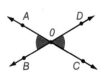

Whole numbers: the set of numbers: {0, 1, 2, 3, 4, ...}.

Selected Answers

Chapter 1

Exercises 1 - 1

1. a) 40¢ b) $4000 **3.** 8 s; 6 s

5. a) 384 cm, 160 cm b) 9 cm

7. 50¢ **9.** a) 26 m b) 22 m²

11. $8.25 **13.** 144 km/h

15. a) 9.2 min b) 14 min c) 17.5 min d) 107.1 min

Exercises 1 - 2

1. a) 20°C b) 80°C c) 176.7°C d) 0°C

3. a) 225 mm² b) 85 mm² c) 86 mm²

5. About 5570 m³ **7.** a) 4 b) 20

9. a) i) $2.30 ii) $7.80 iii) $5.30 b) 1—2 km

11. a) i) 144° ii) 150° b) 6

Exercises 1 - 3

1.

	Expression	Variables	Terms	Coefficients
a)	$6a - 2b$	a, b	$6a, 2b$	6, 2
b)	$a - 2b + 9c$	a, b, c	$a, 2b, 9c$	1, 2, 9
c)	$1.8C + 32$	C	$1.8C, 32$	1.8
d)	$2\pi r$	r	$2\pi r$	2π

3. a) 352 b) $11d$ c) $32r$ d) dr

5. a) 9, 13, 17, 23, 31 b) 23, 18, 13, 8, 3

c) 44 d) 72 e) 21

7. a) 19 b) $\frac{1}{4}$ c) $\frac{17}{30}$ d) $\frac{26}{55}$ e) $\frac{5}{72}$

9. 754 cm³ **11.** a) 114 b) 87

Exercises 1 - 4

1. a) $11a$ c) $71x$ e) $5g$ g) $21p$ i) $5w$

3. a) $m + 5$ c) $10a + 3b$ e) $8u + 14v + 7$

g) $5x + y$ i) $3x + 3y + z$

5. a) 33 c) 497 e) 200 g) 88 i) 360

Exercises 1 - 5

1. a) $3m - 24$ c) $11p + 77$ e) $14p + 42$

g) $36m - 108n$ i) $12s + 66t - 30$

3. a) $34x + y$ c) $87m + 14n$ e) $94k + 5l$

g) $33r + 22s + 14p + 7q$ i) $35a + 15b + 5c$

Exercises 1 - 6

1. a) 250 c) 160 e) 120 g) 22 i) 5

3. a) $3b(c + d)$ c) $10n(m - 1)$ e) $6m(n - 1)$

g) $3(a + b + c)$ i) $5(p - q - r)$ k) $5a(c + d - 1)$

5. $74.86

7. a) $2(3x + y)$ c) $4(3m + 5n)$ e) $6(5a - 4b + 1)$

g) $3a(2b + 5c)$ i) $4m(2a - 3)$ k) $7a(2x + 5y - 1)$

Exercises 1 - 7

1. a) i) 14 ii) 10 c) i) 50 ii) 30 e) i) 7 ii) 15

3. a) $(3 + 5)(4 - 2)$ b) $3 + (5)(4) - 2$

c) $(3 + 5) 4 - 2$

5. a) 13 c) 45 e) $6\frac{2}{5}$ g) 48 i) 49 k) 22

7. a) 59, 27 c) 153, 108 e) 157, 187

9. a) $4x$ c) $5a$ e) $7d$

11. a) $2a + 5b$ c) $16s + 6t$ e) $28p + 6q$

Exercises 1 - 8

1. a) 15 c) 34 e) 36 **3.** a) $8\frac{1}{2}$ c) 3 e) $8\frac{2}{3}$

5. a) 5 c) 7 e) 2 **7.** a) 9 c) 19 e) 17

9. a) 5 c) 5 e) 3

11. a) 15 cm b) 13 cm **13.** a) 16 cm b) 6 cm

15. a) i) 1.6 km ii) 3.2 km iii) 1.12 km

b) i) 25 s ii) 18.75 s

17. a) 250 b) 3

Review Exercises

1. 10.0 L **3.** 1.3 cm **5.** 20%

7. a) $12x$ c) $9x$ e) $8x + 2y$

g) $2x + 10y$ i) $13m + 23n$ k) $11r + 3s$

9. $3.74

11. a) $14a$ b) $36b$ c) $99x$ d) $120 + 52z$

13. a) 4 c) 5 e) 9 g) 16 i) 6

PSS 2. About 8.2 km **4.** a) $7875 b) 875 km

MAU 1. 1.32 m **2.** 0.66 m

5. a) About 2750 A.D. b) About 900 A.D.

Chapter 2

Exercises 2 - 1

1. a) +$9 c) +80° e) +$50

g) −$81 i) −15 s

3. a) debt of $12   c) credit of $15

5. a) 2   c) −1   e) 3   g) 10

7. a) −4, −1, 3, 5   c) −8, −2, −1, 0, 4, 5

9. a) −2,   c) +2   e) +4   g) +5

Exercises 2 - 2

1. a) −4   c) −3   e) +5   g) +8   i) +2
   k) +6   m) +35   o) −50   q) +40   s) −80
   u) −260   w) +2009

3. a) 0   c) +35   e) −6

5. a) 0   c) 0   e) 0
   g) The sum of an integer and its opposite is 0.

7. a) 15 + (+17) + (−5) = +27
   b) 85 + (−29) + (−37) + (+52) + (−66) = +5
   c) 80 + (−7) + (+5) = 78

Exercises 2 - 3

1. +10°C   3. a) +20°C   b) +12°C
   c) −54°C   d) +24°C

5. a) −2   c) −12   e) 7   g) −9
   i) 1   k) −2   m) 2   o) −5

7. a) −19   c) 16   e) −16

9. a) 1   b) 10   c) −3   d) 24

11. a) 6   c) 2   e) −2

13. 56°C   15. a) 15 : 10   b) 09 : 20

17. a) i) 1 p.m.   ii) 9 a.m.   iii) noon
    b) 4 a.m. the next day

19. a) −600 m   c) +2000 m   e) +6900 m
    g) −1200 m   i) −3700 m

MAU 1. a) +99 m   b) +2 m   c) −183 m
    d) −177 m

    3. Lock 3 of the Welland Canal

Exercises 2 - 4

1. a) −30   c) 63   e) −60
   g) −72   i) 25   k) −12

3. a) 70   c) −120   e) 18   g) −5   i) 1080

5. a) 9   c) 25   e) 10

7. a) −8   c) 11   e) −21

9. a) −32   b) 64   c) 28   d) −11

13. a) $4 000 000   c) $6 000 000
    e) $24 000 000   g) $76 000 000

15. a) 4   c) 7   e) −8   g) 14   i) −3

Exercises 2 - 5

1. a) −12   c) −4   e) 5   g) −6   i) 13

3. a) 5   c) 8   e) −5   g) 9   i) −8

5. a) −4   c) −18   e) 6   g) −27   i) 33

9. a) −5   b) −17   c) 26   d) −13

11. d) and f) are positive. a), b), c), e) are negative.

Exercises 2 - 6

1. a) −3   c) −16   e) 80

3. a) −5   c) −21   e) −5

5. a) 4 − x   c) y − x   e) 46 − y

7. a) 1   b) 16   c) −2   d) −16

9. a) −2x + 8y   c) 14a − 14b

Review Exercises

1. a) −9, −7, −5, −4, 3, 6
   b) −9, −7, −6, −5, 6, 9

3. −45°C

5. a) 3   c) 7   e) 9   g) 9   i) 5

7. a) −23   c) −30   e) 80   g) −1   i) −5

9. a) −2   c) −15   e) −6   g) −3

11. a) 72   c) −90   e) 60

13. a) −4   c) 3   e) 8   g) −3   i) 4

15. a) −10   b) 19   c) 10

Chapter 3

Exercises 3 - 1

1. a) −2°   c) $-\dfrac{3}{4}^{\circ}$ e) $-3\dfrac{1}{4}^{\circ}$

3. a) $-\dfrac{1}{2}$ c) $\dfrac{2}{5}$ e) $\dfrac{6}{11}$ g) $\dfrac{2}{7}$ i) $-\dfrac{3}{7}$

5. $\dfrac{28}{72}, \dfrac{16}{72}, -\dfrac{18}{72}, -\dfrac{24}{72}, -\dfrac{26}{72}, -\dfrac{27}{72}$

7. $\dfrac{-3}{4}, \dfrac{15}{-20}, \dfrac{12}{-16}, \dfrac{-6}{-8}, \dfrac{9}{-12}$

Exercises 3 - 2

1. a) $-\dfrac{3}{10}$ c) $-\dfrac{1}{6}$ e) $\dfrac{1}{4}$ g) $\dfrac{14}{5}$ i) 20

3. a) $\dfrac{1}{16}$ c) $-\dfrac{81}{10}$ e) $-\dfrac{2}{3}$ g) −2

5. a) $-\dfrac{12}{5}$ c) $-\dfrac{9}{5}$ e) $\dfrac{3}{8}$ g) $\dfrac{4}{15}$

7. a) 5   c) −2   e) −6   g) −2

Exercises 3 - 3

1. a) $\frac{17}{12}$ c) $-\frac{11}{24}$ e) $\frac{11}{18}$ g) $-\frac{7}{20}$ i) $\frac{13}{8}$

3. a) $-\frac{1}{12}$ c) $\frac{5}{24}$ e) $3\frac{1}{12}$ g) $-9\frac{1}{3}$ i) $2\frac{3}{4}$

5. a) $\frac{4}{5}$ b) $-\frac{13}{24}$ c) $-\frac{35}{24}$ d) $-\frac{31}{28}$

7. a) 4   c) 2   e) -8

9. b) \$17 625   c) 200

   d) $\$10\frac{1}{2}$, $\$7\frac{3}{8}$, $\$8\frac{1}{2}$, \$36, $\$9\frac{3}{4}$, $\$15\frac{3}{4}$, $\$12\frac{1}{2}$, \$17

Exercises 3 - 4

1. a) $\frac{1}{4}$ b) $-\frac{28}{15}$ c) $\frac{1}{24}$ d) $\frac{22}{5}$

3. \$3440, \$2950   **5.** $\frac{21}{55}$, $\frac{8}{21}$

7. a) \$1143   b) \$14 700   c) \$9200   d) \$67.20

9. a) 37.5 m   b) 86.6 m   **11.** a) \$1250   b) \$2000

PSS 1. c)   **3.** d)

Exercises 3 - 5

1. a) 3.232 323 23   c) $-81.466\,666\,66$ e) $-2.651\,351\,35$
   g) 0.069 069 06   i) $-0.007\,474\,74$

3. a) 0.6   c) $0.\overline{4}$ e) $0.\overline{3}$
   g) $0.4\overline{6}$ i) 0.3125   k) $0.91\overline{6}$

5. a) Duffy 0.438, Cobb 0.420, Ruth 0.378, Gehrig 0.373,
     T. Williams 0.406, Carew 0.318, B. Williams 0.322,
     Cash 0.305
   b) Duffy, Cobb, T. Williams, Ruth, Gehrig, B. Williams,
     Carew, Cash

7. $\frac{13}{15}$, $\frac{6}{7}$, $\frac{9}{11}$, $\frac{10}{13}$, $\frac{5}{8}$

9. a) 0.028 901 7   b) 0.289 017 3
   c) 2.890 173 4   d) 28.901 734

Review Exercises

1. $-\frac{5}{2}$, $\frac{1}{8}$, $-\frac{9}{5}$ **3.** $-\frac{31}{-32}$, $-\frac{15}{16}$, $\frac{-43}{48}$, $\frac{2}{3}$, $\frac{5}{6}$, $\frac{11}{12}$

5. a) $-\frac{7}{12}$ c) $-\frac{21}{40}$ e) $\frac{8}{15}$ g) $-\frac{2}{3}$ i) 1

7. a) $\frac{41}{35}$ c) $\frac{29}{36}$ e) $-\frac{5}{6}$ g) $\frac{2}{15}$
   i) $-\frac{79}{63}$ k) $\frac{31}{30}$

9. a) $\frac{11}{18}$ c) $\frac{11}{24}$ e) $-\frac{1}{4}$ g) $\frac{78}{55}$
   i) $8\frac{1}{12}$ k) $-17\frac{1}{18}$

11. a) $-\frac{21}{4}$ c) $\frac{26}{5}$ e) $\frac{2}{3}$ g) $\frac{8}{15}$ i) $-\frac{3}{2}$

13. a) 1   b) 0, -13, $\frac{0}{1}$, 1

   c) $\frac{1}{8}$, 13.25, 0, -13, $\frac{0}{1}$, $-11\frac{3}{8}$, 1, $5.8\overline{5}$, $5\frac{1}{2}$, $5.12\overline{7\,16}$,
     7.692 307 692 …

CP 1. $\frac{17}{20}$ **3.** $\frac{25}{24}$ **5.** $\frac{11}{36}$

Chapter 4

Exercises 4 - 1

1. a) 6   c) -13 e) 8   g) -12 i) -24

3. a) 12   c) 75   e) 145   g) 2   i) 5

Exercises 4 - 2

1. a) 8   c) -2 e) -9 g) -2 i) -1 k) 3

3. a) -9 c) 16   e) -8 g) -6 i) 0.5   k) -30

5. a) $-\frac{3}{2}$ c) 9   e) $\frac{32}{63}$ g) 15   i) 1

7. a) i) -1 iii) 8   v) 2
   b) i) $\frac{2}{3}$ iii) 0   v) -36

9. a) i) \$29.75   ii) \$52.25   iii) \$134.75
   b) i) 108 km   ii) 235 km   iii) 422 km

Exercises 4 - 3

1. a) 2   c) -1 e) 2   g) -2
   i) $\frac{2}{3}$ k) $\frac{1}{6}$ m) $\frac{11}{6}$

3. a) 6   c) -1 e) 1   g) $-\frac{4}{5}$
   i) $\frac{3}{2}$ k) -3 m) 0

Exercises 4 - 4

1. a) 12   c) 14   e) $-\frac{10}{3}$ g) 20   i) $\frac{3}{10}$

3. a) i) $-\frac{1}{4}$ iii) $-\frac{1}{20}$ v) $\frac{19}{20}$
   b) i) $-\frac{5}{12}$ iii) $\frac{5}{12}$ v) $\frac{25}{12}$

5. 24 min

Exercises 4 - 5

1. a) 2   c) 5   e) -1 g) $\frac{1}{2}$ i) 10

5. a) 14   c) No solution   e) No solution
   g) 20   i) $-\frac{1}{2}$

Exercises 4 - 6

1. a) 75 g   b) 22 cm³   **3.** $21.88

5. a) i) 200   ii) 183   iii) 157
    b) i) 50   ii) 28   iii) 79

7. a) 30, 25, 20   b) 60, 70

9. a) i) 9.8 m/s   iii) 78.4 m/s
    b) i) 5.5 s   iii) 35 s

11. a) 2.4 km   c) 1.3 km

13. a) i) 18.7 m   iii) 10.0 m
    b) i) 66.9 m   iii) 171 m

15. a) i) 120 cm   iii) 80 cm   b) 90 cm, 67.5 cm

17. 6

Exercises 4 - 7

1. a) $x + 5$ c) $8x$ e) $8x$ g) $8x - 2$
   i) $3x - 12$ k) $\frac{1}{4}(x + 3)$ m) $4(x + 5)$

Exercises 4 - 8

1. $x + 5x = 12$ **3.** $x + (x + 12) = 42$

5. $x + 2(21 - x) = 30$ **7.** $x + 2(x + 2) = 19$

9. $1.1x - x = 12$ **11.** $x + 2 = \frac{3}{4}[(2x - 4) + 2]$

13. $\frac{1}{2}[(3x + 2) - 5] = x + 5$

Exercises 4 - 9

1. 7 km, 5 km   **3.** 600 km/h, 100 km/h   **5.** 131, 132

7. 57, 58, 59, 60   **9.** 15, 3   **11.** 6 kg, 18 kg

13. 13, 52   **15.** 7, 29   **17.** 3.5, 3.75

19. 11, 16, 48   **21.** $180, $450, $140   **23.** 5

25. 7, 9, 11   **27.** 7.5 g   **29.** 15

Exercises 4 - 10

1. a) $x < 3$ c) $x > -1$ e) $x < -2$

3. a) $x > 6$ c) $z < 2.5$ e) $x < \frac{5}{3}$
   g) $x > -\frac{4}{7}$

Review Exercises

1. a) -16 c) 28   e) -34

3. a) 3   c) -7 e) $\frac{5}{4}$ g) 7   i) 3   k) 1.3

5. a) i) $1.20   ii) $3.10   iii) $5.00   b) 25 min

7. a) 0   c) $-\frac{13}{5}$ e) $\frac{9}{2}$ g) $-\frac{17}{2}$ i) -2 k) $\frac{34}{37}$

9. a) $7x = 56, 8$ c) $x + 16 = 31, 15$
   e) $x - 29 = -2, 27$

11. 14, 21   **13.** $-0.75, 2.25$ **15.** 10, 22

17. $-2, -8$ **19.** Nita 28, Honor 14

21. a) i) 70°   ii) 55°   b) i) 130°   ii) 70°

23. a) $C = 40 + 0.05d$
   b) i) $40   iii) $57.50   c) i) 1200 km   iii) 800 km

25. a) 375 000 km   b) 42 000 000 km (42 Gm)
   c) 9 500 000 000 000 km (9.5 Pm)

MM 1. $\frac{5}{9}$ **3.** $\frac{36}{11}$ **PSS 1.** 37.5 km/h   **3.** 62.5%

CP 1. a) 7   c) 21   e) 12   g) 20   i) 300
   k) -770 m) 100   o) -3.2 q) $9.\overline{216}$ s) 12

Chapter 5

Exercises 5 - 1

1. a) y^4 c) $\left(\frac{2}{5}\right)^5$ e) $(4a)^5$ g) m^4 i) π^6

3. a) 64   c) -32 e) 10 000   g) 0.008
   i) 48   k) 1296

5. a) 25   c) 19   e) 70   g) $6\frac{2}{9}$ i) 6.52

7. a) 10^3 c) 10^2 e) 10^5 g) 10^1 i) 10^{12}

9. a) $3^3, 5^2, 2^4, 3^2, 2^3$
   c) $(1.15)^3, (1.2)^2, (1.1)^3, (1.3)^1, (1.05)^5$
   e) $(0.4)^2, (0.3)^2, (0.2)^2, (0.3)^3, (0.2)^3$

11. a) 3^{25} e) $(0.9)^{11}$

13. a) 3   c) 6   e) 2

15. a) 4^2 c) 2^6 e) $(-2)^4$ g) 7^3
   i) $(-3)^4$ k) 6^5

Exercises 5 - 2

1. a) $(2.5)^2$ c) $\frac{4}{9}x^2$ e) $12.5w^2$

3. a) 25 cm²   c) 2.25 cm²   e) 0.36 cm²
   g) $16a^2$ cm²

5. a) 125 cm³   c) 3.375 cm³   e) 0.216 cm³
   g) $64a^3$ cm³

7. a) i) $233.28   iii) $342.76   b) About 9 years

9. $979.68   **11** 32.7 L   **13.** 7235 cm³

MAU 1. a) 2^1 c) 2^3 e) 2^6 **3.** 5600, 11 200, 1400

Exercises 5 - 3

1. a) 3^{10} c) $(-5)^{25}$ e) $(-8)^6$ g) $\left(\frac{2}{5}\right)^{22}$ i) $\left(-2\frac{1}{4}\right)^{24}$

3. a) 3^5 c) m^{15} e) $-7z^8$ g) 6^6 i) 10^{10}

5. a) $15a^5$ c) $36x^{13}$ e) $30(3)^{12}$ g) $57x^{11}$ i) $84s^{16}$

7. a) m^{20} c) a^{49} e) 12^{35} g) 5^{12} i) 11^{16}

9. a) 625   c) 25   e) 75   g) 15 625   i) 25 000

Exercises 5 - 4

1. a) $\frac{1}{2}$ c) $\frac{1}{9}$ e) $\frac{1}{125}$ g) $\frac{1}{1728}$ i) 2

k) $\frac{1}{100\,000}$ m) 32   o) $\frac{16}{9}$ q) 1000   s) 0.064

3. a) $7\frac{1}{2}$ c) $-6\frac{48}{49}$ e) 17   g) $37\frac{1}{36}$ i) $\frac{7}{144}$

5. $2^7, 2^4, 2^0, 2^{-1}, 2^{-5}, 2^{-6}$

7. a) 7^2 c) $(\frac{1}{7})^3$ e) $(\frac{1}{10})^6$ g) $(0.1)^3$

9. a) $\frac{9}{5}$ c) -5 e) $\frac{1}{216}$ g) $\frac{16}{5}$ i) 1

11. a) $10n^{-21}$ c) $5x^{10}$ e) $4w^{-6}$ g) $27b^{-15}$

13. a) $1\frac{3}{16}$ c) -5

15. a) 0   c) -2 e) -5 g) -2

MAU 1. a) 500   b) 250   **3.** a) 1688

Exercises 5 - 5

1. a) 1×10^3 c) 1×10^2 e) 1.1×10^3
g) 1×10^{-4} i) 1×10^{-6} k) 9.2×10^{-5}
m) 8.5×10 o) 9.9×10^3

3. a) 9 500 000 000 000 km   c) 1×10^{-5} m
e) 120 000 000 000   g) 1.13×10^{-8} cm
i) 360 000 000 km^2 k) 8.5×10^9F

5. a) 1.11×10^{14} c) 4.59×10^9
e) 5.48×10^4 g) 4×10^{-1}

7. $\$1.00 \times 10^9$ **9.** a) About 3×10^4 kg

11. 1.38×10^{21} kg

13. a) 9.46×10^{12} b) 1.54×10^{20}

15. 1×10^{-6} mm   **MAU 1.** 5.6×10^{12}

Exercises 5 - 6

1. a) ±100 c) ±1.3 e) $\pm\frac{1}{10}$ g) $\pm\frac{1}{8}$ i) ±30

3. a) 10   c) 2   e) 40   g) -31 i) 6

5. a) -6 c) 20   e) 35   g) 10   i) 42

7. a) 17 mm, 68 mm   c) 0.08 m, 0.32 m

9. a) 10 mm   c) 1.3 m   e) 0.05 km

Exercises 5 - 7

1. a) 9   c) 6.25   e) 49

3. a) 49   c) 25   e) 289   g) 9   i) 121   k) 4

Exercises 5 - 8

3. a) 7.141   c) 17.916   e) 2.218   g) 0.245   i) 0.348

5. a) 3.146   c) 5   e) 6.922
g) 18.146   i) 14.631

7. a) ±8.06 c) ±1.37 e) ±2.65
g) ±3.61 i) ±6.08

9. a) i) 5.3 cm   iii) 7.1 cm   b) i) 78.3 cm^2 iii) 791.9 cm^2

11. a) 4.2 s   b) 6.7 s   c) 2.1 s   d) 9.5 s

13. a) i) 8.5 cm   ii) 2.6 cm   iii) 12.2 m   iv) 9.2 m
b) 6.9 cm

15. a) i) 4.4 m/s   ii) 3.1 m/s   iii) 1.4 m/s   b) 20.4 cm

17. d) and f)   **MAU 1.** a) i) 2.5 cm   ii) 1.7 cm

Exercises 5 - 9

1. a) 10   c) 0.2   e) $\frac{1}{100}$

3. a) 2.15   b) 4.64   c) 1.07

5. a) 3 cm   b) 6 m   c) 45 mm

7. a) -1.5 b) $\frac{1}{3}$ c) -4

Review Exercises

1. a) 17   c) 2.56   e) 13   g) 125   i) $-\frac{27}{343}$

3. a) ±7 c) ±5 e) ±12

5. a) x^9 c) x^{18} e) $15x^6$ g) $27m^6$ i) $25x^2$
k) $27x^6$ **7.** a) 729   b) 324   c) 1458

9. a) 0   b) 1   c) 4   d) 1

11. a) 1×10^4 b) 7.4×10^5 c) $1. \times 10^{-5}$
d) 5.7×10^{-2}

13. a) 6   c) 120   e) 0   g) 47

15. a) 5.292   b) 4.171   c) 15.811   d) 0.663

17. a) i) 19.2   ii) 0.3   iii) 10.8
b) i) ±3 ii) ±2.5 iii) ±0.91

19. a) i) 1.6 m   ii) 1.02 m   b) i) 2   ii) 4

Chapter 6

Exercises 6 - 1

3. a) $\frac{11}{15}$ c) 6:5   **5.** $\frac{1}{5}, \frac{4}{5}$

7. a) 2.5:1   c) 0.3:1   **9.** a) 11:9   c) 3:55

11. About 159

13. b) 40:28, 40:24, 52:28, 40:20
40:17, 52:20, 40:14, 52:17, 52:14

15. 2.9×10^{32} kg

Exercises 6 - 2

1. a) 11 : 12   b) 2 : 3   c) 3 : 8   **3.** $17 800

5. $22 400, $11 200, $16 800, $33 600

7. a) i) 104   ii) 97   b) i) 9   ii) 9

9. C—264 Hz, D—297 Hz, E—330 Hz, F—352 Hz, G—396 Hz, B—495 Hz, C—528 Hz

11. a) *C*   b) *B*   c) *A*   **13.** 40°, 60°, 80°

Exercises 6 - 3

1. a) 9   c) 5   e) 33   g) 128   i) 49

3. 77 cm   **5.** 192 cm

7. a) 25   b) 500   c) 240

9. a) 325 km   c) 110 km   e) 820 km

11. a) 4 mm   b) 2 mm   c) 4 mm

MAU 1. a) 140 mm   b) 97 mm   c) 521 mm

        3. a) O—30 mm, S—22.5 mm, HO—16.5 mm, TT—12 mm

            b) 1.44 m

Exercises 6 - 4

1. a) 30 km, 75 km   b) 4 L, 18 L

3. a) 23.8 L   b) 2857 km

5. a) 1.85 min, 30.86 h   b) 32 400, 5 443 200

7. 194

9. a) 3.2 h at 50 km/h, 4.0 h at 90 km/h
    b) 160 km at 50 km/h, 360 km at 90 km/h

11. 2.4 h   **13.** 22 km/h   **15.** 1.2 min

Exercises 6 - 5

1. a) 7%   c) 57%   e) 365%   g) 540%   i) 101%

3. a) 38%   c) 81%   e) 3.5%   g) 9.1%
    i) 0.86%   k) 0.07%   m) 306%   l) 3060%

5. a) 13 : 50   c) 16 : 25   e) 5 : 8
    g) 5 : 6   i) 37 : 20   k) 1 : 125

MAU 1. a) 94.7%   b) 324   **3.** a) 226   b) 215

Exercises 6 - 6

1. a) 10   c) 90   e) 75   g) 6.65   i) 67.5

3. $193.38   **5.** $99   **7.** $37.44

9. a) 7.6%   b) 20%

11. a) $1.76, $1.08   b) $1.92, 77¢

13. a) 21.7%   b) About 6253

15. a) $197.95   b) $168.26

17. a) 7.44%
    b) Copper—850, Tin—70, Iron—0.6,
    Lead—2   Phosphorus—3, Zinc—74.4

19. 14.1%   **MAU 1.** 8.7 cm by 7.2 cm, 72.25%

            3. a) 2 and 3   b) 3 and 4

Exercises 6 - 7

1. a) 1.59 cm   c) 3.14 cm

3. a) 60°   b) 240°   c) 225°   **5.** 3.5°

7. Pecans—57.6°, Peanuts—129.6°,
    Hazel nuts—115.2°, Cashews—57.6°

9. 4 cm drain, twice as fast

11. a) 30°   b) 50°   c) 75°   d) 127.5°   **13.** 27

MM 1. 1.618   **3.** 1.625

PSS 1. 29 cm   **3.** 1 min   **5.** 50 m

Review Exercises

1. a) 3 : 1   b) 54 g

3. a) 10.5   c) 9   **5.** Bill $43 750, Laura $81 250

7. 45.7 m   **9.** 305 mm   **11.** a) 29.75 L   b) No

13. 1.2 h   **15.** a) $\frac{2}{5}$ b) 14.4 L   **17.** 40%

19. $199.65/m³   **21.** $15 000 at 16%, $5000 at 12%

23. less than 1   **MM** Best—Chinese: 3.1415929;

Chapter 7

Exercises 7 - 1

1. a) 12   b) 8   **3.** a) 600   b) April   c) 200

5.

7. a)

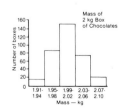

Exercises 7 - 2

1.

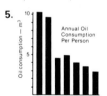

3.

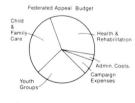

5.

7.

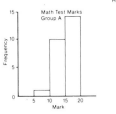

Exercises 7 - 3

1.

	Mean	Median	Mode
a)	11.2	11.5	12
b)	3.3	4	4
c)	3.3	3.4	None
d)	16	16	18
e)	10.25	10	5
f)	8	7	None
g)	$\frac{1}{2}$	$\frac{1}{2}$	$\frac{1}{2}$

3. $8.83   **5.** 18   **7.** Median   **9.** Answers will vary.

Exercises 7 - 5

4. a) 70   b) 290

5. a) Foreign   b) 247   c) $864.50

Exercises 7 - 6

1. 0.502   **5.** 0.53   **8.** About 3125 times

10. a) 0.450   c) 0.249

Exercises 7 - 7

1. a) $\frac{1}{2}$ b) $\frac{1}{4}$ c) $\frac{1}{3}$ **3.** $\frac{1}{12}$ **5.** a) $\frac{1}{4}$ b) $\frac{1}{2}$ c) 0

7. a) $\frac{1}{2}$ b) $\frac{1}{2}$ c) 1 d) 0   **9.** a) 12 b) 50 c) 500

11. a) 6   b) 8   c) 12   d) 4

15. a) 0.02   b) 0.12   c) 0.04

17. a) $\frac{1}{6}$ b) $\frac{7}{30}$ c) $\frac{23}{60}$ d) $\frac{1}{2}$

19. a) $\frac{15}{16}$ b) $\frac{1}{16}$ **21.** a) 0.264 b) 0.085 c) 0.094

23. a) $\frac{3}{4}$ b) $\frac{1}{4}$ c) 0   d) 1

Exercises 7 - 8

1. a)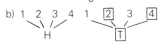

$P(\text{head, 1}) = \frac{1}{8}$

b) 1 2 3 4 1 ②︎ 3 ④︎
H   T

$P(\text{tail, even}) = \frac{2}{8}$, or $\frac{1}{4}$

3. $\frac{1}{32}$ **5.** $\frac{1}{216}$ **7.** $\frac{1}{144}$

9.

	A	*B*	*C*
a)	$\frac{1}{4}$	$\frac{4}{9}$	$\frac{9}{16}$
b)	$\frac{1}{5}$	$\frac{1}{3}$	$\frac{1}{2}$

11. a) $\frac{1}{16}$ b) $\frac{1}{4}$ c) $\frac{1}{169}$ d) $\frac{1}{2704}$

13. b) i) $\frac{1}{2}$ ii) $\frac{1}{2}$ iii) $\frac{1}{3}$ iv) $\frac{1}{6}$

15. a) $\frac{5}{36}$ b) $\frac{25}{216}$ c) $(\frac{5}{6})^9 \times \frac{1}{6}$

MM 1. Break even **3.** Win   **5.** Lose   **7.** Lose

Exercises 7 - 9

1. She should play 4·D more often than 3 S.

3. a)

		Anne	
		4 S	3 H
Bert	5 S	−9	+8
	5 D	+9	−8

+ means Bert wins.

b) i) Each wins $4.25.
ii) Each wins $4.17.
iii) Bert wins $3.78;
Anne wins $4.55.

5. Brenda cannot avoid losing. The game is not fair.

7. a)

		Corinne	
		1	2
John	1	−2	+3
	2	+3	−4

+ means John wins.

i) Each wins $1.50.
ii) Corinne wins $1.75;
John wins $1.50.
iii) Corinne wins $1.13;
John wins $1.88.

b)

		Corinne	
		1	2
John	1	−1	+2
	2	+2	−4

+ means John wins.

i) Corinne wins $1.25;
John wins $1.00.
ii) Corinne wins $1.63;
John wins $1.00
iii) Corinne wins 94¢;
John wins $1.25.

9. a)

		Jeanne	
		H	T
Jacques	H	−1	+1
	T	+1	−1

+ means Jacques wins.

b) i) Each wins 50¢.
ii) Each wins 50¢.
iii) Jeanne wins 55¢;
Jacques wins 45¢.

11. The other player can win by playing heads more often than tails.

Review Exercises

1.

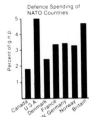

Defence Spending of NATO Countries

3.

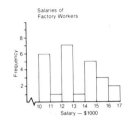

Salaries of Factory Workers

5. 36   **11.** a) $\frac{7}{15}$ b) $\frac{6}{15}$ c) $\frac{3}{15}$ d) $\frac{6}{15}$

13. a) 0.21   b) 0.42   c) 0.36   d) 0.63

15. a) $\frac{9}{38}$ b) $\frac{21}{190}$ c) $\frac{1}{190}$ d) 0

MAU 1.

Number	0	1	2	3	4	5	6	7	8	9
Relative Frequency	0.105	0.12	0.09	0.14	0.085	0.11	0.09	0.07	0.11	0.08

Chapter 8

Exercises 8 - 1

1. a) 14   c) 1   e) -4

3. a) like   c) unlike   e) like
   g) unlike   i) like   k) like

5. a) $14m$ c) $5x$ e) $-4a$ g) $-4y^3$
   i) $-9.4m$ k) $\frac{23}{12}x$

7. a) $4x$ c) $8p$ e) $-5x^3$ g) $-0.8c$
   i) $6b^2$ k) $-m$

9. a) $32m^2 - 22m$ c) $38c^2 + 25c$
   e) $-0.9x^2$ g) $-\frac{5}{12}c$

11. a) $2a^2 - a$ c) —   e) $c - 8$ g) $4x + 2$ i) —

13. a) $-5x^2y - xy + 7xy^2$ c) $4a^2y - 2a^2 - 7ay + ay^2$

Exercises 8 - 2

1. a) 2   c) 8   e) 2.75

3. a) 110   c) 39   **5.** a) iii) 6   c) iii) -5.9

7. a) 2372   b) 1875   **9.** a) 5   b) 9

11. 192.45 m/s, 5787 m   **13.** a) $3400   c) $11 000

Exercises 8 - 3

1. a) $9x + 6$ c) $5 - 6m$ e) $3 - 8t$
   g) $6n^2 - 8n - 4$ i) $7 - 7c - 3c^2$

3. à) $-2 + 3x$ c) $-\frac{1}{2}x + 5$ e) $3 + 2t - t^2$

5. a) $P = 2.5n - 20\,000$ b) i) $5000   ii) $30\,000
   c) i) 12 000   ii) 16 000   iii) 8000

7. a) i) 40   iii) 10   b) 3

9. a) $P = -0.1x^2 + 3x - 1$
   b) i) $11 500 000   ii) $21 500 000

11. a) $-7x^2 - 10x + 25$,   17
   b) $2x^2 - 10x - 5$,   23

Exercises 8 - 4

1. a) $30n^2$ c) $25x^2$ e) $10x^4$
   g) $10x^5$ i) $5x^4$ k) $3m^3$

3. a) $5x - 15$ c) $-6 - 3n$ e) $-2x + 5$
   g) $-30 - 12t$ i) $6 - 10n + 6n^2$ k) $0.6x - 1$

5. a) $10x^2 + 15x$ c) $15c^2 - 6c^2$ e) $-14y^3 + 35y$
   g) $6x^3 - 10x^2$ i) $15s^3 - 10s^2 - 35s$
   k) $-21a^4 + 14a^3 + 28a^2$ m) $-6x^2 + 2.25x^3 + 18x^4$

7. a) 1140   b) 1360   c) 1562.5

9. a) $x + 1$ c) $5a^2 - 4$ e) $2x^2 - 1$
   g) $3x^2 - 6x + 1$ i) $3 - 2y + y^2$

11. 112π cm^2

Exercises 8 - 5

1. a) 3   c) $2x$ e) y^2

3. a) $7(2x^2 + 5x - 1)$ c) $10(2n^2 - 3n + 8)$
   e) $3c(3c^2 + 5)$ g) $4x(1 - 2x + 3x^2)$
   i) $4m(3 + 4m - m^2)$

5. a) $b^3(b - 3)$ c) $a^2(a^2 + 3a - 2)$
   e) $d^4(5d - 1)$ g) $3x(4 + 3x - x^2 - 2x^3)$

7. a) $3xy$ c) $3m$ e) $2xy$

9. a) $(a + 6)(a + 7)$ c) $(1 + y)(8 - 3y)$
   e) $2(x + 3)(x + 2)$

Exercises 8 - 6

1. a) $x^2 + 7x + 12$ c) $a^2 - 8a + 15$
   e) $x^2 + 3x - 10$ g) $a^2 - 2a - 48$
   i) $x^2 + 7x - 60$ k) $n^2 + \frac{3}{4}n + \frac{1}{8}$

3. a) $x^2 + 6x + 9$ c) $4a^2 - 4a + 1$ e) $x^2 - 49$
   g) $9n^2 - 4$ i) $5x^2 - 10x + 5$ k) $-3y^3 + 27y$

5. a) $x^3 + 5x^2 + 3x + 15$ c) $n^4 + 5n^2 + 6$
   e) $x^4 - 5x^2 + 4$ g) $3x^3 + 2x^2 - 15x - 10$
   i) $-3a^6 + 7a^4 - 2a^2$

7. a) $-0.5x^2 + 105x - 1000$ b) i) $3500   ii) $4400

9. a) $x^3 + 7x^2 + 11x + 5$ c) $t^3 - t^2 - 17t + 20$
    e) $x^3 + 6x^2 + 11x + 6$ g) $3x^3 - 33x^2 + 99x - 60$

11. a) $x^2 + 7xy + 10y^2$ c) $6m^2 - 5mn + n^2$
    e) $6r^2 - 17rs - 3s^2$ g) $2p^2 - pq - 15q^2$
    i) $42a^2 + ab - 56b^2$

Exercises 8 - 7

1. a) $(x + 2)(x + 5)$ c) $(m + 4)(m + 6)$
    e) $(x + 3)(x + 3)$ g) $(n + 7)(n + 7)$
    i) $(1 + x)(1 + x)$

3. a) $(x - 2)(x - 4)$ c) $(a - 2)(a - 9)$
    e) $(n - 5)(n - 5)$ g) $(p + 8)(p + 8)$
    i) $(8 + x)(7 + x)$

5. a) $(r - 9)(r + 4)$ c) $(n - 9)(n + 6)$
    e) $(k - 9)(k + 7)$

7. a) $(x - 8)(x + 1)$ c) $(t - 3)(t + 1)$
    e) $(x - 9)(x - 8)$ g) $(m + 11)(m - 5)$
    i) $(s + 5)(s - 4)$ k) $(6 - m)(2 + m)$

9. a) 2750   b) $(x - 9)(x - 4)$ c) 2750

11. a) $2(x + 5)(x + 1)$ c) $10(n + 2)(n - 1)$
    e) $3(x^2 + 5x + 2)$ g) $x(x - 3)(x + 1)$
    i) $2y(y + 3)(y + 4)$ k) $10n(3 + n)(2 + n)$

13. a) $(7 - x)(5 + x)$
    b) i) \$36 000   ii) \$35 000   iii) \$32 000

15. a) \$6.00   b) 37.5%   **17.** a) 3   b) 9   c) 12

Exercises 8 - 8

1. a) $x^2 + 10x + 25$ c) $a^2 - 6a + 9$
    e) $c^2 + 8c + 16$ g) $4a^2 + 12a + 9$
    i) $4 + 4a + a^2$ k) $t^2 + 18t + 81$

3. a) $x^2 + 14x + 49$ c) $n^2 + 8n + 16$
    e) $25c^2 + 30c + 9$ g) $16 + 40x + 25x^2$
    i) $25 + 50s + 25s^2$ k) $25a^2 - 50a + 25$

5. a) i) $x^2 - 6x + 9$ ii) $x^2 - 6x + 9$
    b) i) $9 - 6x + x^2$ ii) $9 - 6x + x^2$

7. a) $(x - 4)^2$ c) $(a + 2)^2$ e) $(x + 3)^2$
    g) $(x + 7)^2$ i) $(b - 6)^2$ k) $(x + 10)^2$

9. a), d), h), j), k), l)

11. a) $(3 - 4w)^2$ c) $(2b - 5)^2$ e) $(0.3x - 0.2)^2$

13. a) $(x - 7y)^2$ c) $(2a + b)^2$ e) $(8x - 3y)^2$

Exercises 8 - 9

1. a) $x^2 - 9$ c) $a^2 - 16$ e) $t^2 - 1$
    g) $s^2 - 625$ i) $x^2 - 2.25$ k) $x^2 - \dfrac{1}{4}$

3. a) $144x^2 - 49$ c) $225a^2 - 121$ e) $x^2 - 2$
    g) $x^4 - 36$ i) $9 - 49y^4$ k) $x^4 - 64y^2$

5. a) $(3a + 2)(3a - 2)$ c) $(4s + 1)(4s - 1)$
    e) $(10x + 11)(10x - 11)$ g) $(\frac{1}{2}x + \frac{2}{3})(\frac{1}{2}x - \frac{2}{3})$
    i) $(1.5 + d)(1.5 - d)$

7. a) $2(x + 3)(x - 3)$ c) $3(a + 4)(a - 4)$
    e) $7(1 + 2y)(1 - 2y)$ g) $x(x + 5)(x - 5)$
    c) $c(2 + 9c)(2 - 9c)$

9. a) $x^4 - 5x^2 + 4$ c) $a^4 - 34a^2 + 225$
    e) $-y^4 + 5y^2 - 4$

11. a) i) $x^2 - 1$ ii) $x^3 - 1$ iii) $x^4 - 1$ b) $x^5 - 1$

Exercises 8 - 10

1. a) $8x + 11$ c) $7x - 38$ e) $3x - 5x^2$
    g) $13 - 2n$ i) $26x^2 + 8x$

3. a) $2x^2 - 4x + 34$ c) $-20x + 8$
    e) $2x^2 + x + 11$ g) $10x^2 - 17$

5. a) i) $V = 9x^3 + 24x^2 + 12x$ ii) $A = 30x^2 + 40x + 8$
    b) i) 45 m³, 78 m²   ii) 495 m³, 398 m²

7. a) -33 b) 8 or -6 c) 5 or -3 d) 9 or -7

9. a) $2.5x - 5$ c) $5 - a$ e) $a - \dfrac{11}{4}$ g) $20x$
    i) $5x^2 - 2x + 4$ **11.** a) -8 b) 41   c) 64

13. a) 48   b) -5 or -7 c) -2 or -10 d) -3 or -9

15. a) 35   b) 7

17. a) $-2x^2 - 4x + 1350$
    b) i) \$1 344 000; \$1 334 000   ii) \$1 352 000; \$1 350 000

PSS 1. 7   **3.** a) 45   b) $\dfrac{n(n - 1)}{2}$
    5. a) odd   b) even   c) $-\dfrac{n}{2}; \dfrac{n + 1}{2}$ **7.** a) 100   b) n^2

CP 1. a) 83   b) 1.625   c) 1462.779

Review Exercises

1. a) $3n$ c) $8y^2$ e) $5b^2$ g) $0.9y^3 + 2y$

3. a) 2   b) 10   c) 44   d) 85.25

5. a) $35a^2$ c) $16y^2$ e) $-8y^3$ g) $-30x^4$

7. a) $3x - 4$ c) $3x + 1$ e) $x - 4$ g) $3 - c + 2c^2$

9. a) $x^2 - 7x + 12$ c) $a^2 + 3a - 10$
    e) $n^2 - 3n - 28$ g) $4x^2 - 44x + 120$
    i) $-21y + 18y^2 + 3y^3$

11. a) $x^2 + 6x + 9$ c) $25 - 10q + q^2$

13. a) $(b + 5)(b - 5)$ c) $(y + 11)(y - 11)$
    e) $(3x + 4)(3x - 4)$ g) $2(x + 4)(x - 4)$

15. a) 9   b) 37   c) 1   d) 21

Chapter 9

Exercises 9 - 1

1. $A(-2, 5)$, $B(-4, 3)$, $C(-3, 1)$, $D(-5, -2)$, $E(-2, -3)$, $F(1, -5)$, $G(3, -3)$, $H(5, -2)$, $I(5, 3)$, $J(3, 5)$

3. a) On y-axis   c) To the left of the y-axis
   e) On a line through (0, 0) bisecting the first and third quadrants

5. c) Rectangle

7. a) $(3, -2)$ b) $(-3, -2)$ c) $(-1, 4)$ d) $(0, 3)$

9. Square

Exercises 9 - 2

1. $\frac{1}{4}$ **3.** a) $\frac{4}{7}$ c) 0   e) $-\frac{10}{11}$ g) $\frac{12}{7}$

4. AB, 7; CD, $-\frac{3}{2}$; EF, $-\frac{1}{2}$; GH, 5; IJ, $-\frac{1}{4}$

7. $\frac{1}{40}$ **9.** b) $\frac{2}{5}$ c) 4 km   d) 5 min

11. Lines appear to be parallel.

13. b) Equal; equal   c) Parallelogram

Exercises 9 - 3

1. a) $\{(1, 0.50), (2, 0.95), (3, 1.35), (4, 1.70), (5, 2.00)\}$

3. a) $\{(11, 1), (10, 2), (9, 3), (8, 4), (7, 5), (6, 6)\}$
   c) $\{(1, 3.90), (2, 4.40), (3, 4.90), (4, 5.40), (5, 5.90), (6, 6.40)\}$

5. a) (2, 30), (6, 12), (10, 20), (13, 25)
   c) $\{(2, 30), (3, 20), (5, 15), (6, 12), (7, 15), (10, 20), (13, 25), (16, 30)\}$

7. a) i) 0   ii) 5   iii) 9   iv) 20   c) 35

Exercises 9 - 4

1. b) 40   c) i) 135   ii) 20

3. b) i) About 115 m   ii) About 240 m
   c) i) About 70 km/h   ii) About 110 km/h

5. b) $71.25   c) 700 km

7. a) $y = x + 2$ c) 7   **9.** a) $\{(x, y) \mid 3x + y = 18\}$

Exercises 9 - 5

1. a), c), d)   **3.** a) $-\frac{2}{3}$,   c) $\frac{7}{4}$ e) 2

5. $-\frac{1}{2}$. As age increases, amount of sleep required decreases 0.5 h/year.

7. a) 180 km   c) 50 km   f) No

9. b) i) About 195°F   ii) About 250°F   iii) About 390°F
   c) $\frac{5}{9}$. An increase of 5°C corresponds to an increase of 9°F.

Exercises 9 - 6

1. a) Parabola c) Hyperbola

3. a) Parabola c) Hyperbola   **7.** b) 200 kPa   c) 1.2 L

Exercises 9 - 7

1. a) Quadratic   c) Other   e) Quadratic

3. b) About 0.5 m   c) Fourth

5. b) About 1945   c) About 8.8 million
   d) About 27 million

7. b) About 4 days   c) About 35%

9. a) i) 5:45 p.m.   ii) 8:45 p.m.   iii) 4:30 p.m.
   c) Beginning and end of daylight-saving time

11. a) $x - y = xy$

MM **2.** b) 81   **3.** a) 1, also 7   b) 17   c) 42

Exercises 9 - 8

1. a) $\{(6, 4), (4, 6), (2, 3), (-1, -1), (-4, -3), (-6, -4)\}$

7. a) $x \to \frac{2x - 18}{3}$ c) $x \to 2x$ e) $x \to x + 2$

Review Exercises

1. Rectangle   **3.** a) -2 c) $\frac{1}{2}$

5. a) (2, 3), (1, 2), (0, 1), (-1, 0), (-2, -1), (-3, -2)
   b) (1, 1,), (1, -1), (3, 3), (3, -3), (5, 4), (5, -4)

7. a) -3 c) 5   **9.** a) Parabola   c) Circle

11. a) $P = 48 - 2x - 2y$ b) $x + y = 8$

Chapter 10

Exercises 10 - 1

3. a) $\angle AOB$, $\angle AOC$, $\angle BOC$
   b) $\angle AOB$, $\angle AOC$, $\angle AOD$, $\angle BOC$, $\angle BOD$, $\angle COD$

5. a) 4   b) 5   **7.** a) 3   b) 4

MAU 1. Yes   **3.** No   **5.** No   **7.** No

Exercises 10 - 2

1. a) $\angle ABE$ and $\angle CBD$; $\angle ABD$ and $\angle CBE$
   b) $\angle SQR$ and $\angle TQP$; $\angle SQP$ and $\angle TQR$

3. a) 90° and 90°   b) 60° and 120°

5. a) Acute   b) Obtuse   c) Right   **7.** 3

9. a) 65°   b) 55°   c) 120° d) 36°   e) 70°   f) 25°

Exercises 10 - 3

1. a) $\angle KLO$ and $\angle LOP$; $\angle MLO$ and $\angle LON$

3. a) 40°, 40°, 140°, 40°   c) 55°, 115°, 55°
  e) 45°, 35°, 100°

5. No   **7.** 180°

Exercises 10 - 4

1. 95°, 44°, 41°; 180°

3. a) Acute   b) Obtuse   c) Right   d) Acute

5. a) 80°   b) 20°   c) 125°

7. a) 20°   c) 100°   e) 80°   g) 70°   i) 45°

11. a) i) 120°   ii) 140°   c) $y - x = 90°$

13. a) 48°   b) 36°   c) 45°   d) 30°

Exercises 10 - 5

1. a) 70°   c) 75°   e) 50°   g) 150°   i) 100°

3. a) i) 80° west of north   iii) 60° east of north
  b) i) 55° east of north   iii) 75° west of north

5. a) 30° and 120° or 75° and 75°
  c) 50° and 50° or 80° and 20°

7. a) i) 40°   ii) 130°   b) i) 50°   ii) 35°
  c) $2x + y = 180°$

9. a) 1   b) 3   c) 4

Exercises 10 - 6

7. c) i) 3.1 cm   ii) 105°   iii) 110°   **9.** 37°, 83°, 8.7 cm

15. b) In any triangle, the sum of the lengths of any two
  sides is greater than the length of the third side.

Exercises 10 - 8

1. a) $MR = CL$ $\angle MRX = \angle CLP$
  $RX = LP$ $\angle RXM = \angle LPC$
  $MX = CP$ $\angle XMR = \angle PCL$

3. a) SAS c) ASA d) SSS f) SAS g) SSS h) SSS

5. a) 4 cm, 5 cm, 6 cm   c) 130°, 6 cm, 9 cm
  e) 7 cm, 9 cm, 120°   g) 90°, 30°, 5 cm

Exercises 10 - 9

5. a) i) 5   ii) 9   b) i) 14   ii) 36   c) i) 30   ii) 100

7. a)

Number of Sides	4	5	6	8	10
Number of Triangles	2	3	4	6	8
Sum of Angles	360°	540°	720°	1080°	1440°

  b) $S = 180(n - 2)$, $n \geq 3$

9. a) 30°   b) 80°   c) 60°

Exercises 10 - 10

1. a) 5   c) 41   e) 9

3. a) 2, 4, 4.47   c) 2, 3, 3.61

5. 51.42 m   **7.** 6710 mm   **9.** 1.2 km

11. a) 5 m, 13 m   c) 5.29 cm, 4.58 cm   e) 2.12 cm

13. 50.70 cm by 42.25 cm

15. a)

s	t	$s^2 - t^2$	$2st$	$s^2 + t^2$
4	2	12	16	20
4	3	7	24	25
5	1	24	10	26
5	2	21	20	29
5	3	16	30	34

CP 1. 4.81 cm   **3.** 38.74 cm

Exercises 10 - 11

3. a) 5.0   c) 7.8   e) 4.4

7. a) 17.0   c) 8.0   e) 7.0

9. a) 5.7 cm   b) 5.7 cm   c) 2.8 cm   d) 8.0 cm

11. a) i) 60°   ii) 72°   b) i) 15°   ii) 33°   c) $x + y = 90$

13. a) 83.7 cm   b) 94.2 cm

Review Exercises

1. a) 116°, 32°   c) 50°, 115°   e) 70°, 70°, 80°
  g) 35°, 37.5°   i) 100°, 130°

7. a) 10.9, 4.9   b) 10.0, 18.0   **9.** No

11. a) 30°, 60°   b) 6.9, 7.4   c) 17, 15, 70°

13. a) 10.0 cm   c) 4.0 cm

Chapter 11

Exercises 11 - 1

1. a) Reflection, Translation, Rotation
  c) Rotation, Translation, Reflection

3. a) Rotation   b) Reflection   c) Reflection

Exercises 11 - 2

3. $A'(-1, -4)$, $B'(-5, -5)$, $C'(0, 0)$

5. $P'(3, -5)$, $Q'(-1, 7)$, $R'(-2, -1)$

7. a) It passes through the turn centre.
  b) They are opposites.

9. a) and c)

11. a) 47°, 133°, 133°   c) 55°, 55°, 125°

Exercises 11 - 3

3. a) $A'(2, -5)$, $B'(6, -5)$, $C'(9, -1)$, $D'(2, -1)$

6. $x = 1$ **7.** $y = -2$

13. a) $A'(-2, -6)$, $B'(4, -3)$, $C'(3, 3)$, $D'(-5, 2)$

Exercises 11 - 5

1. b) i) $(2, 4)$ ii) $(7, 8)$ iii) $(-1, 1)$ iv) $(-2, 4)$

3. a) $120°$, $60°$, $120°$ b) $70°$, $70°$, $110°$

7. b) $P(-6, 6)$, $Q(-4, 10)$, $R(-8, 7)$

Exercises 11 - 6

1. a) Reflection   c) Translation or rotation   e) Reflection

5. a) $135°$ c) $65°$, $115°$, $50°$ **7.** a) 4   b) 3

Exercises 11 - 7

3. 10   **5.** a) i) 3   ii) 4   iii) 6   iv) 8   b) 12

7. Rhombus   **9.** a) 4

Exercises 11 - 8

5. a) $(7.5, 0)$ b) $(4.5, 0)$ and $(0, 3.5)$

Review Exercises

1. a) Translation   b) Reflection   c) Rotation

d) Reflection   e) Reflection   f) Rotation

g) Rotation   h) Translation

3. a) $100°$, $80°$, $80°$ c) $95°$, $95°$ e) $60°$, $140°$, $40°$

5. a) $A'(2, -5)$, $B'(-3, -1)$, $C'(1, 4)$ **7.** $x = -1$

Index